Fundamentals of Machining and Machine Tools

Third Edition

MECHANICAL ENGINEERING
A Series of Textbooks and Reference Books

Founding Editor

L. L. Faulkner

Columbus Division, Battelle Memorial Institute
and Department of Mechanical Engineering
The Ohio State University
Columbus, Ohio

Fundamentals of Machining and Machine Tools

Third Edition

Geoffrey Boothroyd
Winston A. Knight

Taylor & Francis
Taylor & Francis Group
Boca Raton   London   New York

A CRC title, part of the Taylor & Francis imprint, a member of the
Taylor & Francis Group, the academic division of T&F Informa plc.

Published in 2006 by
CRC Press
Taylor & Francis Group
6000 Broken Sound Parkway NW, Suite 300
Boca Raton, FL 33487-2742

International Standard Book Number-10: 1-57444-659-2 (Hardcover)
International Standard Book Number-13: 978-1-57444-659-3 (Hardcover)
Library of Congress Card Number 2005050224

Library of Congress Cataloging-in-Publication Data

Boothroyd, G. (Geoffrey), 1932-
    Fundamentals of machining and machine tools / by Geoffrey Boothroyd and Winston A. Knight.--3rd ed.
         p. cm. -- (Mechanical engineering)
    ISBN 1-57444-659-2 (alk. paper)
    1. Machining. 2. Machine-tools. I. Knight, W.A. (Winston Anthony), 1941- II. Title. III. Mechanical Engineering (Marcel Dekker, Inc.) ; 198.

TJ1185.B7713 2005
671.3'5--dc22
                                                                                            2005050224

Taylor & Francis Group
is the Academic Division of Informa plc.

**Visit the Taylor & Francis Web site at
http://www.taylorandfrancis.com**

**and the CRC Press Web site at
http://www.crcpress.com**

Preface to Third Edition

Similar to the previous editions, this book is intended for those studying and teaching the principles of machine tools and machining in universities and colleges, as well as providing a useful background for those involved in manufacturing in industry.

Since the second edition was published in 1989, there have been some significant developments in the analytical modeling of machining processes, together with changes in machining practice in industry, in particular as a result of developments in modern cutting tool materials. For this reason most of the chapters in the book have been updated to reflect these recent developments.

As in previous editions, the mathematical content of the book has been limited, but it was found appropriate to include some description of modern analytical methods that can be applied to machining, in particular to outline the capabilities and limitations of the various modeling approaches. The chapter on tool materials and tool life has been updated to reflect the recent developments in super hard cutting tool materials, tool geometries, and surface coatings. These new tool materials have enabled cutting speeds to be increased significantly and have led to more effective machining of difficult to cut work materials. For this reason some description of developments in high speed machining and hard machining has been included.

The chapter on cutting fluids has also been updated to include developments in cutting fluid applications, including the trends towards dry and minimum-quantity lubrication machining. Similarly, the chapter on chip control has been modified to reflect developments in tool geometries for chip breaking and chip control.

As in the previous editions, significant emphasis is placed on the economics of machining processes and design for machining. This material has been expanded to include improvements in the cost modeling of machining processes, including the application to grinding processes.

We are indebted to those we have been associated with in recent years and who have contributed both directly and indirectly in the preparation of this book, including colleagues and graduate students whose work has been helpful. The discussion on planning for multispindle automatics in Chapter 11 is derived from the Master of Science thesis of Jeremy Young and the procedures for determining recommended cutting conditions in Chapter 6 are based on the Master of Science thesis of Richard Turner. We are also grateful to Brian Rapoza of Boothroyd Dewhurst Inc. for developing the empirical equations for laser and plasma cutting used in Chapter 14.

Winston Knight

Preface to Second Edition

This book is intended primarily for those studying and teaching the principles of machine tools and metal machining in universities and colleges. It should also prove useful to those concerned with manufacturing in industry.

The mathematical content of the book is deliberately limited. Those who have taken basic courses in statics and dynamics and who have had an introduction to calculus should have no difficulty in comprehending the material.

Many of the present texts dealing with the same material are purely descriptive. In this book, the approach is to illustrate, through fundamentals and analysis, the causes of various phenomena and their effects in practice. Emphasis is given to the economics of machining operations and the design of components for economic machining.

A significant portion of the book is based on a previous text written by one of the authors (Geoffrey Boothroyd) and published by McGraw-Hill. While much of this material has been retained, recent developments have been included where appropriate. Several new chapters have been introduced and others largely rewritten. The section on tool materials has been expanded to include the modern materials that are contributing significantly to increases in productivity in industry. A new chapter on machine tool vibrations has been included, which covers the fundamental aspects of machine tool chatter, the dynamic testing of machine tools, and the practical means of improving machine tool stability. The chapter on grinding has been expanded to include thermal aspects of the process and a description of new grinding processes, including creep feed grinding.

New emphasis in the book has been placed on the utilization of machine tools through the inclusion of chapters on manufacturing systems and automation and on computer-aided manufacturing, together with an expanded chapter on design for machining, which serves as an introduction to an area of growing importance, that of design for manufacturability. Various types of automation in machine tools are outlined and an introduction to cellular plant layouts and flexible manufacturing systems is included. Aspects of the programming of numerical control machine tools are discussed in some detail. Finally, because of their growing importance, the main nonconventional machining processes are described and examples of their application given.

We are indebted to those with whom we have been associated in recent years and who have assisted both directly and indirectly in the preparation of this book, including colleagues and graduate students whose work has been helpful in the preparation of this book. Finally, we would like to thank Kathleen Yorkery for typing the manuscript.

Geoffrey Boothroyd
Winston A. Knight

The Authors

Geoffrey Boothroyd is Emeritus Professor of Industrial and Manufacturing Engineering, The University of Rhode Island and co-founder of Boothroyd Dewhurst, Inc., of Wakefield, Rhode Island. The author or coauthor of over 130 professional papers and articles on manufacturing engineering research and product design techniques, Dr. Boothroyd is also the author, coauthor, or coeditor of several books, including *Product Design for Manufacture and Assembly, Second Edition; Assembly Automation and Product Design;* and *Applied Engineering Mechanics* (all titles, Marcel Dekker, Inc.). A fellow of the Society of Manufacturing Engineers and a recipient of the 1991 National Medal of Technology Award, he is a member of the National Academy of Engineering, among others. Dr Boothroyd received his Ph.D. (1962) and D.Sc. (1974) degrees in engineering from the University of London, England.

Winston Knight is Professor of Industrial and Manufacturing Engineering at the University of Rhode Island. The author of over 120 professional papers and articles, Dr. Knight is also the author or coauthor of several books, including *Product Design for Manufacture and Assembly, Second Edition* (Marcel Dekker, Inc.). Dr. Knight's research interests have focused on various aspects of manufacturing engineering, including product design for manufacture, design for environment and recycling, together with machine tool technology, group technology and aspects of CAD/CAM. A fellow of the Society of Manufacturing Engineers, he received his B.Sc. (1963) and Ph.D. (1967) degrees in mechanical engineering from the University of Birmingham, England, and the M.A. degree (1980) from Oxford University, England.

Contents

Conventions Used
in This Book

STANDARDIZATION

Every attempt is made in this book to follow the International Organization for Standardization (ISO) recommendations for units and definitions. The most important of these is, of course, the International (SI) System of basic and derived units. The basic system is described in ISO Standard 1000, and recommendations for its application are specified in ISO Recommendation R31. The latter document includes suggested symbols for the various derived units.

In addition, numerous standards are becoming available covering various aspects of machining and are followed as closely as possible. Unfortunately, some of these new standards include symbols and definitions that are at variance with the SI; this variance leads to a certain amount of duplication. For example, the recommended symbol for the feed applied in a machining operation is f, which is also the recommended symbol for frequency of vibration. Confusion among measurements and symbols is avoided, however, by the use of appropriate suffixes and subscripts, respectively.

This book is written mainly in SI units because many U.S. manufacturing industries now work in the SI system. However, a nationwide conversion to the metric system is not expected in the foreseeable future. For this reason, in most cases where SI units are specified in the text, the approximate English (U.S. Customary System [USCSI]) equivalents are given in parentheses. Graphs and results from previously published work are reproduced with the original scales in English (USCS) units, and the corresponding SI scales are added at the right and along the top where appropriate. Graphs produced specifically for this book are given in SI units with the equivalent English units added at the right and along the top. In some places English units are used in the text and the appropriate conversion factors are given. This is particularly the case for Chapter 13.

INTRODUCTION TO THE INTERNATIONAL (SI)
SYSTEM OF UNITS

In 1960 the General Conference of Weights and Measures on the International System of Units formally approved the system of units known as the International (SI) System of Units. This system is now adopted for most countries throughout the world, but it is currently not being adopted by the United States, although many international companies have standardized on the system. In practice the system is convenient because it obviates the need for the insertion of conversion factors into equations and eliminates many of the ambiguities present in other systems. The basic SI units encountered in engineering are defined as follows:

1. The unit of length l is the meter, m, which is the length equal to 1, 650,763.73 wavelengths in vacuo of the radiation corresponding to the transition between the levels $2p_{10}$ and $5d_5$ of the krypton-86 atom.
2. The unit of mass m is the kilogram, kg, which is equal to the mass of the international prototype of the kilogram.
3. The unit of time t is the second, s, which is the duration of 9, 192, 631,770 periods of the radiation corresponding to the transition between the two hyperfine levels of the ground state of the cesium-133 atom.
4. The unit of temperature T is the kelvin, K, which is 1/273.16 of the thermodynamic temperature of the triple point of water.
5. The unit of electric current I is the ampere, A, which is that constant current which, if maintained in two straight parallel conductors of infinite length, of negligible, circular cross section and placed 1 meter apart in a vacuum, would produce between these conductors a force equal to 2×10^{-7} newton per meter of length.

To encourage the student or reader to become familiar with SI units, most of the problems presented at the end of each chapter are given in SI units without the English equivalents. It is hoped that the student or reader will thereby discover that the new system is considerably easier to work with than the U.S. Customary System (pounds, feet, etc.).

For those not sufficiently familiar with the SI system, the tables introduce the basic and derived units and their definitions and recommended symbols. Useful conversions are also given.

Table SI.1 presents a selection of derived units.

The most common prefixes for SI units are listed in Table SI.2. It should be noted that:

1. The prefix refers to the whole unit; for example, the m (milli) in m.N/m^2 means m(N/m^2), not mN/m^2. The dot is used to indicate multiplication when confusion could arise.
2. Prefixes in denominators are avoided, except for k in kg (kilogram).

Useful conversion factors are given in Table SI.3.

TABLE SI.1
Units Derived from the Basic SI Units

Unit Being Measured	Symbol	Name of Unit	Symbol of Unit
Angle, plane	α, β, γ, etc.	Radian	rad
Angular velocity	ω	—	rad/s
Angular acceleration	α	—	rad/s^2
Frequency	f	Hertz	Hz
Rotational frequency	n	Reciprocal second	s^{-1}
Area	A	—	m^2
Volume	V	—	m^3
Velocity	v	—	m/s
Acceleration	a	—	m/s^2
Density	ρ	—	kg/m^3
Force	F	Newton	N
Energy, work, heat	W	Joule	N.m or J
Pressure, stress	p	Pascal	N/m^2 or Pa
Power	P	Watt	J/s or W
Temperature difference	θ	Degree Celcius	K-273.15 or °C

TABLE SI.2
Prefixes Used for SI Units

Multiple and Sub-Multiple[a]	Prefix	Symbol
10^9 or $E + 09$	giga	G
10^6 or $E + 06$	mega	M
10^3 or $E + 03$	kilo	k
10^{-3} or $E - 03$	milli	m
10^{-6} or $E - 06$	micro	μ
10^{-9} or $E - 09$	nano	n
10^{-12} or $E - 12$	pico	p

[a] $E \pm ab = 10^{\pm ab}$

TABLE SI.3
USCS — SI Conversion Table

Quantity	To Convert From	To	Multiply By[a]	
Length	inch (in.)	meter (m)	2.54*	E – 02
	foot (ft)	meter (m)	3.048*	E – 01
Mass	pound (lb)	kilogram (kg)	4.535 924	E – 01
Time	minute (min)	second (s)	6.0*	E + 01
	hour (hr)	second (s)	3.6*	E + 03
Force	pound force (lbf)	newton (N)	4.448 222	E + 00
Speed	foot per minute (ft/min)	meter per second (m/s)	5.08*	E – 03
	inch per second (in./s)	meter per second (m/s)	2.54*	E – 02
	inch per minute (in./min)	meter per second (m/s)	4.233*	E – 04
	revolution per minute (rpm)	radian per second (rad/s)	1.047 192	E – 01
Acceleration	foot per second per second (ft/s^2)	meter per second per second (m/s^2)	3.048*	E – 01
	inch per second per second (in./s^2)	meter per second per second (m/s^2)	2.54*	E – 02
Area	square inch (in.^{2})	square meter (m^2)	6.451 6*	E – 04
Volume	cubic inch (in.^{3})	cubic meter (m^3)	1.638 706	E – 05
Volume flow rate	cubic inch per minute (in.^{3}/min)	cubic meter per minute (m^3/min)	2.731 177	E – 07
Density	pound per cubic inch (lb/in.^{3})	kilogram per cubic meter (kg/m^3)	2.767 991	E + 04
Pressure, stress	pound force per square inch (lbf/in.^{2})	newton per square meter (N/m^2)	6.894 757	E + 03
Energy, work, heat	British thermal unit (Btu)	newton-meter (N-m) or (J)	1.055 06	E + 03
Power	horsepower (hp)	joule per second (J/s) or (W)	7.457	E + 02
	foot-pound force per minute (ft-lbf/min)	joule per second (J/s) or (W)	2.259 697	E – 02
Thermal conductivity	Btu in./hr ft^2 °F	joule per second meter kelvin (J/smK), or watt per meter kelvin (W/mK)	1.442 279	E – 01
Specific heat capability	Btu/lb °F	joule per kilogram kelvin (J/kgK)	4.148	E + 03

[a] Values followed by an asterisk are exact conversions.

1 Machine Tools and Machining Operations

1.1 INTRODUCTION

The history of metal cutting dates from the latter part of the eighteenth century. Before that time machine tools did not exist, and the following extract from the diary of an English engineer, Richard Reynolds, dated October 1760, gives some idea of the manufacturing problems that had to be faced. Richard Reynolds was attempting to produce a cylinder for a tire engine for drawing water from a coal pit. The cylinder, of cast brass, had a length of 9 ft and a bore 28 in. in diameter. He wrote:

> Having hewed two balks of deal to a suitable shape for the cylinder to lie therein solidly on the earth in the yard, a plumber was procured to cast a lump of lead of about three hundred weight, which being cast in the cylinder, with a dike of plank and putty either side, did make it of a curve to suit the circumference, by which the scouring was much expedited. I then fashioned two iron bars to go around the lead, whereby ropes might be tied, by which the lead might be pulled to and fro by six sturdy and nimble men harnessed to each rope, and by smearing the cylinder with emery and train oil through which the lead was pulled, the circumference of the cylinder on which the lead lay was presently made of a superior smoothness: after which the cylinder being turned a little, and that part made smooth, and so on, until with exquisite pains and much labour the whole circumference was scoured to such a degree of roundness as to make the longest way across less than the thickness of my little finger greater than the shortest way: which was a matter of much pleasure to me, as being the best that we so far had any knowledge of.

In 1776 James Watt built the first successful steam engine, and one of his greatest difficulties in developing this machine was the boring of the cylinder casting. His first cylinder was manufactured from sheet metal, but it could not be made steam tight. Even attempts to fill the gap between the piston and the cylinder with cloth, leather, or tallow were of no avail. John Wilkinson eventually solved the problem when he invented the horizontal-boring machine. This machine consisted of a cutting tool mounted on a boring bar that was supported on bearings outside the cylinder. The boring bar could be rotated and fed through the cylinder. Thus generating, with the tool points, a cylindrical surface independent of the irregularities of the rough casting (Figure 1.1). This boring machine was the first effective machine tool, and it enabled James Watt to produce a successful steam engine.

1

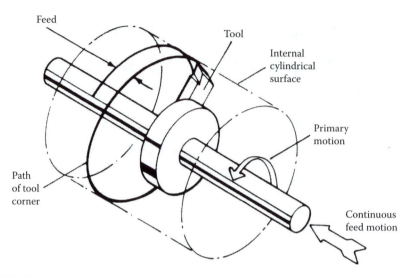

FIGURE 1.1 Generation of a cylindrical surface in horizontal boring.

Metal cutting, as we know it now, started with the introduction of this first machine tool. Today, machine tools form the basis of our industry and are used either directly or indirectly in the manufacture of all the products of modern civilization.

1.2   GENERATING MOTIONS OF MACHINE TOOLS

The principle used in all machine tools is one of generating the surface required by suitable relative motions between the cutting tool and the workpiece. The cutting edge or edges on the cutting tool remove a layer of work material; the removed material is called a shaving or a chip. The simplest surfaces to generate are flat surfaces and internal or external cylindrical surfaces. For example, if a tool is reciprocated backward and forward in a straight line and a workpiece is incrementally fed beneath the tool in a direction at right angles to the motion of the tool, a flat surface will be generated on the workpiece. Similarly, rotating the workpiece and feeding the tool parallel to the axis of workpiece rotation can generate a cylindrical surface. Thus, in general, two kinds of relative motion must be provided by a metal-cutting machine tool. These motions are called primary motion and feed motion and are defined as follows:

> The primary motion is the main motion provided by a machine tool or manually to cause relative motion between the tool and workpiece so that the face of the tool approaches the workpiece material. Usually, the primary motion absorbs most of the total power required to perform a machining operation.

The feed motion is a motion that may be provided to the tool or workpiece by a machine tool which, when added to the primary motion, leads to a repeated or continuous chip removal and the creation of a machined surface with the desired geometric characteristics. This motion may proceed by steps or continuously; in either case it usually absorbs a small proportion of the total power required to perform a machining operation.

To facilitate the descriptions of machine tool motions it is useful to employ a system of machine tool axes recommended by the International Organization for Standards (ISO) [1]. Although this system was established for the purposes of programming numerically controlled machine tools, it is suitable as a general system for all machine tools. The system is based on the right-hand coordinate system shown in Figure 1.2a and refers to the possible motions of a tool in a machine tool. The three coordinate axes X, Y, and Z refer to possible linear motions of the tool, and the motions A, B, and C refer to possible rotary motions of the tool about these axes, respectively. To understand the right-hand screw sign convention used in this system, it is helpful to imagine a right-hand screw thread aligned with each axis. If a screwed nut mounted on any axis is rotated such that it moves along the axis in a positive direction away from the origin, that rotary motion is positive.

A particular machine tool can move the tool in only a few of the possible directions shown in Figure 1.2a. However, before the actual directions of motion in a particular machine tool can be labeled, it is necessary to orient the coordinate system relative to that machine tool. The coordinate system is oriented as follows:

The Z axis of motion is arranged parallel to the axis of the machine spindle, which provides primary motion. If the machine has no spindle, the Z axis is arranged perpendicular to the work-holding surface. Positive Z motion increases the distance between the workpiece and tool holder.

Where possible, the X axis is horizontal and parallel to the work-holding surface. On machines with no main spindle, the X axis is parallel to and positive in the principal direction of cutting (primary motion). On machines with rotating workpieces, the X motion is radial and parallel to the cross slide. (Positive X motion is defined as the motion of a tool when it recedes from the axis of rotation of the workpiece.) On machines with rotating tools:

1. If the Z axis is horizontal, positive X motion is to the right when viewed from the main spindle toward the workpiece.
2. If the Z axis is vertical, positive X motion is to the right when viewed from the main spindle toward the machine column.

Positive Y motion is selected to complete the coordinate system shown in Figure 1.2a.
For motion of the workpiece in a machine tool, a similar set of axes and motions is employed (Figure 1.2b), but the appropriate letter is primed, and the

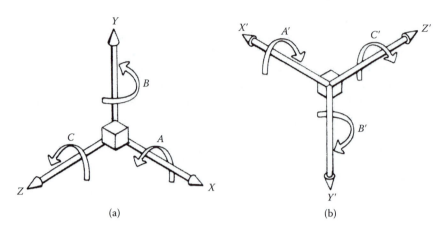

FIGURE 1.2 Coordinate system for specification of machine tool motions. (a) Motions of the tool; (b) motions of the workpiece.

sign convention is reversed. The reason for the prime and reversed sign is that if, for example, a tool can be made to approach a workpiece by motion in the positive X direction, the same effect can be obtained by moving the workpiece in the positive X' direction.

In the following sections some common general-purpose low-production machine tools will be described and their generating motions illustrated. The same basic motions are also used in higher productivity and automatic machine tools and description of machines of this type can be found in Chapter 11. Machine tools in general can be divided into three groups:

1. Those using single-point tools
2. Those using multipoint tools
3. Those using abrasive wheels

Machines using single-point tools will be described first, starting with the most common, the engine lathe. The description of the engine lathe and some of the operations that can be performed on it will be used to introduce numerous terms and definitions that are applicable to all types of machining operations. This description is therefore necessarily quite lengthy, but it is important even for those familiar with machining operations because of the international terminology used.

For the present purposes the term *machining* will be taken to mean any process in which material is removed gradually from a workpiece, including metal cutting with single-point and multipoint tools and grinding with abrasive wheels.

For each machining operation, equations are developed to estimate the time taken to perform the operation and the power required. At the end of the chapter a summary of these equations is presented, together with typical characteristics of the machines described (including speed range, feed range, power, efficiency, size, and accuracy) and characteristics of various work materials.

1.3 MACHINES USING SINGLE-POINT TOOLS

1.3.1 ENGINE LATHE (CENTER LATHE)

An engine lathe is shown diagrammatically in Figure 1.3a; it consists of a horizontal bed supporting the headstock, the tailstock, and the carriage. All machine tools must have a means of supporting or holding the workpiece. In Figure 1.3 the workpiece is gripped at one end by a chuck mounted on the end of the main spindle of the machine and is supported at the other end by a center mounted in the tailstock. The tailstock can be clamped at various positions along the bed to accommodate workpieces of various lengths. Short workpieces need only be gripped by the chuck and, as a rule-of-thumb, parts with a length to diameter ratio of less than three to one need be supported in the chuck only.

Primary motion, the rotation of the workpiece (motion C'), is provided by the movement of a series of gears driving the main spindle, the gears being driven by an electric motor mounted at the rear of the machine. The main spindle and the gears are all mounted in the headstock. Levers on the front of the headstock allow various rotational speeds to be selected.

The single-point cutting tool is held in a toolholder or tool post, which is mounted on a cross slide, which in turn is mounted on the carriage. The carriage is driven along the bed (motion Z) by a lead screw (for screw cutting) or a rack, pinion gear, and feed rod (for turning); both the lead screw and the feed rod are connected to the main spindle through a train of gears. Alternatively the carriage can remain stationary on the bed, and the gear train can be used to drive the toolholder across the carriage (motion X) using a lead screw in the cross slide.

Figure 1.3b shows a cylindrical surface being generated on a workpiece by the rotation of the workpiece (-C' motion) and the movement of the carriage along the lathe bed (-Z motion); this operation is known as cylindrical turning.

The feed motion setting on the lathe is the distance moved by the tool during each revolution of the workpiece. Once chosen, the feed setting remains constant regardless of the spindle speed because of the geared drive between the main spindle and the feed rod. The feed f for all machine tools is defined as the displacement of the tool relative to the workpiece, in the direction of feed motion, per stroke or per revolution of the workpiece or tool. Thus, to turn a cylindrical surface of length l_w, the number of revolutions of the workpiece is l_w/f, and the machining time t_m, is given by

$$t_m = \frac{l_w}{fn_w} \tag{1.1}$$

where t_m, is the rotational frequency of the workpiece.

It should be emphasized at this point that n_w, is the time for one pass of the tool (one cut) along the workpiece. This single pass does not necessarily mean, however, that the machining operation is completed. If the first cut is designed to remove a large amount of material at high feed (roughing cut), the forces generated during the operation will probably have caused significant deflections

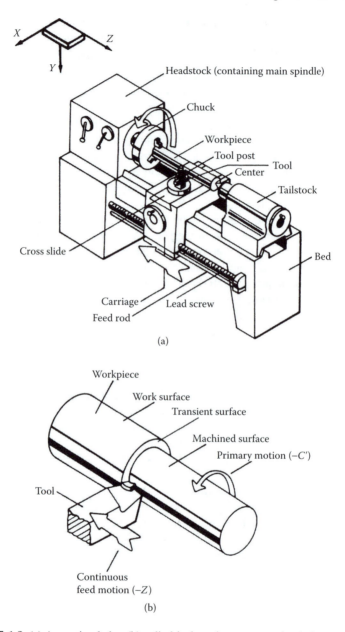

FIGURE 1.3 (a) An engine lathe; (b) cylindrical turning on an engine lathe.

in the machine structure. The resulting loss of accuracy will necessitate a further machining operation at low feed (finish cut) to bring the workpiece diameter within the limits specified and provide a smooth machined surface. For these reasons, it is usual to machine the workpiece oversize deliberately during the roughing cut, leaving a small amount of material that will subsequently be removed during the finishing cut.

Before discussing the operation of the lathe further, it will be helpful to describe some of the features of the single-point tool used in the lathe and in several other types of machine tools.

1.3.2 SINGLE-POINT TOOLS

Single-point tools are cutting tools having one cutting part (or chip-producing element) and one shank. They are commonly used in lathes, turret lathes, planers, shapers, boring mills, and similar machine tools. A typical single-point tool is illustrated in Figure 1.4. The most important features are the cutting edges and adjacent surfaces. These are shown in the figure and defined as follows:

1. The face is the surface or surfaces over which the chip flows.
2. The flank is the tool surface or surfaces over which the surface produced on the workpiece passes.
3. The cutting edge is that edge of the face which is intended to perform cutting. The tool major cutting edge is that entire part of the cutting edge which is intended to be responsible for the transient surface on the workpiece. The tool minor cutting edge is the remainder of the cutting edge.
4. The corner is the relatively small portion of the cutting edge at the junction of the major and minor cutting edges; it may be curved or straight, or it may be the actual intersection of these cutting edges.

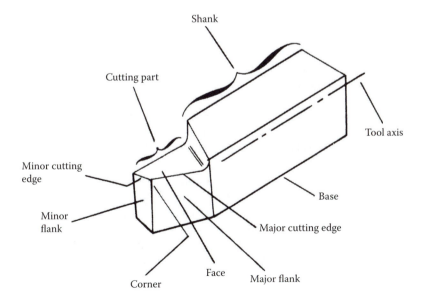

FIGURE 1.4 Typical single-point tool.

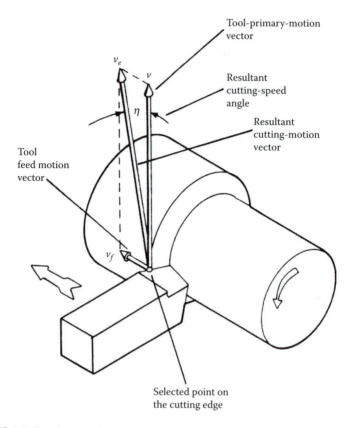

FIGURE 1.5 Resultant cutting motion in cylindrical turning.

In general, when a tool is applied to a workpiece, its motion relative to the workpiece has two components:

1. The motion resulting from the primary motion of the machine tool, which can be called the primary motion of the tool
2. The motion resulting from feed motion of the machine (Figure 1.5)

The resultant of these two tool motions is called the resultant cutting motion and is defined as the motion resulting from simultaneous primary and feed motions.

It should be noted that in machine tools where the feed is applied while the tool is not engaged with the workpiece (as in shaping or planing, for example), the resultant cutting motion is identical to the primary motion. Where the feed motion is applied continuously, the angle between the direction of primary motion and the resultant cutting direction is called the resultant cutting-speed angle η. This angle is usually extremely small and for most practical purposes can be assumed to be zero. Further, the cutting speed v, the instantaneous velocity of the primary motion of the selected point on the cutting edge relative to the workpiece, can vary along the major cutting edge. The feed speed v_f, the instantaneous velocity of the feed motion of the selected point on the cutting edge relative to the workpiece, is constant.

Finally, the resultant cutting speed v_e, the instantaneous velocity of the resultant cutting motion of the selected point on the cutting edge relative to the workpiece, is given by

$$v_e = \frac{v}{\cos \eta} \tag{1.2}$$

but since for most practical operations η is very small, it can generally be assumed that

$$v_e = v \tag{1.3}$$

One of the important tool angles when considering the geometry of a particular machining operation is the angle in Figure 1.6 called the major cutting-edge angle κ_r. The thickness of the layer of material being removed at the selected point on the cutting edge, known as the undeformed chip thickness a_c, significantly affects the power required to perform the operation. Strictly, this dimension should be measured both normal to the cutting edge and normal to the resultant cutting direction. However, for all practical purposes, since η is small, as described above, a_c, can be measured normal to the direction of primary motion; thus in Figure 1.6 and all subsequent figures, a_c will be measured this way. From Figure 1.6, therefore, a_c is given by $a_f \sin \kappa_r$, where a_f is the feed engagement, the instantaneous engagement of the tool cutting edge with the workpiece measured in the direction of feed motion. For single-point cutting operations a_f is equal to the feed f, and therefore

$$a_c = f \sin \kappa_r \tag{1.4}$$

The cross-sectional area A_c of the layer of material being removed (cross-sectional area of the uncut chip) is approximately given by

$$A_c = f a_p \tag{1.5}$$

where a_p is the back engagement, previously known as depth of cut. The back engagement is the instantaneous engagement of the tool with the workpiece, measured perpendicular to the plane containing the directions of primary and feed motion (Figure 1.5). In general the back engagement determines the depth of material removed from the workpiece in a single-point cutting operation.

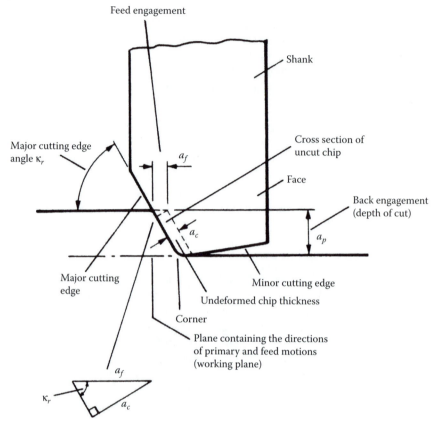

FIGURE 1.6 Single-point tool operation, where $a_c = a_f \sin_r$.

1.3.3 TYPICAL LATHE OPERATIONS

Figure 1.7 illustrates five typical lathe operations: cylindrical turning, facing, boring, external threading, and cutoff. In each case the primary motion and the feed motion, together with certain other terms and dimensions, are indicated. In any machining operation the workpiece has three important surfaces:

1. The work surface, the surface on the workpiece to be removed by machining
2. The machined surface, the desired surface produced by the action of the cutting tool
3. The transient surface, the part of the surface formed on the workpiece by the cutting edge and removed during the following cutting stroke, during the following revolution of the tool or workpiece, or, in other cases (as, for example, in a thread-turning operation) (Figure 1.7d) during the following pass of the tool

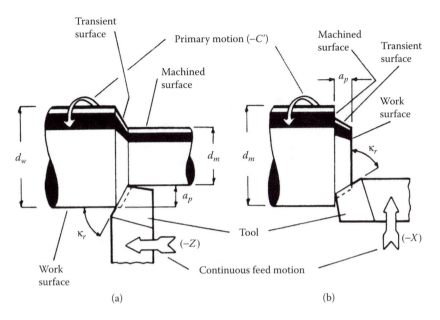

FIGURE 1.7 Lathe operations. (a) Cylindrical turning; (b) facing; (c) boring; (d) external threading; (e) parting and cutoff.

In Figure 1.7a, which shows the geometry of a cylindrical-turning operation, the cutting speed at the tool corner is given by $d_m n_w$, where n_w is the rotational frequency of the workpiece, and d_m is the diameter of the machined surface. Thus the average, or mean, cutting speed v_{av} is given by

$$v_{av} = \frac{\pi n_w \left(d_w + d_m \right)}{2} \tag{1.6}$$

The metal removal rate Z_w is the product of the mean cutting speed and the cross sectional of the uncut chip, A_c. Thus

$$Z_w = A_c v_{av}$$

$$= \frac{\pi a_f a_p n_w \left(d_w + d_m \right)}{2} \tag{1.7}$$

$$= \pi f a_p n_w \left(d_m + a_p \right)$$

The same result could have been obtained by dividing the total volume of metal removed by the machining time t_m.

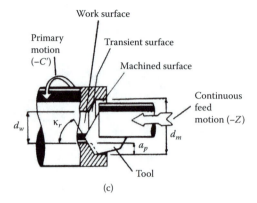

(c)

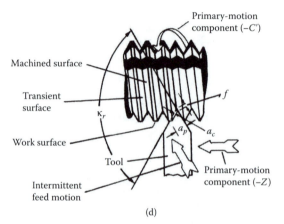

(d)

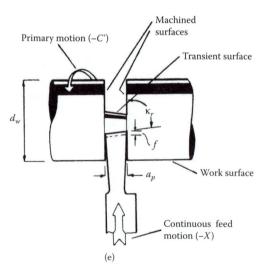

(e)

FIGURE 1.7 (continued).

For operations where the back engagement a_p is small compared to the diameter of the machined surface d_m, the metal removal rate is approximately given by

$$Z_w \cong \pi f a_p n_w d_m \qquad (1.8)$$

For a given work material machined under given conditions, the energy required to remove a unit volume of material, p_s, can be measured. This factor is mainly dependent on the work material. The power P_m required to perform any machining operation can be obtained from

$$P_m = p_s Z_w \qquad (1.9)$$

Finally, if the overall efficiency of the machine tool motor and drive systems is denoted by η_m, the electrical power P_e consumed by the machine tool is given by

$$P_e = \frac{P_m}{\eta_m} \qquad (1.10)$$

Approximate values of p_s, for various work materials and various values of the undeformed chip thickness are presented at the end of this chapter.

An operation in which a flat surface is generated by an engine lathe is shown in Figure 1.7b and can be performed by feeding a tool in a direction at right angles (-X motion) to the axis of workpiece rotation. In this operation, known as *facing*, the cutting speed at the tool corner varies from a maximum at the beginning of the cut to zero when the tool reaches the center of the workpiece. Clearly this operation can only be performed when the workpiece does not require support by the tailstock center.

The undeformed chip thickness a_c and the cross-sectional area of the uncut chip A_c are given, as in cylindrical turning, by Equation 1.4 and Equation 1.5, respectively.

The machining time t_m is given by

$$t_m = \frac{d_m}{2 f n_w} \qquad (1.11)$$

The maximum cutting speed v_{max} and the maximum metal removal rate $Z_{w_{max}}$ are given by

$$v_{max} = \pi n_w d_m \qquad (1.12)$$

$$Z_{w_{max}} = \pi f a_p n_w d_m \qquad (1.13)$$

Figure 1.7c shows an internal cylindrical surface being generated on a lathe. This operation is termed *boring* and can only be used to enlarge an existing hole in the workpiece. If the diameter of the work surface is d_w and the diameter of the machined surface is d_m, the mean cutting speed is given by Equation 1.6 and the metal-removal rate by

$$Z_w = \pi f a_p n_w \left(d_m - a_p \right) \tag{1.14}$$

Finally the machining time t_m is given by Equation 1.1 if l_w is taken as the length of the hole to be bored.

The lathe operation illustrated in Figure 1.7d is known as external threading, or screw cutting. In this operation the primary motion of the tool is considered to be a combination of the motions -C' and -Z. This motion generates a helix on the workpiece and is obtained by setting the gears that drive the lead screw to give the required pitch of the machined threads. The machining of a thread necessitates several passes of the tool along the workpiece, each pass removing a thin layer of metal from one side of the thread. The feed is applied in increments, after each pass of the tool, in a direction parallel to the machined surface. In calculating the production time, allowance must be included for the time taken to return the tool to the beginning of the cut, to increment the feed, and to engage the lathe carriage with the lead screw.

The last lathe operation to be illustrated (Figure 1.7e) is used when the finished workpiece is to be separated from the bar of material gripped in the chuck. It is known as a parting, or cutoff, operation and produces two machined surfaces simultaneously. As with a facing operation, the cutting speed and hence the metal removal rate varies from a maximum at the beginning of the cut to zero at the center of the workpiece. The machining time is given by Equation 1.11 and the maximum metal removal rate by Equation 1.13.

1.3.4 WORK AND TOOL HOLDING IN AN ENGINE LATHE

In the most common method of work holding, the chuck has either three or four jaws (Figure 1.8) and is mounted on the end of the main spindle. A three-jaw chuck is used for gripping cylindrical workpieces when the operations to be performed are such that the machined surface is concentric with the work surfaces. The jaws have a series of teeth that mesh with spiral grooves on a circular plate within the chuck. This plate can be rotated by the key inserted in the square socket, resulting in simultaneous radial motion of the jaws. Since the jaws maintain an equal distance from the chuck axis, cylindrical workpieces are automatically centered when gripped.

With the four-jaw chuck, each jaw can be adjusted independently by rotation of the radially mounted threaded screws. Although accurate mounting of a workpiece can be quite time-consuming, a four-jaw chuck is often necessary for noncylindrical workpieces.

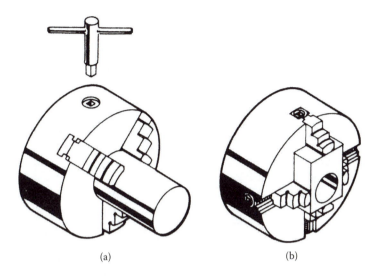

(a)   (b)

FIGURE 1.8 Lathe work holding using chucks. (a) Three-jaw (self-centering) chuck; (b) independent four-jaw chuck.

For very complicated shapes, a circular faceplate can be used. The faceplate has radial slots that provide a means of bolting the workpiece to the faceplate.

For small lathes employed extensively for work on material provided in bar form, collets are often used. These collets are effectively split sleeves that fit snugly over the workpiece and have a taper on their outer surface. Drawing the collet into a matching tapered hole in the end of the spindle has the effect of squeezing the collet and gripping the workpiece. In general collets are not usually suitable for gripping bar materials larger than 50 mm in diameter. The range of standard collets available includes those with gripping surfaces suitable for non-round bar materials, such as hexagonal and square section bar stock materials.

For accurate turning operations or in cases where the work surface is not truly cylindrical, the workpiece can be turned between centers. This form of work holding is illustrated in Figure 1.9. Initially the workpiece has a conical center hole drilled at each end to provide location for the lathe centers. Before supporting the workpiece between the centers (one in the headstock and one in the tailstock), a dog (a clamping device) is secured at the headstock end. The dog is arranged so that the tip is inserted in a slot in the drive plate mounted on the main spindle, ensuring that the workpiece will rotate with the spindle.

The simplest form of tool holder is illustrated in Figure 1.10 and is suitable for holding one single-point tool. Immediately below the tool is a curved block resting on a concave spherical surface; this method of support provides an easy way of inclining the tool so that its corner is at the correct height for the machining operation. In the figure the tool post is shown mounted on a compound rest. The rest is a small slideway that can be clamped in any angular position in the horizontal plane and is mounted on the cross slide of the lathe. The compound

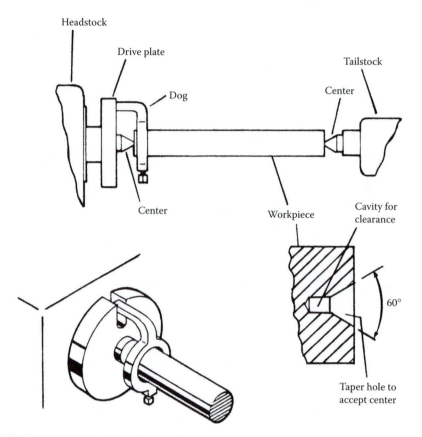

FIGURE 1.9 Work holding when turning between centers.

rest allows the tool to be hand-fed at an oblique angle to the lathe bed and is required, for example, in screw-threading operations and the machining of short tapers or chamfers.

Another common form of tool post is shown in Figure 1.11; it also would be mounted on the compound rest. This four-way tool post can, as its name suggests, accommodate as many as four cutting tools. Any cutting tool can be quickly brought into position by unlocking the tool post with the lever provided, rotating the tool post, and, finally, reclamping with the lever. Variations on this type of tool post with quick-change toolholders are commonly used and a number of proprietary tooling systems are available.

1.3.5 OTHER TYPES OF LATHES

The engine lathe is the most common machine tool used in low-production work. Several other kinds of lathes are available: for medium-production work, turret lathes are widely employed. In a turret lathe the tailstock is replaced with a rotary-indexing, hexagonal turret that can be driven along the bed by the lead screw. A

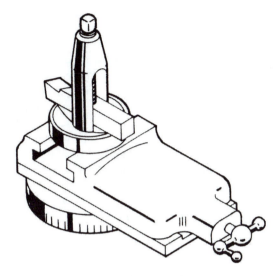

FIGURE 1.10 Typical lathe tool post and compound rest.

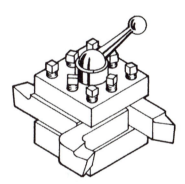

FIGURE 1.11 Four-way tool post.

variety of tools can be mounted on the faces of the turret (in some turret lathes two or more tools can be mounted on the one face) and are brought into operation by manually indexing the turret and engaging the feed drive.

Single-spindle and multi-spindle automatic lathes are suitable for high-volume or mass production of small components machined from work material in bar form. Specially machined cams control the various motions of these lathes, and the operations are completely automatic, including the gradual feeding of the workpiece through the hollow spindle to the collet. The machine needs attention only when a new bar of material is required. Items such as small screws needed in large quantities are manufactured on this type of lathe. More descriptions of multi-spindle automatic lathes and their operation can be found in Chapter 11.

A wide range of programmable numerically controlled (nc) lathes and turning centers are also available. The basic cutting motions are the same as above and additional descriptions of these machines can be found in Chapter 11.

1.3.6   VERTICAL-BORING MACHINE (VERTICAL BORER)

A horizontal-spindle lathe is not suitable for turning heavy, large-diameter work-pieces. The axis of the machine spindle would have to be so elevated that a machinist could not easily reach the tool- and work-holding devices. In addition, it would be difficult to mount the workpiece on a vertical faceplate or support it between centers; for this reason a machine that operates on the same principle as a lathe but has a vertical axis is used and is known as a vertical-boring machine (Figure 1.12). Like the lathe, this machine rotates the workpiece (motion C′) and applies continuous, linear feed motion to the tool. This motion can be either at right angles to the axis of workpiece rotation (motion X) or parallel to it (motion Z).

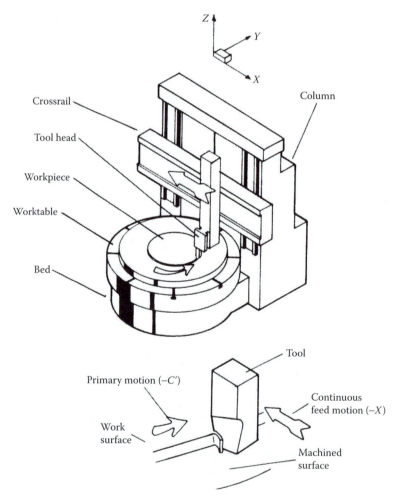

FIGURE 1.12 Facing on a vertical-boring machine.

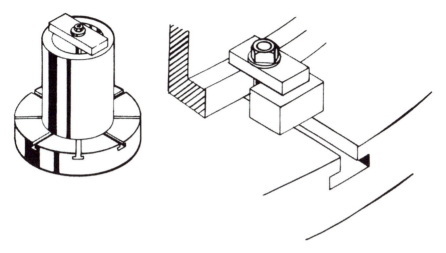

FIGURE 1.13 Work holding in vertical boring machines.

Single-point tools are employed and mounted in a toolholder similar to the four-way tool post shown in Figure 1.11 but without the rapid-indexing feature. The operations carried out are generally limited to turning (-Z motion), facing (-X motion), and boring (-Z motion). These operations were illustrated in Figure l.7a to c; the geometry described and equations developed still apply.

The horizontal work-holding surface, which facilitates the positioning of large workpieces, consists of a rotary table having radial T slots for clamping purposes. Two typical work-holding methods are illustrated in Figure 1.13.

1.3.7 HORIZONTAL-BORING MACHINE (HORIZONTAL BORER)

The last type of machine described here that uses single-point tools and has a rotary primary motion is a horizontal-boring machine (Figure 1.14). This machine is needed mostly for heavy noncylindrical workpieces in which an internal cylindrical surface is to be machined. In general, the words *horizontal* or *vertical* are used when describing a machine tool to refer to the orientation of the machine spindle that provides primary motion (main spindle). Thus, in the horizontal borer the main spindle is horizontal.

The principal feature of the machine is that the workpiece remains stationary during machining, and all the generating motions are applied to the tool. The most common machining process is boring and is shown in the figure; boring is achieved by rotating the tool, which is mounted on a boring bar connected to the spindle (motion C), and feeding the spindle, boring bar, and tool along the axis of rotation (motion Z). The machine tool motions that can be used to move the workpiece are for positioning of the workpiece and are not generally employed while machining is taking place. A facing operation is carried out by using a special toolholder (Figure 1.15) that feeds the tool radially as it rotates.

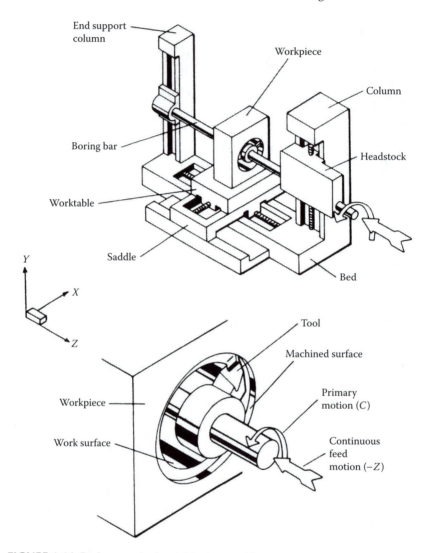

FIGURE 1.14 Boring on a horizontal-boring machine.

The worktable has T slots to facilitate clamping of the workpiece. Several common methods of using these T slots are shown in Figure 1.16.

Again, the equations developed earlier for the undeformed chip thickness, the machining time, and the metal removal rate in boring and facing apply.

1.3.8 SHAPING MACHINE (SHAPER)

The shaper is a small machine on which the primary motion is linear (Figure 1.17). The single-point tool is gripped in a toolhead mounted on the end of a ram. The ram is made to move backward and forward (X motion) either by a mechanical

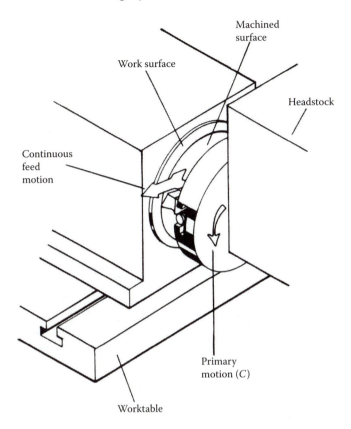

FIGURE 1.15 Facing on a horizontal-boring machine.

drive system or a hydraulic piston and cylinder. The cutting stroke is the forward stroke, and with either mechanical or hydraulic shapers, the forward ram speed is slower than the speed on the return stroke, causing the production time to be reduced as much as possible. A quick-return mechanism often used in mechanical shapers is shown in Figure 1.18. In all shapers the length of the stroke can be adjusted to suit the particular workpiece being machined. The feed is applied to the workpiece in increments at the end of the return stroke of the ram by a ratchet and-pawl mechanism driving the lead screw in the crossrail.

Shapers are most commonly used to machine flat surfaces on small components and are only suitable for low-batch quantities. For the machining of a horizontal surface (Figure 1.17), the workpiece is fed horizontally (Y motion); for vertical surfaces, the workpiece is fed vertically (Z motion).

Typical tool- and work-holding methods are illustrated in Figure 1.19, where it can be seen that the toolhead is similar to the tool post and compound rest used on a lathe. The toolhead can be rotated and clamped in various positions in the vertical plane to allow inclined surfaces to be generated by hand feeding of the tool. An essential feature of the shaper toolhead is the clapper box. Effectively the

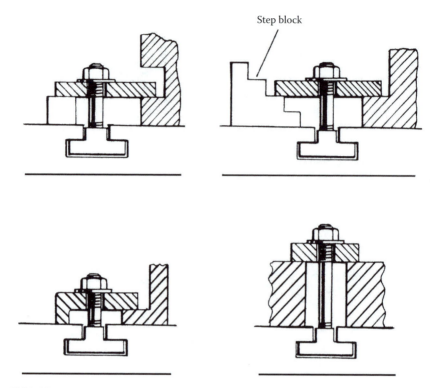

FIGURE 1.16 Work holding methods for machine tables having T slots.

tool post is pivoted on a horizontal pin and the tool is free to rotate forward, thereby lifting the tool corner during the return stroke of the ram. This action is brought about simply by the force exerted by the workpiece on the tool as the tool is dragged backward across the work surface. The workpiece is generally conveniently held in a vise bolted to the surface of a table provided with T slots for this purpose.

The geometries when shaping horizontal, vertical, and inclined flat surfaces are shown in Figure 1.20. For a surface of width b_w the machining time t_m is given by

$$t_m = \frac{b_w}{fn_r} \tag{1.15}$$

where n_r is the frequency of reciprocation or cutting strokes, and f is the feed.

The metal removal rate Z_w during cutting will be given by

$$Z_w = A_c v = fa_p v \tag{1.16}$$

where v is the cutting speed, and a_p is the back engagement.

The undeformed chip thickness a_c will be given by

$$a_c = f \sin \kappa_r \tag{1.17}$$

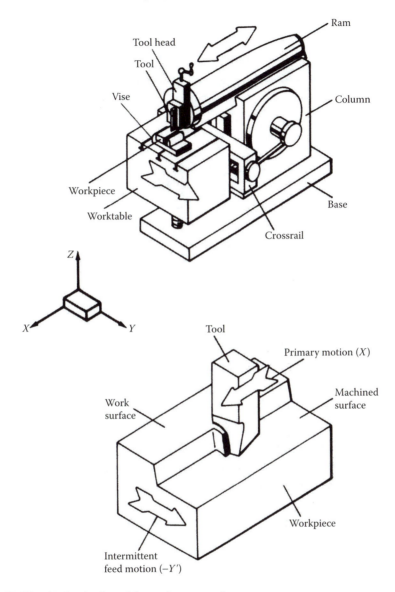

FIGURE 1.17 Production of flat surfaces on a shaper.

1.3.9 PLANING MACHINE (PLANER)

The shaper is unsuitable for generating flat surfaces on very large parts because of limitations on the stroke and overhang of the ram. This problem is solved in the planer (Figure 1.21) by applying the linear primary motion to the workpiece (motion X′) and feeding the tool at right angles to this motion (motion Y or Z). The primary motion is normally accomplished by a rack-and-pinion drive using

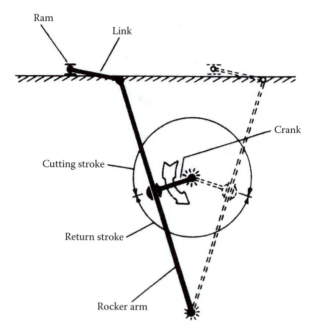

FIGURE 1.18 Quick-return mechanism for a mechanical shaper.

a variable-speed motor. As with the shaper, the tool posts are mounted on clapper boxes to prevent interference between the tools and workpiece on the return stroke, and the feed motion is intermittent. The work is held on the machine table using the T slots provided. The machining time, metal removal rate, and undeformed chip thickness can be estimated using the equations developed for the shaper.

The planer is the last of the machine tools employing single-point cutting tools to be described here. The next section describes machines designed to apply multipoint cutting tools to the workpiece.

1.4 MACHINES USING MULTIPOINT TOOLS

1.4.1 MULTIPOINT TOOLS

A multipoint tool can be regarded as a series of two or more cutting parts (chip-producing elements) secured to a common body. The majority of multipoint tools (milling cutters, drills, etc.) are intended to be rotated and have either a taper (conical) or parallel (cylindrical) shank for holding purposes or a bore through which a spindle or arbor can be inserted. The terms such as *face*, *flank*, and *cutting edges*, defined earlier for single-point tools, are still applicable, and the cutting action at a selected point on one of the cutting edges is the same.

The common multipoint tools (drills, reamers, milling cutters, and broaches) will be introduced along with descriptions of the machine tools on which they are most often used.

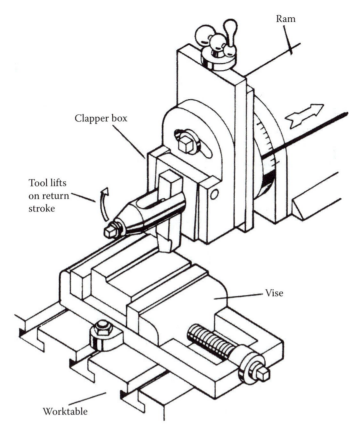

Ram

Clapper box

Tool lifts
on return
stroke

Vise

Worktable

FIGURE 1.19 Tool and work holding on a shaper.

1.4.2 DRILLING MACHINE (DRILL PRESS)

A drill press (Figure 1.22) can perform only those operations where the tool is rotated (motion C) and fed along its axis of rotation (motion Z). The workpiece always remains stationary during the machining process. On many drill presses the tool is fed by the manual operation of a lever to the right of the head. Both the worktable and the head can be raised and lowered to accommodate workpieces of different heights.

The most common operation performed on this machine is drilling with a twist drill to generate an internal cylindrical surface. A twist drill with a taper shank, together with the geometry of the drilling operation, is shown in Figure 1.23. This tool has two cutting edges, each of which is expected to remove its share of the work material. Thus the feed per tooth (the depth of material removed by each tooth and measured parallel to the direction of feed motion) is the feed engagement a_f and is equal to ½ of the feed f. By geometry therefore

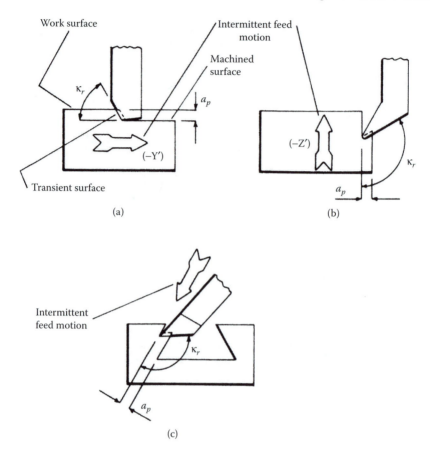

FIGURE 1.20 Shaping operations. (a) Horizontal surface; (b) vertical surface; (c) inclined surface.

$$a_c = \frac{f}{2} \sin \kappa_r \qquad (1.18)$$

where κ_r is the major cutting-edge angle.

The machining time t_m is given by

$$t_m = \frac{l_w}{fn_t} \qquad (1.19)$$

where l_w is the length of the hole produced, and n_t is the rotational frequency of the tool. For blind holes l_w is the depth of holes, but for through holes an allowance must be made for the drill point to clear the exit side of the workpiece. In this case l_w is given by

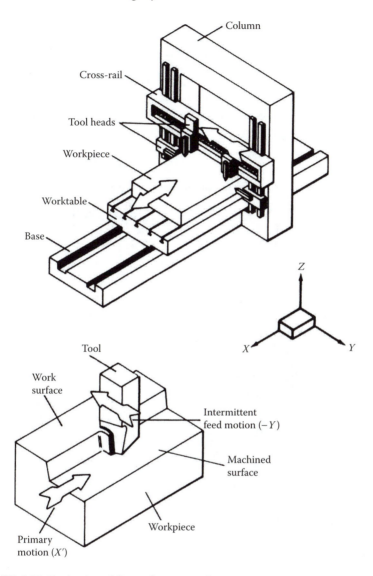

FIGURE 1.21 Production of flat surfaces on a planer.

$$l_w = t_w + \frac{a_p}{2 \tan \kappa_r} = t_w + \frac{d_m}{2 \tan \kappa_r} \qquad (1.20)$$

where

t_w = thickness of the workpiece

a_p = the work engagement (in this case equal to the drill diameter, d_m)

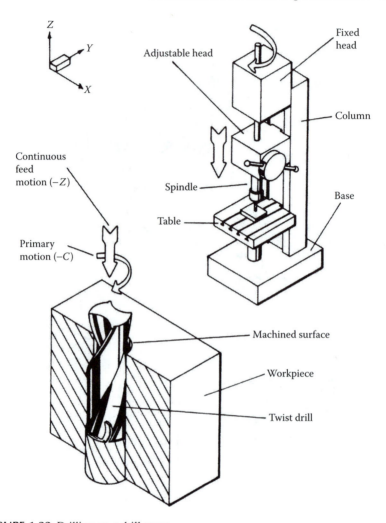

FIGURE 1.22 Drilling on a drill press.

The metal removal rate Z_w may be obtained by multiplying the feed speed v_f by the cross-sectional area of the hole produced. Thus

$$Z_w = \frac{\pi}{4} d_m^2 v_f = \frac{\pi f d_m^2 n_t}{4} \tag{1.21}$$

where d_m is the diameter of the machined surface. If an existing hole of diameter d_w is being enlarged,

$$Z_w = \frac{\pi f \left(d_m^2 - d_w^2 \right) n_t}{4} \tag{1.22}$$

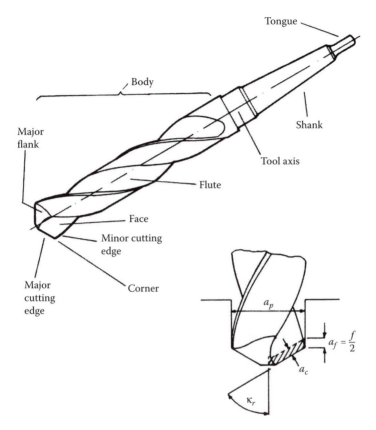

FIGURE 1.23 Twist drill, where $a_c = a_f \sin \kappa_r$.

Clearly the cutting speed will be a maximum at either corner (the outer radius of the major cutting edges) and will be zero at the tip of the drill, which is in the form of a short chisel edge. In drilling a new hole, this chisel edge must force its way through the work material, effectively pushing the material outward to be removed by the cutting edges. The poor cutting conditions in this region have little effect on the quality of the hole produced, which is mainly determined by the conditions at the minor cutting edges. The chips removed by the cutting edges take a helical form and travel up the drill flutes.

Twist drills are usually considered suitable for machining holes having a length less than five times their diameter. Special drills requiring special drilling machines are available for drilling deeper holes and are not described here.

The workpiece is often held in a vise bolted to the machine worktable. The drilling of a concentric hole in a cylindrical workpiece, however, is often carried out on an engine lathe with the drill held either in the tailstock (for hand feeding) or in a special drill holder mounted on the carriage.

Large twist drills are usually provided with a taper shank, as shown in Figure 1.23. This shank is designed to be inserted in a corresponding taper hole

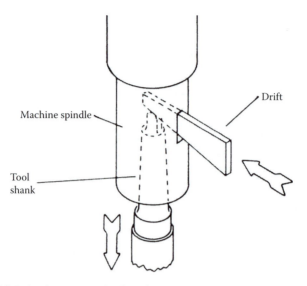

FIGURE 1.24 Releasing a taper-shank tool.

in the end of the machine spindle. The tongue on the end of the shank is inserted in a slot at the bottom of the hole in the spindle and prevents rotation of the drill relative to the spindle during machining. The drill is removed by tapping a wedge (drift) into the slot in the machine spindle, as illustrated in Figure 1.24.

Small twist drills have a parallel shank and are held in a three-jaw chuck of the familiar type used in hand drills. These chucks are provided with a taper shank for location in the drill-press spindle or in the tailstock of a lathe.

Several other machining operations can be performed on a drill press, and some of the more common ones are illustrated in Figure 1.25. The center-drilling operation produces a shallow, conical hole with clearance at the bottom. This center hole can provide a location for a lathe center when the workpiece is subsequently to be supported between centers (Figure 1.9) or can provide a guide for a subsequent drilling operation to prevent the drill point from "wandering" as the hole is started. The reaming operation is intended for finishing a previously drilled hole. The reamer is similar to a drill but has several cutting edges and straight flutes. It is intended to remove a small amount of work material only, but it considerably improves the accuracy and surface finish of a hole. The spot-facing operation is designed to provide a flat surface around the end of a hole and perpendicular to its axis; this flat surface provides a suitable seating for a washer and nut.

For large workpieces a radial-arm drilling machine is used (Figure 1.26). In this machine the drilling head and motor can be positioned along an arm that is free to swing in a horizontal plane about the column, allowing large areas to be covered. The radial-arm drilling machine is particularly suitable for drilling large numbers of holes in heavy workpieces.

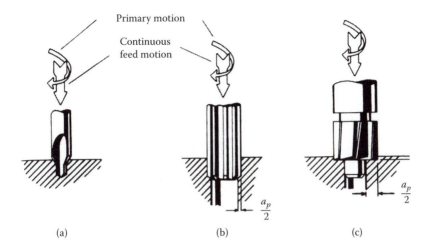

Primary motion

Continuous
feed motion

$\frac{a_p}{2}$

$\frac{a_p}{2}$

(a)   (b)   (c)

FIGURE 1.25 Some drill-press operations. (a) Center drilling; (b) reaming; (c) spot-facing.

1.4.3 HORIZONTAL-MILLING MACHINE (HORIZONTAL MILLER)

There are two main types of milling machines: horizontal and vertical. These words again refer to the orientation of the main spindle. In the horizontal-milling machine shown in Figure 1.27, the milling cutter is mounted on a horizontal arbor driven by the main spindle. The tools are therefore rotated (motion C) and the work fed continuously (motions X' or Y').

The simplest operation, slab milling, is used to generate a horizontal surface on the workpiece, as shown in Figure 1.27. The figure shows the conventional slab-milling operation; if the workpiece had been fed in the opposite direction, it would tend to climb onto the work surface, and this type of slab milling is called *climb* milling. Some evidence suggests that the forces and power consumption are less in climb milling than in conventional milling. However, high rigidity of the machine tool and work-and tool-holding devices is required for this operation.

The geometry in conventional slab milling is shown in Figure 1.28. The feed *f* which is equal to the distance moved by the workpiece during one revolution of the tool, is given by

$$f = \frac{v_f}{n_t} \tag{1.23}$$

where v_f is the feed speed of the workpiece, and n_t is the rotational frequency of the cutter. The feed engagement a_f, which is equal to the thickness of the chip removed by one tooth and measured parallel to the feed direction (feed per tooth), is given by f/N, where N is the number of teeth on the cutter. The maximum undeformed chip thickness, $a_{c_{\max}}$ (measured normal to the direction of primary motion) is therefore given by

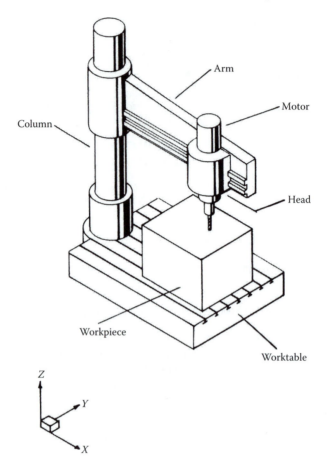

FIGURE 1.26 Radial-arm drill.

$$a_{c_{max}} = \frac{v_f \sin \theta}{N n_t} \tag{1.24}$$

where θ is given by (Figure 1.28).

$$\cos \theta = 1 - \frac{2a_e}{d_t} \tag{1.25}$$

where d_t is the diameter of the cutter, and a_e is the working engagement (the instantaneous engagement of the tool with the workpiece), measured in the plane containing the directions of primary and feed motions and perpendicular to the feed motion. In slab milling the working engagement is known as depth of cut.

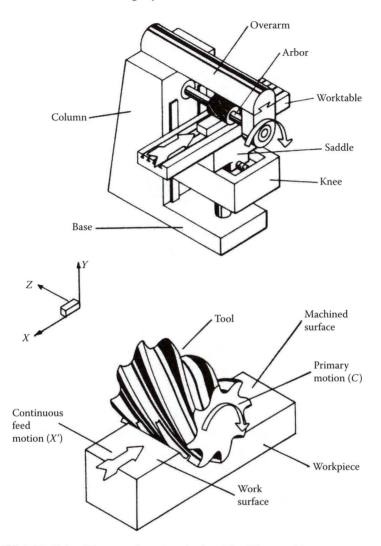

FIGURE 1.27 Slab milling on a knee-type horizontal-milling machine.

Now

$$\sin \theta = \sqrt{1 - \cos^2 \theta} = 2\sqrt{\frac{a_e}{d_t} - \left(\frac{a_e}{d_t}\right)^2} \qquad (1.26)$$

and substitution of Equation 1.26 in Equation 1.24 gives

$$a_{c_{max}} = \frac{2v_f}{Nn_t}\sqrt{\frac{a_e}{d_t}\left(1 - \frac{a_e}{d_t}\right)} \qquad (1.27)$$

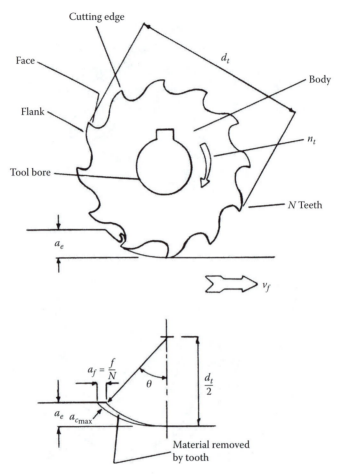

FIGURE 1.28 Geometry of a slab-milling operation, where $a_{c_{max}} = (2v_f/Nn_t) \sqrt{a_c/d_t}$.

Finally, if a_e/d_t is small,

$$a_{c_{max}} = \frac{2v_f}{Nn_t} \sqrt{\frac{a_e}{d_t}} \qquad (1.28)$$

When estimating the machining time t_m in a milling operation, it should be remembered that the distance traveled by the cutter would be larger than the length of the workpiece. This extended distance is illustrated in Figure 1.29 in which it can be seen that the cutter travel distance is given by $l_w + \sqrt{a_e(d_t - a_e)}$ where l_w is the length of the workpiece. Thus the machining time is given by

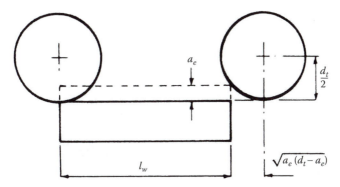

FIGURE 1.29 Relative motion between a slab-milling cutter and the workpiece during machining time.

$$t_m = \frac{l_w + \sqrt{a_e\left(d_t - a_e\right)}}{v_f} \tag{1.29}$$

The metal removal rate Z_w will be equal to the product of the feed speed and the cross-sectional area of the metal removed, measured in the direction of feed motion. Thus, since the back engagement a_p is equal to the workpiece width,

$$Z_w = a_c a_p v_f \tag{1.30}$$

Figure 1.30 shows further horizontal-milling operations. In form cutting, the special cutter has edges shaped to form the cross section required on the workpiece. Because these cutters are generally expensive to manufacture, form milling is used only when the quantity of production is sufficiently large. In slotting, a standard cutter with minor cutting edges at each end of the major cutting edges is used to produce a rectangular slot in a workpiece. Similarly, in angular milling a standard cutter machines a triangular slot. The straddle-milling operation shown in the figure is only one of an infinite variety of operations that can be carried out by mounting more than one cutter on the machine arbor. In this way combinations of cutters can machine a wide variety of cross-sectional shapes. When cutters are used in combination, the operation is often called *gang milling*.

In horizontal milling the tools are mounted on a spindle or arbor (Figure 1.31) and constrained by a key. The arbor has a tapered section at one end that is designed to fit into a corresponding tapered bore or socket in the end of the main spindle. A drawbar passed through the hollow main spindle from the rear of the machine is screwed into the end of the arbor, which results in the arbor being drawn tightly into the spindle socket. Sleeves are used as spacers to locate the tool in its required axial position on the arbor. The free end of the arbor is supported in a bearing mounted at the end of the overarm, as illustrated in Figure 1.27.

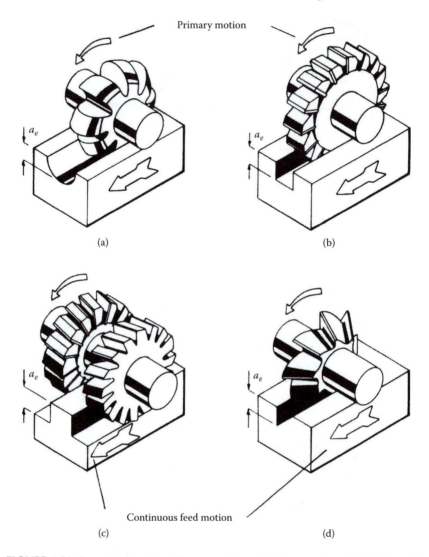

FIGURE 1.30 Some horizontal milling operations. (a) Form cutting; (b) slotting; (c) straddle milling; (d) angular milling.

Work holding is accomplished by either using a machine vise bolted to the worktable or by bolting the workpiece directly onto the worktable using the T slots provided.

1.4.4 VERTICAL-MILLING MACHINE (VERTICAL MILLER)

A wide variety of operations involving the machining of horizontal, vertical, and inclined surfaces can be performed on a vertical-milling machine. Vertical milling machines are very versatile machines. A variety of end milling tools can be used

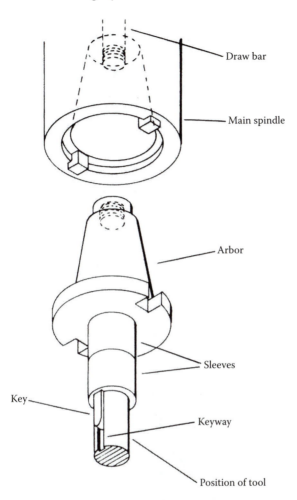

Draw bar

Main spindle

Arbor

Sleeves

Key

Keyway

Position of tool

FIGURE 1.31 Tool holding in horizontal milling.

to machine slots, pockets, peripheral steps, and profiles. In addition drill, tapping, reaming, and boring operations can be carried out. As the name of the machine implies, the spindle is vertical. In the knee-type machine illustrated in Figure 1.32 the workpiece can be fed either:

1. Along the vertical axis (Z' motion) by raising or lowering the knee
2. Along a horizontal axis (Y' motion) by moving the saddle along the knee
3. Along a horizontal axis (X' motion) by moving the table across the saddle

In larger vertical-milling machines the saddle is mounted directly on the bed, and relative motion between the tool and workpiece along the vertical axis is achieved by motion of the head up or down the column (Z motion); these machines are called bed-type, vertical-milling machines.

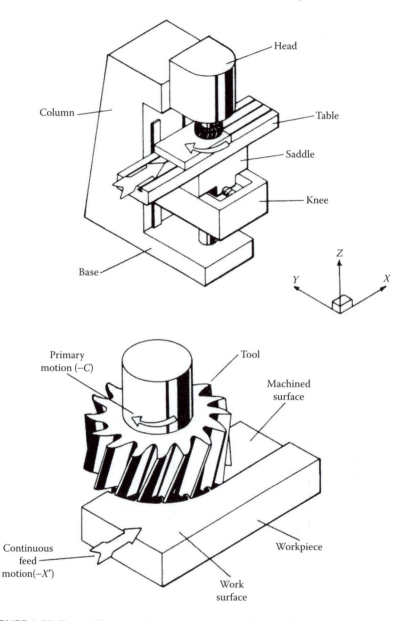

FIGURE 1.32 Face milling on a knee-type vertical milling machine.

A typical face-milling operation, where a horizontal flat surface is being machined, is shown in Figure 1.32. The cutter employed, known as a face-milling cutter, is shown in Figure 1.33, which also illustrates the geometry of the operation.

The feed f is the distance the cutter advances across the workpiece during one revolution. Thus

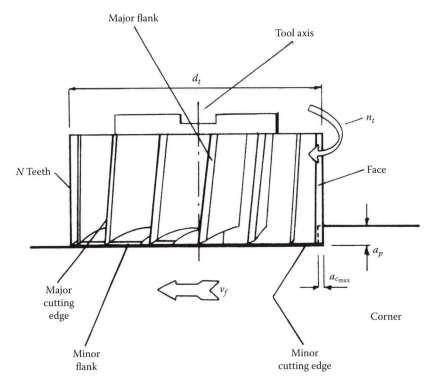

FIGURE 1.33 Geometry of face milling, where $a_{c_{max}} = (v_f/Nn_t)$.

$$f = \frac{v_f}{n_t} \qquad (1.31)$$

where v_f is the feed speed, and n_t is the rotational speed of the cutter.

If the tool axis passes over the workpiece, the undeformed chip thickness increases to a maximum value and then decreases during the time each tooth is engaged with the workpiece; its maximum value, $a_{c_{max}}$, is equal to the feed engagement, which is equal to f/N where N is the number of teeth on the cutter. Thus

$$a_{c_{max}} = \frac{v_f}{Nn_t} \qquad (1.32)$$

In estimating the machining time t_m allowance should again be made for the additional relative motion between the cutting tool and workpiece. As can be seen in Figure 1.34, the total motion when the path of the tool axis passes over the workpiece is given by $(l_w + d_t)$, and therefore the machining time is given by

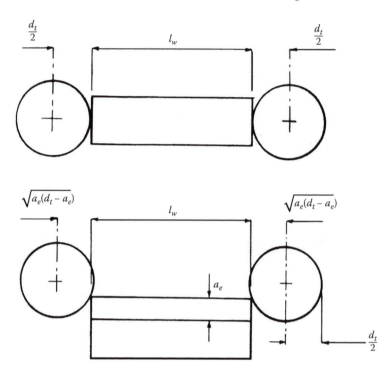

FIGURE 1.34 Relative motion between face-milling cutter and the workpiece during machining time.

$$t_m = \frac{\left(l_w + d_t\right)}{v_f} \qquad (1.33)$$

where l_w is the length of the workpiece, and d_t is the diameter of the cutter. When the path of the tool axis does not pass over the workpiece,

$$t_m = \frac{l_w + 2\sqrt{a_e\left(d_t - a_e\right)}}{v_f} \qquad (1.34)$$

where a_e is the working engagement. In this case the operation is similar to slab milling with a large working engagement, and the maximum value of the unde-formed chip thickness is given by Equation 1.26.

The metal removal rate Z_w in both cases is given by Equation 1.29. A variety of vertical-milling machine operations is illustrated in Figure 1.35. It can be seen that in one pass of the tool, several combinations of machined surfaces can be produced.

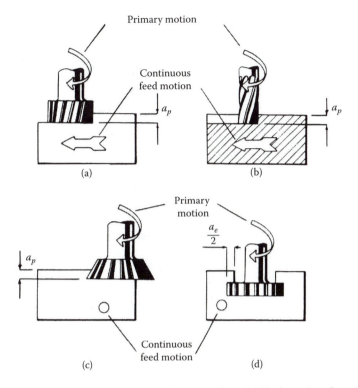

FIGURE 1.35 Some vertical-milling machine operations. (a) Horizontal surface; (b) slot; (c) dovetail; (d) T slot.

Milling cutters for vertical-milling machines generally have either a bore or a straight shank. Those having a bore are called shell end mills and are secured to an arbor (Figure 1.36) held in a socket in the machine spindle with a draw bar. Those having a straight shank are either gripped in a chuck or held in the spindle by a screw bearing on a flat surface machined into the shank.

Work holding is again accomplished by a machine vise or by using the T slots in the machine table.

1.4.5 BROACHING MACHINE (BROACHER)

The last machine using multipoint tools to be described here is the broaching machine. A vertical broaching machine suitable for machining shaped slots in the workpiece is shown in Figure 1.37. In broaching, the machine provides the primary motion (usually hydraulically powered) between the tool and workpiece (motion X), and the feed is provided by staggering of the teeth on the broach, each tooth removing a thin layer of material. Since the machined surface is usually produced during one pass of the tool, the machining time t_m is given by

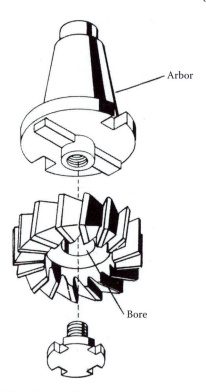

FIGURE 1.36 Tool holding for a shell end mill.

$$t_m = \frac{l_t}{v} \tag{1.34}$$

where l_t is the length of the broach, and v is the cutting speed. The feed f in broaching is defined as the motion an imaginary single cutting edge would have to be given by the machine tool to produce the same result as the array of cutting edges with which the tool or machine is actually provided. The undeformed chip thickness a_c at a selected point on the cutting edge is therefore equal to the feed or the feed engagement; thus

$$a_c = a_f = f \tag{1.36}$$

The average metal removal rate Z_w can be estimated by dividing the total volume of metal removed by the machining time.

Broaching is widely used to produce noncircular holes. In these cases the broach can be either pulled or pushed through a circular hole to enlarge the hole

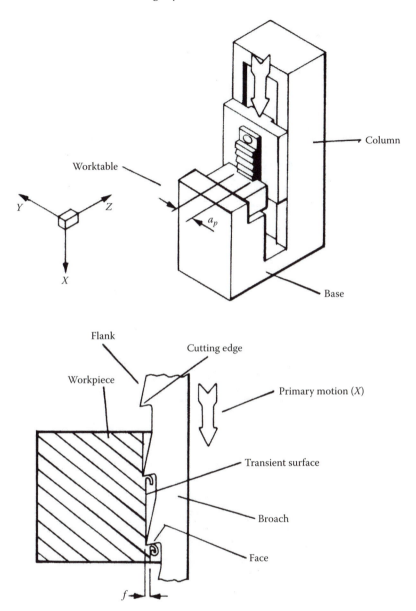

FIGURE 1.37 Broaching on a vertical-broaching machine, where $a_c = f$.

to the shape required or to machine a keyway, for example (Figure 1.38). Broaches must be designed individually for the particular job and are expensive to manufacture. It follows that broaching can be justified only when a very large batch of components (100,000 to 200,000) are to be machined.

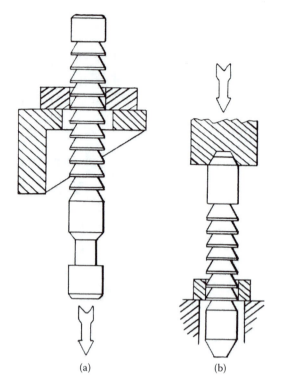

(a)   (b)

FIGURE 1.38 Methods of broaching a hole. (a) Pull broach; (b) push broach.

1.4.6 TAPS AND DIES

The production of internal and external screw threads can be accomplished by the use of taps and dies. These multipoint tools can be thought of as helical broaches, and because the amount of material removed is quite small, they can often be operated by hand.

In Figure 1.39 a tap is fed into a prepared hole and rotated at low speed. The relative motion between a selected point on a cutting edge and the workpiece is therefore helical; this motion is the primary motion. All the machining is done by the lower end of the tap, where each cutting edge removes a small layer of metal (Figure 1.39, inset) to form the thread shape; the fully shaped thread on the tap serves to clear away fragments of chips which may collect. A die has the same cutting action as a tap but is designed to produce an external thread.

Internal threading using taps can be carried out on turret lathes, drill presses, and vertical milling machines. External threading using dies can be carried out on turret lathes and special screw-cutting machines.

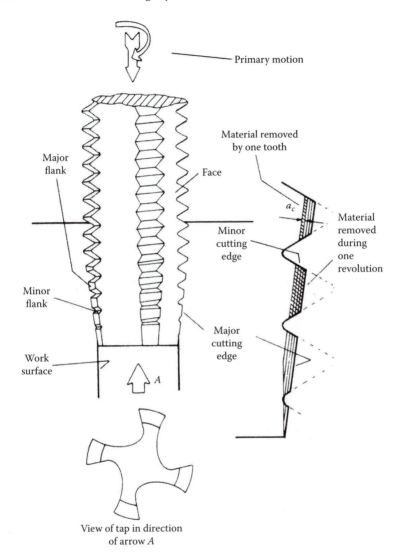

Primary motion

Material removed
by one tooth

Major
flank

Face

a_c

Material
removed
during
one
revolution

Minor
cutting
edge

Minor
flank

Major
cutting
edge

Work
surface

A

View of tap in direction
of arrow A

FIGURE 1.39 Tapping.

1.5 MACHINES USING ABRASIVE WHEELS

1.5.1 Abrasive Wheels

Abrasive wheels (or grinding wheels) are generally cylindrical, disc-shaped, or cup-shaped (Figure 1.40). The machines on which they are used are called grinding machines, or grinders; they all have a spindle that can be rotated at high

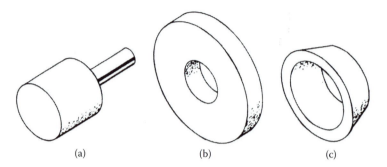

(a)   (b)   (c)

FIGURE 1.40 Common shapes of abrasive wheels. (a) Cylindrical; (b) disk; (c) cup.

speed and on which the grinding wheel is mounted. The spindle is supported by bearings and mounted in a housing; this assembly is called the *wheel head*. A belt drive from an electric motor usually provides power for the spindle. The abrasive wheel consists of individual grains of very hard material (usually silicon carbide or aluminum oxide) bonded in the form required. More discussion on abrasive machining processes can be found in Chapter 10.

Abrasive wheels are sometimes used in rough grinding where material removal is the important factor; more commonly, abrasive wheels are used in finishing operations where the resulting surface finish is the criterion.

In the metal-cutting machine tools described earlier, generation of a surface is usually obtained by applying a primary motion to either the tool or workpiece and a feed motion to either the tool or the workpiece. In grinding machines, however, the primary motion is always the rotation of the abrasive wheels, but often two or more generating (feed) motions are applied to the workpiece to produce the desired surface shape.

It is difficult to estimate the undeformed chip thickness in grinding because the individual grains in the wheel are so irregular. As a rough guide, values of 0.00025 to 0.025 mm (0.00001 to 0.001 in.) can be assumed if it is remembered that a closely packed grain structure, small grains, a high cutting speed, a low feed or feed rate, and a small depth of cut will all tend to result in a low value of the undeformed chip thickness. Further, because of the relatively inefficient cutting conditions in grinding, the specific cutting energy p_s is considerably higher in grinding than in metal-cutting processes.

1.5.2 HORIZONTAL-SPINDLE SURFACE-GRINDING MACHINE (HORIZONTAL-SPINDLE SURFACE GRINDER)

The horizontal-spindle surface grinder (Figure 1.41) has a horizontal spindle that provides primary motion to the wheel (motion C). The principal feed motion is the reciprocation of the worktable on which the work is mounted (motion X′); this motion is known as the traverse and is hydraulically operated. Further feed motions may be applied either to the wheel head, by moving it down the column (motion -Y) (known as *infeed*), or to the table, by moving it parallel to the machine

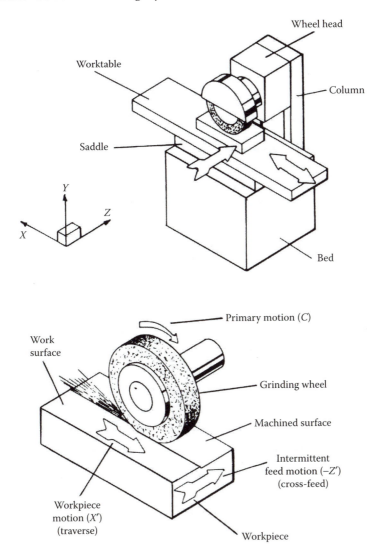

FIGURE 1.41 Surface grinding on a horizontal-spindle surface grinder.

spindle (motion Z′) (known as *cross-feed*). In Figure 1.41 a horizontal surface is being generated on a workpiece by a cross-feed motion (motion -Z′). This feed motion, which is intermittent, is usually hydraulically operated and applied after each stroke or pass of the table. The amount of cross-feed f may therefore be defined as the distance the tool advances across the workpiece between each cutting stroke. The operation is known as *traverse grinding*.

Figure 1.42 shows the geometries of both traverse grinding and plunge grinding on a horizontal-surface grinder. From Figure 1.42a, the metal removal rate in traverse grinding is given by

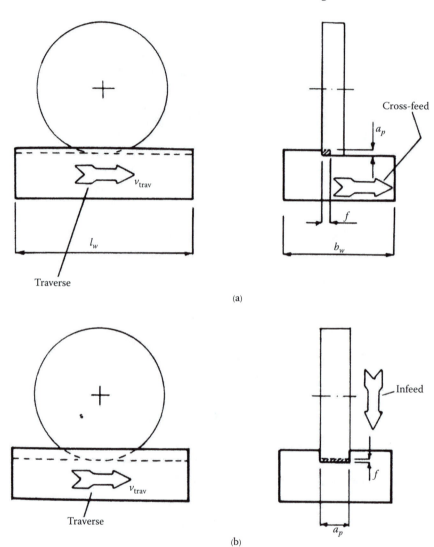

FIGURE 1.42 Horizontal-spindle surface-grinding operations. (a) Traverse grinding; (b) plunge grinding.

$$Z_w = f a_p v_{trav} \tag{1.37}$$

where

f = cross-feed per cutting stroke
a_p = back engagement
v_{trav} = traverse speed

The machining time t_m is given by

$$t_m = \frac{b_w}{2 f n_r} \tag{1.38}$$

where n_r is the frequency of reciprocation, and b_w is the width of the workpiece. The frequency of reciprocation depends on the traverse speed setting and on the stroke setting. Typical relationships between these parameters will be presented later. In a similar way, for the plunge-grinding operation (Figure 1.42b), the metal removal rate is given by Equation (1.37).

Before estimating the machining time in the plunge-grinding operation, it is necessary to describe a phenomenon known as *sparking-out*. In any grinding operation where the wheel is fed in a direction normal to the work surface (infeed), the feed *f*, which is the depth of the layer of material removed during one cutting stroke, will initially be less than the nominal feed setting on the machine. This feed differential results from the deflection of the machine tool elements and workpiece under the forces generated during the operation. Thus, on completion of the theoretical number of cutting strokes required, some work material will still have to be removed. The operation of removing this material, called sparking-out, is achieved by continuing the original grinding operation with no further application of feed until metal removal becomes insignificant (no further sparks appear). If the time for sparking-out is denoted by t_s, the machining time in plunge grinding is given by

$$t_m = \frac{a_t}{2 f n_r} + t_s \tag{1.39}$$

where a_t is the total depth of work material to be removed.

Typical tool and work holding in a surface-grinding machine are illustrated in Figure 1.43. Tool holding is generally accomplished by mounting the wheel on the end of the machine spindle between two flanges. Work holding is often achieved by use of a magnetic vise controlled by a lever that can be placed in the "on" or "off" position. When "off," the magnetic field is short-circuited. Motion of the lever to the "on" position slides the magnet into a position where the magnetic field passes through the workpiece. Clearly, this type of vise is suitable only for workpieces of ferromagnetic material and for finishing operations where the forces generated during grinding are small. In other situations, a mechanical vise or other direct clamping means would be employed.

1.5.3 Vertical-Spindle Surface-Grinding Machine (Vertical-Spindle Surface Grinder)

The vertical-spindle surface grinder (Figure 1.44) employs a cup-shaped abrasive wheel and performs an operation similar to face milling. The worktable is reciprocated (X motion) and the tool fed intermittently downward (-Z motion); these

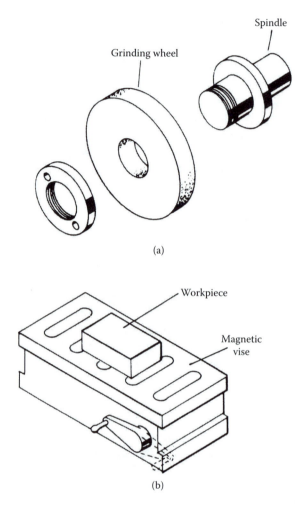

(a)

(b)

FIGURE 1.43 Tool and work holding in surface grinding. (a) Tool holding; (b) work holding.

motions are known as *traverse* and *infeed*, respectively. A horizontal surface is generated on the workpiece, and because of the deflection of the machine structure, the feed f will initially be less than the feed setting on the machine tool. This means that sparking-out is necessary, as in plunge grinding on a horizontal-spindle machine.

The metal removal rate is given by

$$Z_w = f a_p v_{trav} \qquad (1.40)$$

where a_p, the back engagement, is equal to the width of the workpiece, and v_{trav} is the traverse speed.

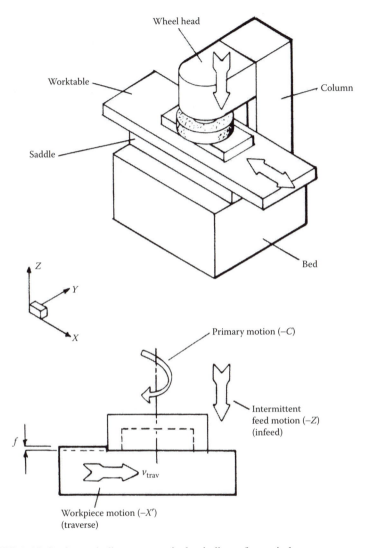

FIGURE 1.44 Surface grinding on a vertical-spindle surface grinder.

The machining time is given, as in plunge grinding on a horizontal-spindle machine, by Equation 1.39.

For vertical-spindle surface grinders with a horizontal, rotary worktable on which several workpieces can be mounted the machining time is given by

$$t_m = \frac{a_t}{fn_w} + t_s \qquad (1.41)$$

where n_w is the rotational frequency of the worktable.

1.5.4 CYLINDRICAL-GRINDING MACHINE (CYLINDRICAL GRINDER)

In the cylindrical-grinding machine (Figure 1.45) the workpiece is supported and rotated between centers. The headstock provides the low-speed rotational drive (C′ motion) to the workpiece and is mounted, together with the tailstock, on a worktable that is reciprocated horizontally (Z′ motion) using a hydraulic drive. The grinding-wheel spindle is horizontal and parallel to the axis of workpiece rotation, and horizontal, hydraulic feed can be applied to the wheel head in a direction normal to the axis of workpiece rotation (-X motion); this motion is known as infeed.

Figure 1.45 shows a cylindrical surface being generated using the traverse motion; an operation that can be likened to cylindrical turning where the single-point cutting tool is replaced by a grinding wheel. In fact, grinding attachments are available that allow this operation to be performed on an engine lathe.

The geometries of traverse and plunge grinding on a cylindrical grinder are shown in Figure 1.46. In traverse grinding, the maximum metal removal rate is closely given by

$$Z_{w_{max}} \cong \pi f d_w v_{trav} \tag{1.42}$$

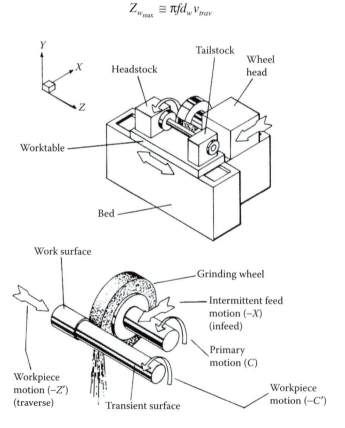

FIGURE 1.45 Cylindrical grinding.

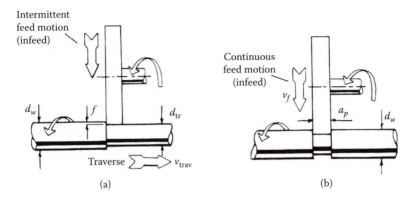

FIGURE 1.46 Cylindrical-grinding operations. (a) Traverse grinding; (b) plunge grinding.

where

d_w = diameter of the work surface

v_{trav} = traverse speed

f = feed per stroke of the machine table (usually extremely small compared to d_w)

The machining time will be given by Equation 1.39.

In the plunge-grinding operation shown in Figure 1.46b, the wheel is fed into the workpiece, without traverse motion applied, to form a groove. If v_f is the feed speed of the grinding wheel, d_w the diameter of the work surface, and a_p the back engagement (the width of the grinding wheel), the maximum metal removal rate is given by

$$Z_{w_{max}} = \pi f d_m v_f \tag{1.43}$$

and the machining time will be

$$t_m = \frac{a_t}{v_f} + t_s \tag{1.44}$$

where a_t is the total depth of material to be removed, and t_s is the sparking-out time.

1.5.5 INTERNAL-GRINDING MACHINE (INTERNAL GRINDER)

The last machine to be described here is the internal grinder (Figure 1.47), which is designed to produce an internal cylindrical surface. The wheel head supports a horizontal spindle and can be reciprocated (traversed) in a direction parallel to

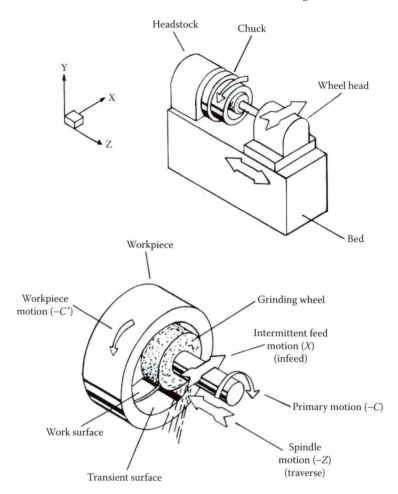

FIGURE 1.47 Internal grinding.

the spindle axis (Z motion). A small cylindrical grinding wheel is used and is rotated at very high speed. The workpiece is mounted in a chuck or on a magnetic faceplate and rotated (motion C′). Horizontal feed is applied to the wheel head in a direction normal to the wheel spindle (motion X); this motion is known as in feed. Again, traverse and plunge grinding can be performed, the geometries of which are shown in Figure 1.48.

Traverse grinding is shown in Figure 1.48a, and the maximum removal rate, which occurs at the end of the operation, is given by

$$Z_{w_{max}} = \pi f d_m v_{trav} \qquad (1.45)$$

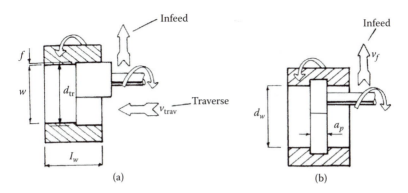

FIGURE 1.48 Internal-grinding operations. (a) Transverse grinding; (b) plunge grinding.

where

f = feed

v_{trav} = traverse speed

d_m = diameter of the machined surface

The machining time is again given by Equation 1.38.

Finally, in plunge grinding (Figure 1.48b) the maximum removal rate is given by

$$Z_{w_{max}} = \pi a_p d\, v_f \tag{1.46}$$

and the machining time in Equation 1.44.

1.5.6 CENTERLESS GRINDING MACHINES

A special class of cylindrical grinding machines exist called *centerless grinding machines*. These machines are used mainly for machining the outside diameters of short cylindrical parts, but an adaptation of the process can be used for machining the inside diameters of tubular parts. Figure 1.49 shows the general configuration of the external centerless grinding process. As the name implies the workpiece is not held between centers as is generally the case for conventional cylindrical grinding. Two wheels contact the workpiece, while being supported by a rest blade situated between the two wheels. Material is removed by the grinding wheel, which rotates at similar speeds to conventional cylindrical grinding. The regulating wheel, which is usually rubber bonded, rotates at slow speeds and causes the workpiece to rotate slowly so that the whole of the outside cylindrical surface is machined.

The axis of the regulating wheel is inclined at a small angle, α_i, to the axis of the grinding wheel. This induces an axial feed velocity, v_f, onto the part and the magnitude of v_f is given by

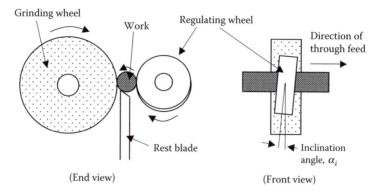

FIGURE 1.49 External centerless grinding.

$$v_f = \pi d_r n_r \sin \alpha_i \qquad (1.47)$$

where

d_r = the diameter of the regulating wheel
n_r = the rotational frequency of the regulating wheel
α_i = the inclination angle

In this manner the part is fed automatically through the grinding process. This is particularly useful for the continuous machining of short cylindrical parts fed by gravity feed track into the process and then out the other side. Parts such as rollers for bearings and piston pins are final ground in this way, with high productivity.

The time to machine each part is given by

$$t_m = \frac{l_w}{v_f} \qquad (1.48)$$

where l_w is the length of the workpiece.

Centerless grinding can also be used for internal grinding, and the typical configuration is shown in Figure 1.50. In this case the workpiece is supported by two supporting rolls, with the grinding wheel machining the inside diameter and the regulating wheel contacting the outside diameter of the part. Automatic through feed is not achievable with internal centerless grinding because of the need to support the workpiece. However, the main advantage is the close concentricity achieved between the inner and external surfaces and this is particularly useful for some tubular parts such as roller bearing races.

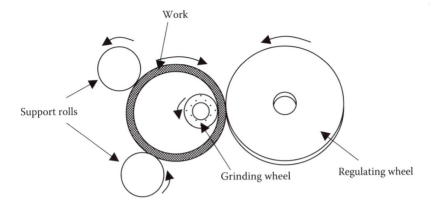

FIGURE 1.50 Internal centerless grinding.

1.6 SUMMARY OF MACHINE TOOL CHARACTERISTICS AND MACHINING EQUATIONS

In this section a summary of the equations derived for each machining operation is presented with empirical information that allow rough estimates to be made of machining times and power consumption.

First, however, Table 1.1, Table 1.2, and Table 1.3 present characteristics and capabilities of typical general-purpose machine tools. Table 1.1 lists the common machines that employ single-point tools and shows the nature of the primary motion and its typical speed range; the nature of the feed motion and the range of feed f; the power available P_e; the overall efficiency η_m of the machine drive system; the capacity of the machine in terms of the maximum workpiece dimensions that can be accommodated; and, finally, an estimate of the machining accuracy that can be achieved without special techniques.

Table 1.2 and Table 1.3 present similar information for machines employing multipoint tools and grinding machines, respectively. It should be noted that for each type of machine a wide range of sizes is generally available: the tables give details of only one typical machine within the range.

Table 1.4 summarizes the equations developed earlier for the machining time t_m, the metal removal rate Z_w, and the mean undeformed chip thickness $a_{c_{av}}$ for metal-cutting operations. Table 1.5 summarizes the equations for the machining time and the metal removal rate for grinding operations.

The machining time is an important factor when considering the economics and production rate of an operation, but it should be remembered that the time spent by the operator on loading and unloading workpieces, changing tools, and positioning the workpiece or tool before engaging the feed is not included. To estimate the machining time for those machines having a reciprocating motion,

TABLE 1.1
Typical Characteristics of Machines Using Single-Point Tools

Machine	Primary Motion				Feed Motion			Power P_e	Overall Efficiency η_m	Maximum Size of Workpiece	Machining Accuracy
	Type	Axis	Applied to	Range of Speed	Type	Applied to	Range of Feed f				
Engine lathe	Rotary	Horizontal	Workpiece	0.3–27 s^{-1} (20–1600 rpm)	Continuous	Tool	0.05–2.5 mm (0.002–0.2 in.)	4 kW (5 hp)	70	l_w < 760 mm (30 in.) d_w < 360 mm (14 in.)	0.013 mm (0.0005 in.)
Vertical borer	Rotary	Vertical	Workpiece	0.07–3.3 s^{-1} (4–200 rpm)	Continuous	Tool	0.05–2.5 mm (0.002–0.2 in.)	15 kW (20 hp)	70	d_w < 1.2 m (48 in.)	0.025 mm (0.001 in.)
Horizontal borer	Rotary	Horizontal	Tool	0.3–27 s^{-1} (20–1600 rpm)	Continuous	Tool	0.05–0.75 mm (0.002–0.03 in.)	7.5 kW (10 hp)	70	l_w < 1.2 m (48 in.)	0.025 mm (0.001 in.)
Shaper	Linear	Horizontal	Tool	To 0.75 m/s (150 ft/min)	Intermittent	Workpiece	0.05–0.75 mm (0.002–0.03 in.)	4 kW (5 hp)	70	l_w < 410 mm (16 in.) b_w < 410 mm (16 in.)	0.075 mm (0.003 in.)
Planer	Linear	Horizontal	Workpiece	To 1.5 m/s (300 ft/min)	Intermittent	Tool	0.05–2.5 mm (0.002–0.1 in.)	30 kW (40 hp)	80	l_w < 7.6 m (25 ft) b_w < 1.2 m (48 in.)	0.064 mm (0.0025 in.)

Key: b_w = width of workpiece; d_w = diameter of workpiece; l_w = length of workpiece

TABLE 1.2
Typical Characteristics of Machines Using Multipoint Tools

Machine	Primary Motion				Feed Motion			Power P_e	Overall Efficiency η_m	Maximum Size of Workpiece	Maximum Size of Tool	Machining Accuracy
	Type	Axis	Applied to	Range of Speed	Type	Applied to	Range of Feed v_f or f					
Upright drill press	Rotary	Vertical	Tool	1–10 s⁻¹ (60–600 rpm)	Continuous	Tool	0.15–0.64 mm (0.006–0.025 in.)	3.7 kW (5hp)	70	600 × 600 mm (24 × 24 in.)	25 mm (1 in.) dia. drill	0.13mm (0.005 in.)
Radial-arm drill	Rotary	Vertical	Tool	1–10 s⁻¹ (60–600 rpm)	Continuous	Tool	0.15–0.64 mm (0.006–0.025 in.)	3.7 kW (5hp)	70	1.2 × 1.2 m (48 × 48 in.)	25 mm (1 in.) dia. drill	0.13mm (0.005 in.)
Bench drill press	Rotary	Vertical	Tool	12–83 s⁻¹ (700–5000 rpm)	Hand operated	Tool	—	250 W (0.33 hp)	85	250 × 250 mm (10 × 10 in.)	8 mm (5/16 in.) dia drill	0.05 mm (0.002 in.)
Horizontal miller	Rotary	Horizontal	Tool	0.3–8.3 s⁻¹ (20–500 rpm)	Continuous	Workpiece	0.2–8.5 mm/s (0.5–20 in./min)	3.7 kW (5hp)	50	l_w < 915 mm (36 in.) b_w < 250 mm (10 in.)	200 mm dia (8 in.)	0.076 mm (0.003 in.)
Vertical knee-type miller	Rotary	Vertical	Tool	0.3–8.3 s⁻¹ (20–500 rpm)	Continuous	Workpiece	0.2–8.5 mm/s (0.5–20 in./min)	3.7 kW (5hp)	50	l_w < 915 mm (36 in.) b_w < 250 mm (10 in.)	150 mm dia (6 in.)	0.076 mm (0.003 in.)
Vertical bed-type miller	Rotary	Vertical	Tool	0.3–8.3 s⁻¹ (20–500 rpm)	Continuous	Workpiece	0.2–8.5 mm/s (0.5–20 in./min)	3.7 kW (5hp)	50	760 × 760 mm (30 × 30 in.)	250 mm dia (10 in.)	0.076 mm (0.003 in.)
Broacher	Linear	Horizontal or vertical	Tool	To 0.5 m/s (100 ft/min)	Provided by cutting edges	Tool	0.013–0.13 mm (0.0005–0.005 in.)	19 kW (25hp)	80	l_w < 460 mm (18 in.)	1500 mm long (60 in.)	0.13mm (0.005 in.)

Key: b_w = width of workpiece; d_w = diameter of workpiece; l_w = length of workpiece

TABLE 1.3
Typical Characteristics of Machines Using Multipoint Tools

Machine	Rotary Primary Motion Applied to Wheel		Traverse			Rotary Motion of Workpiece		Intermittent feed		Power P_e	Overall Efficiency η_m	Maximum Size of Workpiece	Machining Accuracy
	Axis	Speed	Applied to	Axis	Maximum Feed	Axis	Speed	Applied to	Maximum Feed f				
Horizontal-spindle surface grinder	Horizontal	33 s⁻¹ (2000 rpm)	Workpiece	Horizontal	0.75 m/s (150 ft/min)	—	—	Workpiece	6.4 mm (0.25 in.)	1 kW (1.5 hp)	85	l_w < 600 mm (24 in.) b_w < 200 mm (8 in.)	0.0025 mm (0.0001 in.)
Vertical-spindle surface grinder	Vertical	33 s⁻¹ (2000 rpm)	Workpiece	Horizontal	0.75 m/s (150 ft/min)	—	—	Wheel	1.5 mm (0.06 in.)	3.7 kW (5 hp)	85	l_w < 760 mm (30 in.) b_w < 250 mm (10 in.)	0.025 mm (0.001 in.)
Vertical-spindle surface grinder with rotary table	Vertical	33 s⁻¹ (2000 rpm)	—	—	—	Vertical	1.7 s⁻¹ (100 rpm)	Wheel	0.25 mm (0.01 in.)	1 kW (1.5 hp)	85	400 mm (16 in.) dia. table	0.0025 mm (0.0001 in.)
808 Cylindrical grinder	Horizontal	33 s⁻¹ (2000 rpm)	Workpiece	Horizontal	0.5 m/s (100 ft/min)	Horizontal	1.7 s⁻¹ (100 rpm)	Wheel	0.4 mm (0.015 in.)	1 kW (1.5 hp)	85	l_w < 600 mm (24 in.) d_w < 250 mm (10 in.)	0.0025 mm (0.0001 in.)
Internal grinder	Horizontal	1700 s⁻¹ (100000 rpm)	Wheel	Horizontal	0.75 m/s (150 ft/min)	Horizontal	3.3 s⁻¹ (200 rpm)	Wheel	0.025 mm (0.001 in.)	1 kW (1.5 hp)	85	l_w < 200 mm (8 in.) d_w < 150mm (6 in.)	0.00025 mm (0.00001 in.)

Key: b_w = width of workpiece; d_w = diameter of workpiece; l_w = length of workpiece

TABLE 1.4
Summary of Equations for Cutting Operations

Operation	Figure	Machining Time t_m	Metal Removal Rate Z_w or $Z_{w_{max}}$	Mean Undeformed Chip Thickness, $a_{c_{av}}$
Turning	1.7a	$\dfrac{l_w}{fn_w}$	$\pi f a_p n_w \left(d_m + a_p\right)$	$f \sin\kappa_r$
Boring	1.7c		$\pi f a_p n_w \left(d_m - a_p\right)$	
Facing / Parting	1.7b / 1.7e	$\dfrac{d_m}{2fn_w}$	$\pi f a_p n_w d_m$	
Shaping and planing	1.20	$\dfrac{b_w}{fn_r}$	$f a_p v$	
Drilling	1.23	$\dfrac{l_w}{fn_t}$	$\dfrac{\pi f_m^2 n_t}{4}$	$\dfrac{f}{2}\sin\kappa_r$
Slab milling	1.28 / 1.29	$\dfrac{l_w + \sqrt{a_e\left(d_t - a_e\right)}}{v_f}$		$\dfrac{v_f}{Nn_t}\sqrt{\dfrac{a_e}{d_t}}$
Face milling	1.33 / 1.34	$\dfrac{l_w + d_t}{v_f}$	$a_e a_p v_f$	$\dfrac{v_f}{Nn_t}$
Side and face milling	1.33 / 1.34	$\dfrac{l_w + 2\sqrt{a_e\left(d_t - a_e\right)}}{v_f}$		$\dfrac{v_f}{Nn_t}\sqrt{\dfrac{a_e}{d_t}\left(1 - \dfrac{a_e}{d_t}\right)}$
Broaching	1.37	$\dfrac{l_t}{v}$	$f a_p v$ per engaged cutting edge	f

Key: a_e = working engagement; a_p = back engagement; b_w = width of surface to be machined; d_m = diameter of machined surface; d_t = diameter of tool; f = feed; l_t = length of tool; l_w = length of surface or hole to be machined; n_r = frequency of reciprocation; n_t = rotational frequency of tool; n_w = rotational frequency of workpiece; N = number of teeth on cutter; v = cutting speed; v_f = feed speed; κ_r = major cutting edge angle.

it is necessary to know the relationships between the frequency of reciprocation n_r, the cutting speed v (or traverse speed v_{trav} in grinding machines), and the length of the workpiece l_w. Typical relationships presented in Figure 1.51 allow a rough estimate to be made of the rate of reciprocation n_r for use in the equations in Table 1.4 and Table 1.5.

Although the cost of the power consumed in a machining operation is not usually an important economic factor, it is necessary to be able to estimate the quantity of power required because of the limitations that may be imposed by the power available to operate a particular machine tool. The ability to estimate power requirements is important because, for rough-machining operations, it is

TABLE 1.5
Summary of Equations for Cutting Operations

Machine		Operation	Figure	Machining Time t_m	Metal Removal Rate Z_w or $Z_{w_{max}}$
Horizontal-spindle surface grinder		Traverse	1.42a	$\dfrac{b_w}{2fn_r}$	$fa_p v_{trav}$
		Plunge	1.42b		
Vertical spindle surface grinder	Reciprocating worktable	Traverse	1.44	$\dfrac{a_t}{2fn_r}+t_s$	
	Rotary worktable	Traverse		$\dfrac{a_t}{fn_w}+t_s$	
Cylindrical grinder		Traverse	1.46a	$\dfrac{a_t}{2fn_r}+t_s$	$\pi f d_w v_{trav}$
		Plunge	1.46b	$\dfrac{a_t}{v}+t_s$	$\pi a_p d_w v_f$
Internal grinder		Traverse	1.48a	$\dfrac{a_t}{2fn_r}+t_s$	$\pi f d_m v_{trav}$
		Plunge	1.48b	$\dfrac{a_t}{v}+t_s$	$\pi a_p d_m v_f$

Key: a_p = back engagement; a_t = total depth of material to be removed; b_w = width of workpiece to be machined; d_m = diameter of machined surface; f = feed; n_r = frequency of reciprocation; n_w = rotational frequency of workpiece; t_s = spark-out time; v_f = feed speed; v_{trav} = traverse speed.

usually desirable to remove metal at the highest rate possible. It was shown earlier (Equation 1.19) that the power required for machining P_m is found from

$$P_m = p_s Z_w \qquad (1.49)$$

In practice the specific cutting energy p_s depends mainly on the hardness of the work material and the mean value of the undeformed chip thickness $a_{c_{av}}$. A guide to the relationships between p_s and $a_{c_{av}}$ for various work materials is presented in Figure 1.52 and can be used with the equations in Table 1.4 and Table 1.5 to estimate P_m. It should be noted that for slab milling and face milling the mean value of a_c has been approximated roughly by $a_{c_{max}}/2$ ($a_{c_{max}}$ is the maximum value of a_c), except in face milling, where the workpiece width is

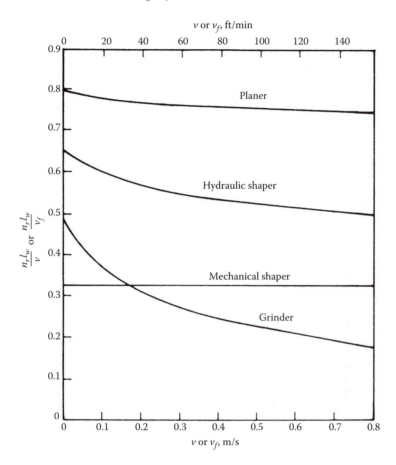

FIGURE 1.51 Typical characteristics of machine tools having reciprocating motion, where n_r = frequency of reciprocation, l_w = length of workpiece, v = cutting speed, and v_{trav} = traverse speed (for grinders).

smaller than the tool diameter. In face milling the mean value of $a_{c_{max}}$ is approximately equal to $a_{c_{av}}$. For all other cases $a_{c_{av}}$ is equal to a_c.

For a particular machine, the power available for machining can be estimated from

$$P_m = P_e \eta_m \qquad (1.50)$$

where P_e is the electrical power available, and η_m is the overall efficiency of the drive system.

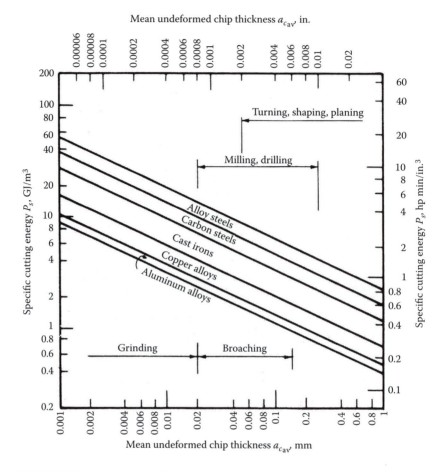

FIGURE 1.52 Approximate values of the specific cutting energy p_s for various materials and operations.

PROBLEMS

1. Two thousand bars 80 mm in diameter and 300 mm long must be turned down to 65 mm diameter for 150 mm of their length. The surface-finish and accuracy requirements are such that a heavy roughing cut (removing most of the material) followed by a light-finishing cut are needed. The roughing cut is to be taken at maximum power. The light-finishing cut is to be taken at a feed of 0.13 mm, a cutting speed of 1.5 m/s, and at maximum power.

   If the lathe has a 2-kW motor and an efficiency of 50%, calculate the total production time in kiloseconds (ks) for the batch of work. Assume that the specific cutting energy for the work material is 2.73 GJ/m³, the time taken to return the tool to the beginning of the cut is 15 s, and the time taken to load and unload a workpiece is 120 s.

2. A workpiece in the form of a bar 100 mm in diameter is to be turned down to 70 mm diameter for 50 mm of its length. A roughing cut using maximum power and a depth of cut of 12 mm is to be followed by a finishing cut using a feed of 0.1 mm and a cutting speed of 1.5 m/s.

   It takes 20 s to load and unload the workpiece and 30 s to set the cutting conditions, set the tool at the beginning of the cut and engage the feed.

   The specific cutting energy for the material is 2.3 GJ/m³, and the lathe has a 3-kW motor and a 70% efficiency. Estimate:
   a. The machining time for the rough cut
   b. The machining time for the finish cut
   c. The total production time for each workpiece

3. A 50 mm diameter bar is to be screw-threaded, for 250 mm of its length, on a lathe. The included angle of the thread is 60 degrees, the pitch is 2.5 mm, and the outside diameter of the threaded portion is to be 50 mm.
   a. If the undeformed chip thickness is limited to 0.13 mm, how many passes of the tool will be required to complete the operation?
   b. If the rotational frequency of the workpiece is 0.8 s⁻¹ and it takes 20 s to return the tool and engage the carriage with the lead screw after each pass, what will be the total production time?

4. A 1.5 m diameter disc with a 600 mm diameter hole in the center is to be faced, starting at the outside, on a vertical-boring machine. The rotational frequency of the table is 0.5 s⁻¹, the feed is 0.25 mm, and the back engagement (depth of cut) is 6 mm. The specific cutting energy for the work material and the particular cutting conditions is 3.5 GJ/m. Calculate:
   a. The machining time, in kiloseconds
   b. The power consumption (in kW) at the beginning of the operation
   c. The power consumption just before the end of the operation

5. In a pipe flange facing operation carried out on a lathe, the depth of cut is 12 mm, and the inside and outside diameters of the flange are 250 mm and 500 mm, respectively. The specific cutting energy of the work material is 2.6 GJ/m³, the maximum rotational frequency of the lathe spindle is 6 s⁻¹, and the maximum feed that can be applied is 0.25 mm. The major cutting-edge angle of the tool is 60 degrees. Find:
   a. The maximum power required for machining when the machine is operating at maximum capacity
   b. The cutting force
   c. The machining time

6. Fifty 1 m diameter discs having an 80 mm diameter hole in the center are to be faced on a vertical-boring machine with a feed of 0.25 mm and a back engagement (depth of cut) of 5 mm. The machine has an automatic control device by which the cutting speed is continuously adjusted to allow maximum power utilization at the cutting tool of 3 kW. However, the maximum rotational frequency of the spindle is limited to 0.7 s^{-1}.

    If the specific cutting energy for the work material is 2.27 GJ/m^3 and it takes 600 s to unload a machined disc, load an unmachined disc, and return the tool to the beginning of the cut, calculate the total production time for the batch, in kiloseconds (ks).

7. In a facing operation carried out on a lathe, the end of a carbon steel cylindrical workpiece 300 mm diameter is to be faced with a high-speed steel tool having a major cutting-edge angle of 60 degrees. The depth of cut (back engagement) is 10 mm and the feed 0.1 mm.
    a. What is the maximum rotational frequency for the lathe if the power available for machining is 4 kW? (Find the specific cutting energy for the work material from Figure 1.50.)
    b. What is the machining time?

8. In a particular mechanical shaper mechanism, the length of the rocker arm is 1.4 m, the rocker-arm pivot is 1.4 m below the connection between the link and the ram, the radius of the crank is 150 mm, and the crank pivot is 700 mm above the rocker-arm pivot. If the crank is rotated at 0.3 s^{-1}, calculate:
    a. The time for a forward stroke of the ram
    b. The time for a return stroke of the ram
    c. The maximum cutting speed, in meters per second (rn/s). during the forward stroke

9. The cross-feed on a shaper consists of a lead screw having 0.2 threads per millimeter. A ratchet and pawl on the end of the lead screw is driven from the shaper crank such that the pawl indexes the ratchet by one tooth during each return stroke of the ram. If the shaper is operating at one stroke per second and the ratchet on the cross-feed lead screw has 20 teeth, calculate:
    a. The feed, in millimeters (mm)
    b. The shortest time to machine the surface of a rectangular workpiece 130 mm long by 100 mm wide

10. A shaper is operated at two cutting strokes per second and is used to machine a workpiece 150 mm in length at a cutting speed of 0.5 m/s using a feed of 0.4 mm and a back engagement (depth of cut) of 6 mm. Calculate:

   a. The total machining time to produce 800 components each 100 mm in width

   b. The percentage of this time during which the tool is not contacting the workpiece

   c. The metal removal rate during actual machining

11. A shaper is operated at a frequency of reciprocation of 1.5 s^{-1} and is used to machine a 4 mm layer of material from the top of a rectangular low-carbon steel workpiece. The width of the workpiece is 150 mm and its length is 200 mm. The tool has a major cutting-edge angle of 30 degrees and is contacting the workpiece for 72% of the time. The feed is 0.25 mm. Find:

   a. The undeformed chip thickness

   b. The total time to complete the operation neglecting setup, loading, and unloading

   c. The metal removal rate during machining

   d. The power required during machining (use Figure 1.50 to obtain the specific cutting energy)

12. In a drilling operation using a twist drill, the rotational frequency of a drill is 5 s^{-1}, the feed 0.25 mm, the major cutting-edge angle 60 degrees, and the drill diameter 12 mm. Assuming that the specific cutting energy for the work material is 2 GJ/m^3, calculate:

   a. The maximum metal removal rate

   b. The undeformed chip thickness

   c. The drill torque, in Newton-meters (N-m)

13. In a slab-milling operation, the width of the medium carbon steel workpiece is 75 mm, its length is 200 mm, and a 5-mm layer is to be removed in one pass.

   a. What feed speed (mm/s) could be used if the power available for cutting is 3 kW and the specific cutting energy for the work material is 3.6 GJ/m^3?

   b. If the cutter diameter is 100 mm and the ideal surface roughness is 1.5 micrometers (μm), what should be the rotational frequency of the cutter?

   c. What is the cutting speed?

   d. What is the machining time?

14. In a slab-milling operation, the cutter has 20 teeth and is 100 mm in diameter. The rotational frequency of the cutter is 5 s^{-1}, the workpiece feed speed is 1.3 mm/s, the working engagement (depth of cut) is 6 mm, and the back engagement (width of the workpiece) is 50 mm. The relationship between the maximum undeformed chip thickness $a_{c_{max}}$ and the specific cutting energy p_s in gigajoules per cubic meter (GJ/m^3), for the work material is:

$$p_s = 1.4\left[1 + \left(25 \times 10^{-6}\right) / a_{c_{\text{max}}}\right]$$

Estimate:
a. The maximum metal removal rate
b. The maximum power, in kilowatts (kW), required at the cutter

15. In a slab-milling operation, the length of the workpiece is 150 mm, its width is 50 mm, and a layer 10 mm in thickness is to be removed from its upper surface. The diameter of the cutter is 40 mm, and it has 10 teeth. The workpiece is of medium carbon steel, the feed speed selected is 2 mm/s, and the cutter speed is 2.5 rev/s. Estimate:
    a. The power required (in kW)
    b. The machining time for the operation
    Note: Use Figure 1.52 to obtain the specific cutting energy.

16. In a face-milling operation the back engagement (depth of cut) is 5 mm, the work feed speed is 0.65 mm/s, and the working engagement (width of the workpiece) is 50 mm. The cutter has a diameter of 100 mm and has 20 teeth. If the cutting speed is to be 1 m/s, calculate:
    a. The rotational frequency of the cutter
    b. The maximum metal removal rate
    c. The time taken to machine 1000 workpieces of length 150 mm if it takes 180 s to load and unload a workpiece and return the cutter to the beginning of the cut

17. In a finish-surface-grinding operation on a horizontal-spindle surface grinder, the length of the workpiece is 100 mm and its width 50 mm. The cross-feed is applied every stroke of the worktable and is set at 0.25 mm, the back engagement (depth of cut) is 0.1 mm, and the maximum traverse speed is 250 mm/s. The frequency of worktable reciprocation is $1 s^{-1}$. Calculate:
    a. The machining time
    b. The maximum metal removal rate
    c. The maximum power consumption in watts (W) if the specific cutting energy for the conditions employed is 25 GJ/m^3
    d. The maximum tangential force on the grinding wheel if its diameter is 150 mm and has a rotational frequency of 60 s^{-1}

REFERENCE

1. International Standards Organization, *Axis and Motion Nomenclature for Numerically Controlled Machine Tools*, ISO Recommendation R841, 1st ed., October, 1968.

2 Mechanics of Metal Cutting

2.1 INTRODUCTION

Research in metal cutting did not start until approximately 70 years after the introduction of the first machine tool. It is not proposed to give here a complete history of research in metal cutting but to indicate some of the more important steps that have been made.

According to Finnie, who published a historical review of work in metal cutting [1], early research in metal cutting started with Cocquilhat in 1851 [2] and was mainly directed toward measuring the work required to remove a given volume of material in drilling. In 1873 tabulations of the work required in cutting metal were presented by Hartig [3] in a book that seems to have been the authoritative work on the subject for several years.

The shavings, or swarf, removed during the cutting of metal are called chips, and the first attempts to explain how chips are formed were made by Time [4] in 1870 and the famous French scientist Tresca [5] in 1873.

Some years later, in 1881, Mallock [6] suggested correctly that the cutting process was basically one of shearing the work material to form the chip and emphasized the importance of the effect of friction occurring on the cutting-tool face as the chip was removed (Figure 2.la). He produced drawings of chip formation from polished and etched specimens of partly formed chips; in addition, he observed the effect of cutting lubricants, the effect of the tool sharpness on the cutting process, and the reasons for instability in the cutting process that leads to undesirable vibrations, or "chatter." Many of these observations are surprisingly close to the accepted modern theories and are still being repeated more than 100 years later.

Finnie [1] reports that a step backward in the understanding of the metal-cutting process was taken in 1900 when Reuleaux [7] suggested that a crack occurred ahead of the tool and that the process could be likened to the splitting of wood. This theory suggests a model of cutting similar to that shown in Figure 2.lb and is a misconception that found popular support for many years.

About this time the now-famous paper by Taylor [8] was published, which reported the results of 26 years of research investigations and experience. Taylor was interested in the application of piecework systems in machine shops, where a time allowance was set for a particular job and a bonus was given to the workman performing his task in the allocated time. To assist in the application of such a system, Taylor investigated the effect of tool material and cutting conditions on

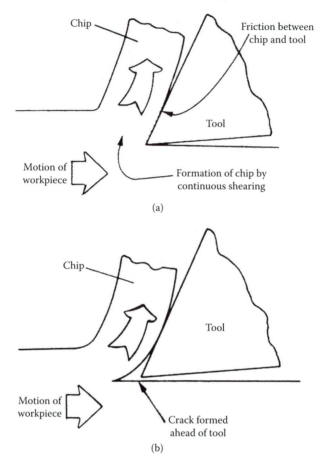

FIGURE 2.1 Models of the cutting process. (a) Present-day model; (b) earlier misconception.

tool life during roughing operations. The principal object was to determine the empirical laws that allow optimum cutting conditions to be established. In a biographical note at the beginning of the paper it was reported that with a combination of the development of a heat-treatment process to produce high-speed-steel cutting tools (known as the Taylor-White process of heat treatment) and the results of empirical research to improve shop methods, Taylor was able to increase the production of the Bethlehem Steel Company's machine shop by 500%. It is of interest to note that the empirical law governing the relationship between cutting speed and tool life suggested by Taylor is still used today and employed as the basis for many recent studies of machining economics. One fundamental discovery made by Taylor was that the temperature existing at the tool cutting edge controlled the tool wear rate.

Since 1906 when Taylor's paper was published, empirical and fundamental work on metal cutting has gradually increased in volume. Most of the fundamental work has been carried out since the publication of the well-known paper by Ernst and Merchant [9] in 1941, dealing with the mechanics of the process. There is now a large volume of published work on the machining process. Chapter 2, Chapter 3, Chapter 4, Chapter 5, and Chapter 6 of this book cover fundamental aspects of the mechanics of machining, largely based on the shear plane models of machining developed initially by Ernst and Merchant [9]. Some of the more advanced methods of analysis and results are outlined at the end of this chapter.

2.2 TERMS AND DEFINITIONS

All metal-cutting operations can be likened to the process shown in Figure 2.2, where the tool is wedge-shaped, has a straight cutting edge, and is constrained to move relative to the workpiece in such a way that a layer of metal is removed in the form of a chip. Figure 2.2b depicts the general case of cutting known as *oblique cutting*. A special case of cutting, where the cutting edge of the tool is arranged to be perpendicular to the direction of relative work-tool motion (Figure 2.2a), is known as *orthogonal cutting*. Since orthogonal cutting represents a two-dimensional rather than a three-dimensional problem, it lends itself to research investigations where it is desirable to eliminate as many of the independent variables as possible. The relatively simple arrangement of orthogonal cutting is therefore widely used in theoretical and experimental work.

The wedge-shaped cutting tool basically consists of two surfaces intersecting to form the cutting edge (Figure 2.3). The surface along which the chip flows is known as the *rake face*, or more simply as the face, and that surface ground back to clear the new or machined workpiece surface is known as the *flank*. Thus, during cutting a wedge-shaped "clearance crevice" exists between the tool flank and the new workpiece surface.

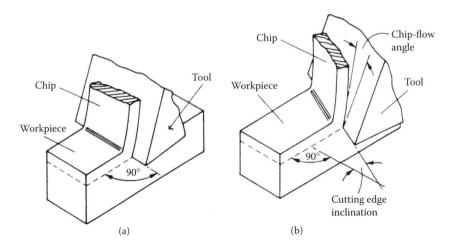

FIGURE 2.2 Orthogonal and oblique cutting. (a) Orthogonal cutting; (b) oblique cutting.

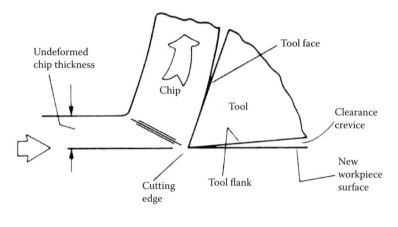

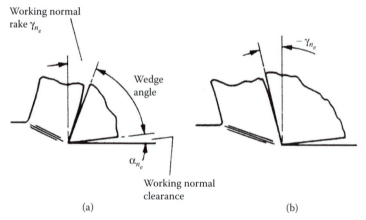

FIGURE 2.3 Terms used in metal cutting. (a) Positive rake; (b) negative rake.

The depth of the individual layer of material removed by the action of the tool is known as the undeformed chip thickness (Figure 2.3), and although in practical cutting operations this dimension often varies as cutting proceeds, for simplicity in much of the research work it is arranged to be constant.

One of the most important variables in metal cutting is the slope of the tool face, and this slope, or angle, is specified in orthogonal cutting by the angle between the tool face and a line perpendicular to the new work surface (Figure 2.3). This angle is known as the rake angle or, in accordance with ISO terminology, the *working normal rake*, and Figure 2.3 illustrates how the sign of the angle is defined. The rake angle is measured from the normal to the cut surface, with the positive direction such that the wedge angle is decreased. Negative rake angles result in a stronger cutting edge and consequently are often preferred for rough machining operations. In addition, negative rake angles usually result in more usable cutting edges for replaceable tool inserts (see Chapter 4).

The tool flank plays no part in the process of chip removal; however, the angle between the flank and the new workpiece surface can significantly affect the rate at which the cutting tool wears and is defined as the clearance angle or, more precisely, the working normal clearance.

Thus from Figure 2.3 the sum of the rake, clearance, and wedge angles is equal to 90 degrees, where the wedge angle is the included angle between the face and the flank.

2.3 CHIP FORMATION

The type of chip produced during metal cutting depends on the material being machined and the cutting conditions used. One of the most useful techniques for a study of chip formation is the quick-stopping device. With this device it is possible to "freeze" or suddenly stop the cutting action and allow subsequent microscopic examination of the chip removal process. A typical quick-stopping device, designed for use in a shaping machine, is depicted in Figure 2.4. Here the workpiece is gripped in a vise that is free to slide in the guide block. During cutting the vise is constrained by the holding ring, which is held in position by shear pins passing through both the guide block and the holding ring. The shear pins are designed to carry the force required to remove the chip. When the cut is partly completed, a tongue on the toolholder contacts the vise, shears the pins, and pushes the vise and holding ring forward. This action of the tongue stops the cutting action by accelerating the speed of the workpiece rapidly to the speed of the cutting tool. With carefully designed devices the effective stopping time is extremely small. For example, in the device described above, the stopping time was estimated [10] to be 0.00017 s when the cutting speed is 0.8 m/s. A requirement of quick-stopping devices is that the tool is removed from the cutting process as rapidly as possible so that the "frozen" chip is representative of the conditions existing during cutting. The device shown in Figure 2.4 is useful for relatively slow cutting speeds. Several other quick-stop mechanisms have been devised, including the use of explosive action, such as in humane killer guns, to propel a lathe tool away from the cutting position at very high speeds [11,12] and these are more suitable for higher cutting speeds.

After the use of a quick-stop device, a segment of the workpiece, with the chip attached, can be cut out and examined in detail at any required magnification. The specimens may be mounted, polished, and etched to reveal the internal action of cutting. In some cases the photographic deposition of a fine grid on the outside surface can aid the determination of the extent of the deformation zone and the material strain during cutting [12]. Since each quick-stop specimen illustrates the cutting action at only one instant in time, several specimens must be prepared to cover a range of cutting conditions.

Examples of some specimens obtained with the quick-stopping device shown in Figure 2.4 are depicted in Figure 2.5, Figure 2.6, and Figure 2.7. The specimens shown in Figure 2.5, Figure 2.6, and Figure 2.7 were chosen because they depict

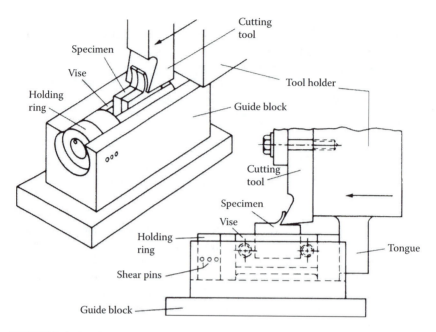

FIGURE 2.4 A quick-stopping device.

three basic types of chip formation commonly occurring in practice: the continuous chip; the continuous chip with built-up edge; and the discontinuous chip. In each case the cutting speed is 0.8 m/s, which means that during the retardation of the cutting action, the tool travels a distance of only 0.07 mm relative to the specimen, which, in Figure 2.5, is equal to about one-tenth of the undeformed chip thickness. These calculations support the assumption that the specimens obtained with this technique are representative of chip formation during cutting.

2.3.1 CONTINUOUS CHIP

Figure 2.5 shows the formation of a continuous chip. This type of chip is common when most ductile materials, such as wrought iron, mild steel, copper, and aluminum, are machined at relatively high cutting speeds. It can be seen that cutting under these conditions is essentially a steady-state process. For this reason much of the research conducted into metal cutting has dealt with continuous-chip formation and most of the discussion in this chapter is related to this type of chip-formation machining. Basically this operation is one of shearing the work material to form the chip and sliding of the chip along the face of the cutting tool. The chip remains in contact with the tool face for a short distance (the contact length) before curling away. The formation of the chip takes place in the zone extending from the tool cutting edge to the junction between the surfaces of the chip and workpiece; this zone is known as the primary deformation zone (Figure 2.5). To deform the material in this manner the forces that must be

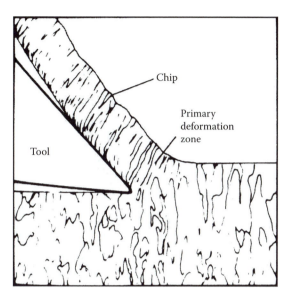

FIGURE 2.5 Continuous chip.

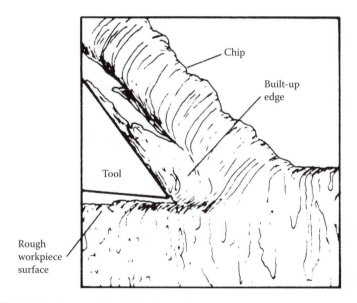

FIGURE 2.6 Continuous chip with built-up edge.

transmitted to the chip across the interface between the chip and the tool are sufficient to deform the lower layers of the chip as it slides along the tool face (secondary deformation zone). Continuous chip formation in ductile materials is associated with reduced cutting forces and generally good surface finish.

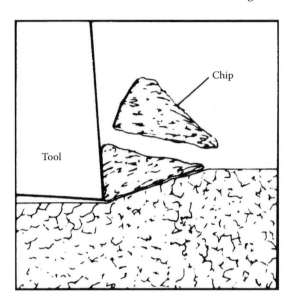

FIGURE 2.7 Discontinuous chip.

2.3.2 CONTINUOUS CHIP WITH BUILT-UP EDGE

Under some conditions, usually at relatively low cutting speeds, the friction between the chip and the tool is so great that the chip material welds itself to the tool face. The presence of this welded material further increases the friction, and this friction leads to the building up of layer upon layer of chip material. The resulting pile of material is referred to as a *built-up edge* (Figure 2.6). Continuous chip formation with a built-up edge usually occurs when machining ductile materials at low cutting speeds, and the size of the built-up edge tends to reduce as cutting speed is increased. In effect machining takes place with a blunt cutting edge and consequently cutting forces are increased relative to continuous chip formation without a built up edge present.

Often the built-up edge continues to grow and then breaks down when it becomes unstable, the broken pieces being carried away by the underside of the chip and the new workpiece surface. Figure 2.6 shows the rough workpiece surface obtained under these conditions. A study of built-up-edge formation in metal cutting is most important as it is one of the principal factors affecting surface finish and can have a considerable influence on cutting-tool wear. These effects will be discussed later.

2.3.3 DISCONTINUOUS CHIP

During the formation of a chip, the material undergoes severe strain, and if the work material is brittle, fracture will occur in the primary deformation zone when the chip is only partly formed, with a crack propagating from the cutting edge to the free surface. Under these conditions the chip breaks into small segments in a cyclic manner (Figure 2.7), and the condition is referred to as *discontinuous-chip formation*. Discontinuous chips are always produced when machining such materials as cast iron or cast brass but may also be produced when machining ductile materials at very low speeds and high feeds. Discontinuous chip formation is not necessarily detrimental to machining performance, but is characteristic of the processing of brittle materials.

2.3.4 OTHER TYPES OF CHIP FORMATION

The three types of chips shown in Figure 2.5, Figure 2.6, and Figure 2.7 occur most commonly in the general machining of engineering materials. However, other types of chip formation can be observed. In particular, the development of more advanced tool materials has led to considerable interest in machining at much higher cutting speeds (so called high speed machining) and the machining of more difficult to machine materials, including materials in their hardened state (hard machining). High speed and hard machining developments are discussed further in Chapter 4. Under these conditions the process of chip formation tends to become periodic with corresponding variations in chip thickness [13–15]. The two basic types of chips observed are *wavy chips*, associated with periodic variations in the underlying shearing process, and *segmented chips*. Wavy chips are sometimes observed as a transition from continuous chip formation to segmented chips as cutting speeds are increased.

Figure 2.8a shows segmented chips, and these are characterized by regions of intense shear separated by regions of material with relatively little deformation. Such chips can be seen in most materials as cutting speeds are increased to high levels, but for materials with low thermal conductivity and low heat capacity such as titanium alloys, segmented chip formation occurs at more moderate speeds. As speed increases, the separation of the segments becomes more severe until complete separation of the segments occurs (Figure 2.8b). This type of chip formation is the result of a thermo-plastic instability and is accompanied by fluctuations in cutting forces. As the severe deformation takes place, temperature increases and this results in a decrease in shear strength. If the heat generated is not dissipated quickly enough, the catastrophic shear becomes very localized and segments are formed separated by relatively undeformed regions.

For most materials there is a transition from one type of chip formation to another as speed is increased, and Figure 2.9 illustrates approximately the speed ranges for different types of chip formation for alloy steel, such as AISI 4340, at different hardness values [13]. It can be seen that as hardness of the workpiece is increased, the cutting speed at which segmental chips are formed becomes less. This transition can also be observed from the separation of the peaks in chip thickness as the segmentation develops, as illustrated in Figure 2.10 [16].

The fact that different types of chip formation can be observed for machining the same material under different cutting conditions makes the analytical modeling of machining difficult because individual approaches tend to be aimed at

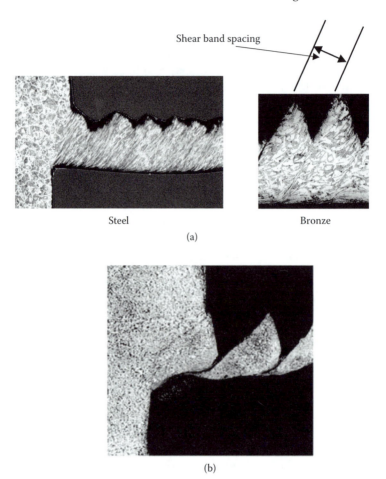

Shear band spacing

Steel   Bronze

(a)

(b)

FIGURE 2.8 Segmented chip formation. (a) Partial segmentation; (b) complete segmentation.

one type of chip formation, for example continuous chip formation. The prediction of when each type of chip formation will occur is a difficult complication to overcome.

2.4 THE FORCES ACTING ON THE CUTTING TOOL AND THEIR MEASUREMENT

In orthogonal cutting the resultant force F_r applied to the chip by the tool lies in a plane normal to the tool cutting edge (Figure 2.11). This force is usually determined, in experimental work, from the measurement of two orthogonal components: one in the direction of cutting (known as the cutting force F_c), the other normal to the direction of cutting (known as the thrust force F_t). The accurate

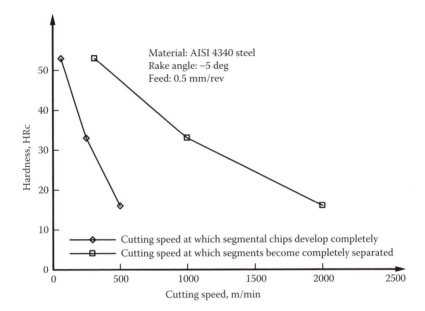

FIGURE 2.9 Speed ranges for segmented chip formation for AISI 4340 steel with different hardness values. (Adapted from Komandhuri, R., et al., *Trans. ASME, J. Eng. Ind.*, Vol. 104, 121–131, 1982.)

measurement of these two components of the resultant tool force has been the subject of considerable effort in the past, and several types of cutting-force dynamometers have been developed.

In most metal-cutting-force dynamometers the tool force is determined by measuring the deflections or strains in the elements supporting the cutting tool. It is essential that the instrument should have high rigidity and high natural frequencies so that the dimensional accuracy of the cutting operation is maintained and the tendency for chatter, or vibrations, to occur during cutting is minimized. The dynamometer must, however, give strains or displacements large enough to be measured accurately. The design of the dynamometer depends, to a large extent, on the strain- or deflection-measuring device and instrumentation employed. Figure 2.12 shows a simple type of two-component, cutting-force dynamometer, where the cutting tool is supported at the end of a cantilever. The vertical and horizontal components of the deflection of the cantilever under the action of the resultant tool force are taken as a measure of the two force components. Many effective dynamometer designs have been developed using a range of deflection-measuring methods, including strain gauges and piezoelectric load cells, for increased resolution and stiffness.

The two components F_c and F_t, of the resultant tool force F_r measured with a dynamometer may be used to calculate many important variables in the process of continuous-chip formation.

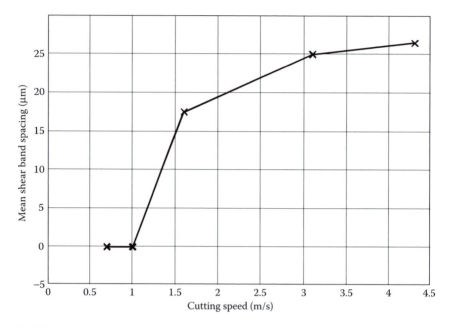

FIGURE 2.10 Effect of cutting speed on the separation of chip segments. (Adapted from Davies, M.A. and Burns, T.J., *Phyl. Trans. Royal Society*, London, Vol. 359, 821–846, 2001.)

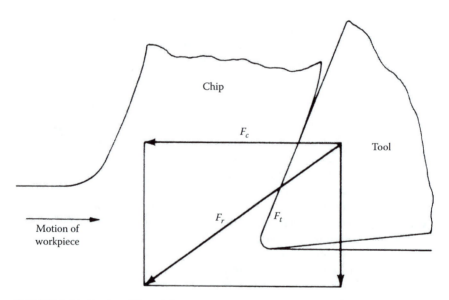

FIGURE 2.11 Cutting (F_c) and thrust (F_t) components of resultant tool force (F_r).

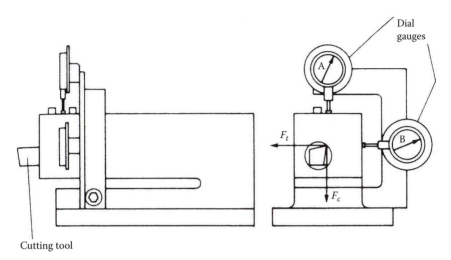

Dial
gauges

FIGURE 2.12 Simple two-component cutting force dynamometer for use on a lathe, where dial gauge A gives a measure of F_c, and dial gauge B gives a measure of F_t.

2.5 SPECIFIC CUTTING ENERGY

The rate of energy consumption during machining P_m is the product of the cutting speed v and the cutting force F_c. Thus,

$$P_m = F_c v \qquad (2.1)$$

Both the rate of energy consumption and the metal removal rate are proportional to the cutting speed. A parameter giving an indication of the efficiency of the process, independent of the cutting speed, is therefore the energy consumed per unit volume of metal removed, referred to as the specific cutting energy p_s. This parameter is given by

$$p_s = \frac{P_m}{Z_w} = \frac{F_c}{A_c} \qquad (2.2)$$

where Z_w is the metal removal rate, and A_c is the cross-sectional area of the uncut chip.

The specific cutting energy can vary considerably for a given material and is affected by changes in cutting speed, feed, tool rake, and so on. However, for a given tool rake at high cutting speeds and large feeds, the specific cutting energy tends to become constant, as can be seen in Figure 2.13. This constant value can be a useful guide, in practice, to the forces required to cut a given material at large speeds and feeds, and for the results presented in Figure 2.13 for low-carbon steel, p_s approaches 1 GN/m².

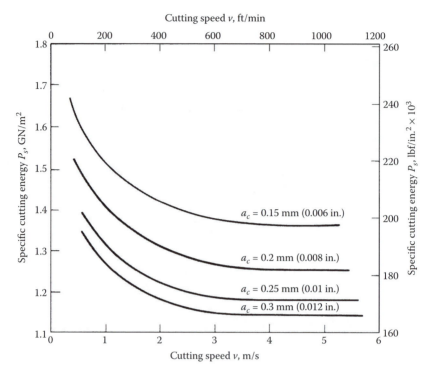

FIGURE 2.13 Effect of cutting speed and undeformed chip thickness on specific cutting energy, where p_s = specific cutting energy, v = cutting speed, and a_c = undeformed chip thickness and the material is mild steel, the normal rake is 10 degrees. and the width of the chip is 1.25 mm.

2.6 PLOWING FORCE AND THE "SIZE EFFECT"

The resultant tool force in metal cutting is distributed over the areas of the tool that contact the chip and workpiece. No cutting tool is perfectly sharp and in the idealized picture shown in Figure 2.14, the cutting edge is represented by a cylindrical surface joining the tool flank and the tool face. Observations in the past [17] have shown that the radius of this edge varies from 0.005 to 0.03 mm for a freshly ground high-speed steel tool. As the tool edge "plows" its way through the work material, the force that acts on the tool cutting edge forms only a small proportion of the cutting force at large values of the undeformed chip thickness a_c. At small values of a_c, however, the force that acts on the tool edge is proportionately large and cannot be neglected.

Because of the high stresses acting near the tool cutting edge, deformation of the tool material may occur in this region. This deformation would cause contact between the tool and the new workpiece surface over a small area of the tool flank. Thus when sharp cutting tools are used, a frictional force may act in

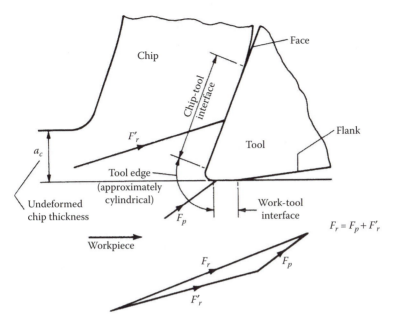

FIGURE 2.14 Contact regions on a cutting tool, where F_r = resultant tool force, F_r' = force required to remove chip, and F_p = plowing force acting on the tool edge and work-tool interface region.

the tool-flank region, this force again forming a small proportion of the cutting force at high feeds.

The force acting on the tool edge and the force that may act on the tool flank do not contribute to removal of the chip. These forces will be referred to collectively as the plowing force F_p.

The existence of the plowing force results in certain important effects and can explain the so-called *size effect*. This term refers to the increase in specific cutting energy (the energy required to remove a unit volume of metal) at low values of undeformed chip thickness. For example, in Figure 2.15 the mean specific cutting energy is plotted against undeformed chip thickness for a slab-milling operation. At the relatively small values of chip thickness occurring in this process, the specific cutting energy p_s increases rapidly with decreasing chip thickness. It is thought that the plowing force F_r is constant and therefore becomes a greater proportion of the total cutting force as the chip thickness decreases. When the total cutting force is divided by the undeformed-chip cross-sectional area to give p_s, the portion of p_s contributing to chip removal will remain constant, and the portion resulting from the plowing force F_p will increase as the chip thickness decreases. The increase in specific cutting energy, p_s as the chip thickness decreases explains to some extent why (as shown in Figure 1.52) those processes such as grinding that produce very thin chips require greater power to remove a given volume of metal.

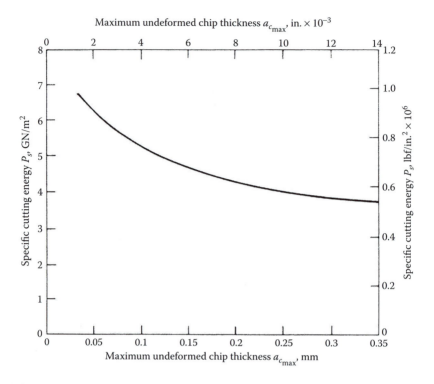

FIGURE 2.15 Effect of maximum undeformed chip thickness $a_{c_{max}}$ on specific cutting energy p_s during slab milling, where the material is steel, 57 ton/in.^{2}, UTS.

2.7 THE APPARENT MEAN SHEAR STRENGTH OF THE WORK MATERIAL

Many analyses of the metal-cutting processes have been developed. In general, assumptions have to be made which are often valid only for a restricted range of conditions. A particular difficulty in formulating an analysis of the metal-cutting process is the lack of constraint in the process that reduces the range of boundary conditions that can be applied. It has been suggested [18,19] that a unique solution does not exist for a particular set of cutting conditions and further that parameters such as the cutting forces, cutting ratio, and so on that occur in a particular case, depend on the conditions that exist when the tool first contacts the workpiece. A further complication is that different forms of chip formation mechanism occur at various cutting conditions. However, the simplified analyses of metal cutting developed do predict many of the general trends observed in the process. Figure 2.16 shows the idealized model of continuous-chip formation employed in much of the previous work on the mechanics of the cutting process. Two of the earliest workers to employ this model were Ernst and Merchant [9], who suggested that the shear zone, or primary deformation zone, could be reasonably

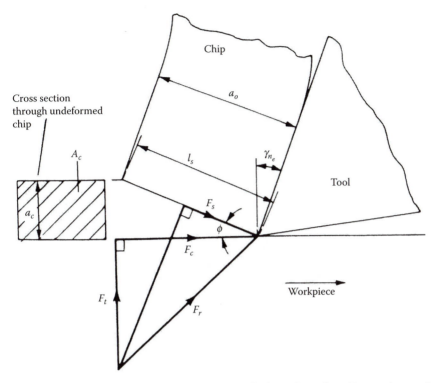

FIGURE 2.16 Shear-plane model of continuous chip formation, where F_r = resultant tool force, F_c = cutting force, F_t = thrust force, F_s = shear force on shear plane, ϕ = shear angle, γ_{n_e} = working normal rake, a_c = undeformed chip thickness, a_o = chip thickness, A_c = cross-sectional area of uncut chip, and l_s = length of shear plane.

represented by a plane, called the *shear plane*. The angle of inclination of the shear plane to the direction of cutting, called the shear angle ϕ, could be determined as shown in Equation 2.3, Equation 2.4, and Equation 2.5.

From Figure 2.16, the length of the shear plane is given by

$$l_s = \frac{a_c}{\sin \phi} = \frac{a_0}{\cos\left(\phi - \gamma_{n_e}\right)} \qquad (2.3)$$

$$\frac{a_c}{a_0} \cos\left(\phi - \gamma_{n_e}\right) = \sin \phi$$

and after rearrangement

$$\tan \phi = \frac{\left(a_c / a_0\right) \cos \gamma_{n_e}}{1 - \left(a_c / a_0\right) \sin \gamma_{n_e}} \qquad (2.4)$$

The ratio a_c/a_0 is known as the cutting ratio and is denoted by r_c. Thus,

$$\tan \phi = \frac{r_c \cos \gamma_{n_e}}{1 - r_c \sin \gamma_{n_e}} \qquad (2.5)$$

where

ϕ = shear angle
r_c = cutting ratio (given by a_c/a_0)
a_c = undeformed chip thickness
a_0 = chip thickness
γ_{n_e} = working normal rake

In experimental work, the working normal rake and the undeformed chip thickness are known, and the chip thickness can be measured either directly with a ball-ended micrometer or obtained from the weight of a known length of chip as follows:

$$a_0 = \frac{m_c}{l_c a_w \rho} \qquad (2.6)$$

where

m_c = mass of the chip specimen
l_c = length of the chip specimen
a_w = width of the chip
ρ = density of the workpiece material

If the resultant tool force is resolved in a direction parallel to the shear plane, the force F_s required to shear the work material and form the chip is obtained. As shown in Figure 2.16, this force may be expressed in terms of the cutting (F_c) and thrust (F_t) components of the resultant tool force:

$$F_s = \left(F_c \cos \phi \right) - \left(F_t \sin \phi \right) \qquad (2.7)$$

The area of shear A_s is given by

$$A_s = \frac{A_c}{\sin \phi} \qquad (2.8)$$

and thus the apparent shear strength of the material τ_s on the shear plane may be obtained:

$$\tau_s = \frac{F_s}{A_s} = \frac{\left[\left(F_c \cos\phi\right) - \left(F_t \sin\phi\right)\right]\sin\phi}{A_c} \qquad (2.9)$$

Experimental work [20] has shown that τ_s calculated in this way, remains constant for a given work material over a wide variety of cutting conditions. It has been observed, however, that at small feeds τ_s increases with a decrease in feed (or undeformed chip thickness). This exception to the constancy of τ_s can be explained by the existence of a constant plowing force F_p. If F_p is subtracted from the resultant cutting force F_r then F_r' the force required to remove the chip and acting on the tool face, is obtained (see Figure 2.14):

$$F_r' = F_r - F_p \qquad (2.10)$$

It has been shown [20] that if the components of F_r' are used to calculate the apparent shear strength of the work material, this apparent shear strength remains constant with respect to changes in feed. Thus,

$$\tau_s' = \left[\left(F_c'\cos\phi\right) - \left(F_t'\sin\phi\right)\right]\frac{\sin\phi}{A_c} \qquad (2.11)$$

where

F_c' = cutting component of F_r'
F_t' = thrust component of F_r'
τ_s' = constant property of the work material

Studies of the deformation of metals at high strain rates have shown that a material deforms at a relatively constant stress when the strain rate is sufficiently large. In metal cutting, the strain rates are on the order of 10^3 to 10^5 s^{-1}, and under these conditions the shear strength of the metal could be expected to be constant and independent of strain rate, strain, and temperature. These shear-strength characteristics, it is suggested, explain why in metal cutting the value of τ_s', the mean shear strength of the work material, is constant and independent of the cutting speed and rake angle under the range of these conditions normally encountered in metal cutting. However, the trend towards the use of higher cutting speeds means that accurate modeling of the constitutive behavior materials taking into account strain, strain rate, and temperature is important, and this is discussed further below.

2.8 CHIP THICKNESS

The chip thickness a_0 in metal cutting is not only governed by the geometry of the cutting tool and the undeformed-chip thickness a_c, but, as will be seen, it can also be affected by the frictional conditions existing at the chip-tool interface. Because the cutting process is affected by these factors, it differs fundamentally from other metal deformation processes, where the final shape of the deformed material is determined by the shape or setting of the tool used. In metal cutting, before predictions of cutting forces can be made, the chip thickness a_0 must be determined in order that the geometry of the process will be known.

It has been shown (Equation 2.5) that knowledge of the shear angle ϕ allows a_0 to be estimated for a given set of cutting conditions. Experimentally, ϕ, and hence the cutting ratio r_c, depends on the work and tool materials and the cutting conditions. A number of attempts have been made, in the past, to establish a theoretical law that predicts the shear angle ϕ, and two of these attempts are considered below.

2.8.1 THEORY OF ERNST AND MERCHANT

Although Piispanen attempted to solve this problem in 1937, the first complete analysis resulting in a so-called shear-angle solution was presented by Ernst and Merchant [9]. In their analysis the chip is assumed to behave as a rigid body held in equilibrium by the action of the forces transmitted across the chip-tool interface and across the shear plane. It is further assumed that the chip remains straight and has infinite contact length with the tool face.

For convenience, in Figure 2.17 the resultant tool force F_r is shown acting at the tool cutting edge and is resolved into components F_n and F_f in directions normal to and along the tool face, respectively, and into components F_{n_s} and F_s normal to and along the shear plane, respectively. The cutting (F_c) and thrust (F_t) components of the resultant tool force are also shown.

It is assumed that the whole of the resultant tool force is transmitted across the chip-tool interface and that no additional force acts on the tool edge or flank (i.e., the plowing force is zero).

The basis of Ernst and Merchants' theory was the suggestion that the shear angle ϕ would take up such a value as to reduce the work done in cutting to a minimum. Since, for given cutting conditions, the work done in cutting was proportional to F_c, it was necessary to develop an expression for F_c in terms of ϕ and then to obtain the value of ϕ for which F_c is a minimum.

From Figure 2.17:

$$F_s = F_r \cos\left(\phi + \beta - \gamma_{n_e}\right) \tag{2.12}$$

and

$$F_s = \tau_s A_s = \frac{\tau_s A_c}{\sin\phi} \tag{2.13}$$

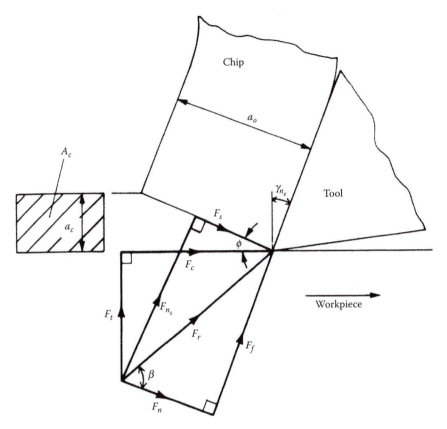

FIGURE 2.17 Force diagram for orthogonal cutting, where F_r = resultant tool force, F_c = cutting force, F_t = thrust force, F_s = shear force on shear plane, F_f = frictional force on tool face, F_n = normal force on tool face, ϕ = shear angle, γ_{n_e} = working normal rake, β = mean friction angle on tool face, a_c = undeformed chip thickness, and a_o = chip thickness.

where

τ_s = shear strength of the work material on the shear plane

A_s = area of the shear plane

A_c = cross-sectional area of the uncut chip

β = mean angle of friction between chip and tool (given by arctan F_f/F_n)

γ_{n_e} = working normal rake

From Equation 2.12 and Equation 2.13:

$$F_r = \frac{\tau_s A_c}{\sin\phi} \frac{1}{\cos\left(\phi + \beta - \gamma_{n_e}\right)} \tag{2.14}$$

Now by geometry:

$$F_c = F_r \cos\left(\beta - \gamma_{n_e}\right) \tag{2.15}$$

Hence from Equation 2.14 and Equation 2.15:

$$F_c = \frac{\tau_s A_c}{\sin\phi} \frac{\cos\left(\beta - \gamma_{n_e}\right)}{\cos\left(\phi + \beta - \gamma_{n_e}\right)} \tag{2.16}$$

Equation 2.16 may now be differentiated with respect to ϕ and equated to zero to find the value of ϕ for which F_c is a minimum. The required value is given by

$$2\phi + \beta - \gamma_{n_e} = \frac{\pi}{2} \tag{2.17}$$

Merchant [21] found that this theory agreed well with experimental results obtained when cutting synthetic plastics but agreed poorly with experimental results obtained for steel machined with a sintered carbide tool.

It should be noted that in differentiating Equation 2.16 with respect to ϕ, it was assumed that A_c, γ_{n_e}, and τ_s would be independent of ϕ. On reconsidering these assumptions, Merchant decided to include in a modified analysis the relationship

$$\tau_s = \tau_{s_0} + k\sigma_s \tag{2.18}$$

which indicates that the shear strength of the material τ_s increases linearly with increase in normal stress, σ_s on the shear plane (Figure 2.18); at zero normal stress τ_s is equal to τ_{s_0}. This assumption agreed with the work of Bridgman [22], where, in experiments on polycrystalline metals, the shear strength was shown to be dependent on the normal stress on the plane of shear.

Now from Figure 2.17:

$$F_{n_s} = F_r \sin\left(\phi + \beta - \gamma_{n_e}\right) \tag{2.19}$$

and

$$F_{n_s} = \sigma_s A_s = \frac{\sigma_s A_c}{\sin\phi} \tag{2.20}$$

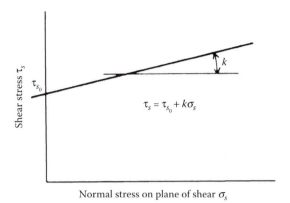

Normal stress on plane of shear σ_s

FIGURE 2.18 Dependence of τ_s and σ_s assumed in Merchant's second theory.

From Equation 2.19 and Equation 2.20:

$$\sigma_s = \frac{\sin\phi}{A_c} F_r \sin\left(\phi + \beta - \gamma_{n_e}\right) \tag{2.21}$$

Combining Equation 2.14 and Equation 2.21,

$$\tau_s = \sigma_s \cot\left(\phi + \beta - \gamma_{n_e}\right) \tag{2.22}$$

and from Equation 2.18 and Equation 2.22

$$\tau_s = \frac{\tau_{s_0}}{1 - k\tan\left(\phi + \beta - \gamma_{n_e}\right)} \tag{2.23}$$

This equation shows how the value of τ_s may be affected by changes in ϕ and is now inserted in Equation 2.16 to give a new equation for F_c in terms of ϕ:

$$F_c = \frac{\tau_{s_0} A_c \cos\left(\beta - \gamma_{n_e}\right)}{\sin\phi\cos\left(\phi + \beta - \gamma_{n_e}\right)\left[1 - k\tan\left(\phi + \beta - \gamma_{n_e}\right)\right]} \tag{2.24}$$

It is now assumed that k and τ_{s_0} are constants for the work material and that A_c and γ_{n_e} are constants for the cutting operation. Thus Equation 2.24 may be differentiated to give the new value of ϕ. The resulting expression is

$$2\phi + \beta - \gamma_{n_e} = C \tag{2.25}$$

where C is given by arccot k and is a constant for the work material. (As mentioned earlier, however, more recent experimental work indicates that τ_s remains constant for a given material over a wide range of cutting conditions, and therefore k would be expected to be zero.)

2.8.2 THEORY OF LEE AND SHAFFER

The theory of Lee and Shaffer [23] was the result of an attempt to apply simplified plasticity analysis to the problem of orthogonal metal cutting. Certain assumptions regarding the behavior of the work material under stress were made as follows:

1. The material is rigid plastic, which means that the elastic strain is negligible during deformation and that once the yield point is exceeded deformation takes place at constant stress. The stress-strain curve for such a material is shown schematically in Figure 2.19, where it can be seen that the material does not work-harden.
2. The behavior of the material is independent of the rate of deformation.
3. The effects of temperature increases during deformation are neglected.
4. The inertia effects resulting from acceleration of the material during deformation are neglected.

These assumptions have been commonly made and have led to useful solutions of many plasticity problems. They should closely approximate the actual behavior of the material during cutting because of the very high strains and strain rates that occur in the cutting process. It is known that the rate of work hardening of most metals decreases rapidly with increasing strain and that the effect of a high strain rate is to raise the yield strength of the metal with respect to its ultimate stress. Also, with the high strains encountered, the elastic strain would form a negligible proportion of the total strain. Thus, the stress-strain curve for the work material would correspond approximately to the ideal case, as shown in Figure 2.19.

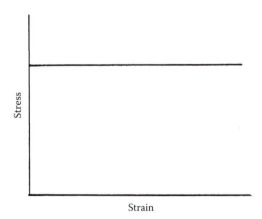

FIGURE 2.19 Stress-strain curve for a rigid-plastic material.

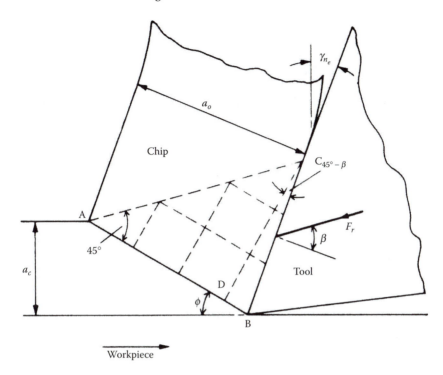

FIGURE 2.20 Lee and Shaffer's slip-line field for orthogonal cutting, where F_r = resultant tool force, ϕ = shear angle, γ_{n_e} = working normal rake, β = mean friction angle on tool face, a_c = undeformed chip thickness, and a_o = chip thickness.

One approach to solving problems in plasticity is to construct a slip line field; this field consists of two orthogonal families of lines (called slip lines), indicating, at each point in the plastic zone, the two orthogonal directions of maximum shear stress. Any field constructed must be compatible with boundary conditions and velocity fields within the deformation zone.

The slip-line field proposed by Lee and Shaffer for the orthogonal cutting of a continuous chip is shown in Figure 2.20. It can be seen that Lee and Shaffer have employed the idealized shear-plane model of cutting, where all the deformation takes place in a plane extending from the tool cutting edge to the point of intersection of the free surfaces of the work and chip. Consideration is given, however, to the manner in which the cutting forces applied by the tool are transmitted through the chip to the shear plane. This transmission of forces results in the triangular plastic zone ABC, where no deformation occurs but the material is stressed to its yield point. Thus, the maximum shear stress throughout this zone is τ_s, the shear stress on the shear plane, and the two directions of this maximum shear stress are indicated by the two orthogonal sets of straight lines (slip lines).

If the boundaries of this triangular zone are considered, it is clear that the shear plane *AB* must give the direction of one set of slip lines since the maximum shear stress must occur along the shear plane. Also, since no forces act on the chip after it has passed through the boundary *AC*, no stresses can be transmitted across this boundary. Thus, *AC* can be regarded as a free surface, and since the directions of maximum shear stress always meet a free surface at $\pi/4$, the angle *CAB* is equal to $\pi/4$. Finally, assuming that the stresses acting at the chip-tool interface are uniform (an unreasonable assumption, as shown later), the principal stresses at the boundary *BC* will meet this boundary at angles β and $[\beta + \pi/2]$ (where β is given by arctan F_f/F_n and is the mean chip-tool friction angle). Directions of maximum shear stress lie at $\pi/4$ to the directions of principal stress, and thus the angle *BCD* is given by $[\pi/4] - \beta$.

It now follows from Figure 2.20 that

$$\phi + \beta + \frac{\pi}{4} - \gamma_{n_e} = \frac{\pi}{2}$$

Thus,

$$\phi + \beta - \gamma_{n_e} = \frac{\pi}{4} \tag{2.26}$$

which is the required shear-angle solution.

Lee and Shaffer realized that Equation 2.26 could not apply where β is equal to $\pi/4$ and γ_{n_e} is zero since with these values, ϕ would be zero. They considered, however, that such conditions of high friction and low rake angle were just those conditions that lead to the formation of a built-up edge in practice. To support this point, a second solution was presented for the new geometry where a built-up edge is present on the tool face [23].

2.8.3 EXPERIMENTAL EVIDENCE

The shear angle relationships outlined above have been compared with the results of independently conducted experiments [24,25]. Figure 2.21 and Figure 2.22 show the most convenient way of making such comparisons, that is, to plot the shear angle ϕ against $\beta - \gamma_{n_e}$. On such a graph, the relationships obtained by the Ernst and Merchant and the Lee and Shaffer theories both form straight lines. It can be seen that neither of these theories approach quantitative agreement with any of the experimental relationships for the various materials tested. However, if these theories are compared qualitatively with the experimental results, both the theories and the experimental results indicate that an approximate linear relationship exists between ϕ and $\beta - \gamma_{n_e}$ and that a decrease $\beta - \gamma_{n_e}$ always results in an increase in ϕ. Thus, for a given rake angle γ_{n_e} a decrease in β, the mean friction angle on the tool face, results in an increase in ϕ, the shear angle, with a corresponding decrease in the area of shear. Since the mean shear strength of the work material in the shear zone remains constant, the force required to form the chip will be reduced. This agrees with general observations of machining that

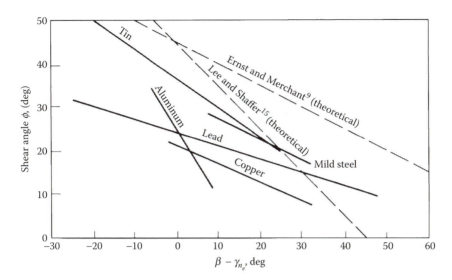

FIGURE 2.21 Comparison of theoretical and experimental shear-angle relationships for orthogonal metal cutting, where ϕ = shear angle, γ_{n_e} = working normal rake, and β = mean friction angle on tool face. (After Pugh., H.D., Proc. Inst. Mech. Eng. Conf. Technol. Eng. Manuf., London, 1958, 235.)

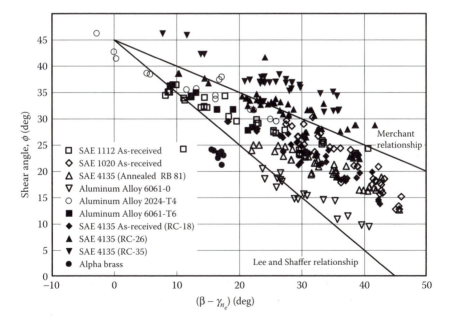

FIGURE 2.22 Experimental shear angle results; top line represents Merchant's relationship and the bottom line represents Lee and Shaffer's relationship. (Adapted from Kobayashi, S., et al., *Trans. ASME, J. Eng. Ind.*, Vol.82B/4, 333–347, 1960.)

the addition of lubricant results in thinner chips and lower cutting forces. An increase in rake γ_{n_e} always results in an increase in the shear angle in ϕ and hence a reduction in the cutting forces. This increase in shear angle, however, is usually accompanied by a relatively small increase in the friction angle β, an effect that will be discussed later. Again this agrees with experimental observations that for a given work material increasing the rake angle produces thinner chips and lower cutting forces.

The comparison in Figure 2.21 and Figure 2.22 shows that no unique relationship of the kind predicted by the Ernst and Merchant and the Lee and Shaffer theories could possibly agree with all the experimental results. Even the modified Merchant theory in which the shear stress on the shear plane is assumed to be linearly dependent on the normal stress could not agree with all the results. The modified Merchant theory yielded the relationship

$$2\phi + \beta - \gamma_{n_e} = C \qquad (2.27)$$

where C is a constant depending on the work material. Substituting various values of C in Equation 2.27 could only give a series of parallel lines on the graph in Figure 2.21. Clearly, the experimental lines are not parallel and could not be represented by Equation 2.27.

It is not difficult to produce reasons why the theories of Ernst and Merchant and of Lee and Shaffer do not find agreement with experimental results. The assumption that the cutting tool is perfectly sharp (i.e., the plowing force F_p is negligible) could be a very rough approximation to actual conditions, particularly at small values of undeformed chip thickness. Further, it has been shown that the primary deformation zone cannot be regarded as a shear plane under a wide variety of cutting conditions. For example, Palmer and Oxley [26], using cinephotography to observe the flow of grains in a steel workpiece during slow-speed cutting, found that the primary deformation zone had the form shown in Figure 2.23. Nakayama [27] showed that this wide deformation zone had constant proportions for cutting speeds as high as 2.5 m/s. In Nakayama's experiments the side of the specimen workpiece was coated with lampblack, and a series of lines was inscribed on the prepared surface parallel to the cutting direction. During cutting, these lines formed stationary material flow lines and could be photographed. It was therefore possible to plot the boundaries of the primary deformation zone. Similar observations have been made using a variety of techniques, including quick-stop tests using photo deposited grids on the work surface to determine the extent of the deformation zone [28].

The shear-angle theories of Ernst and Merchant and of Lee and Shaffer and experimental results such as those in Figure 2.21 and Figure 2.22 indicate that the friction on the tool face is a most important factor in metal cutting. In many of the metal-cutting theories produced, the simplifying assumption has been made that chip-tool frictional behavior can be represented by a single value of the mean

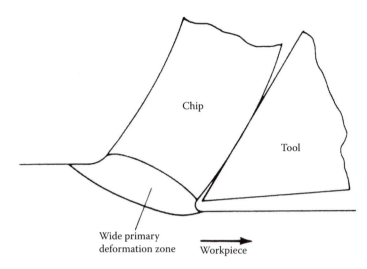

FIGURE 2.23 Model of an orthogonal cutting process. (After Palmer, W.B. and Oxley, P.L.B., *Proc. Inst. Mech. Eng.*, London, Vol. 173, 623–634, 1959.)

coefficient of friction on the tool face. It is therefore important to consider in detail the nature of the frictional conditions between the chip and the tool in metal cutting.

2.9 FRICTION IN METAL CUTTING

It is clear from the preceding discussion that, by some mechanism not completely understood, the frictional behavior on the tool face affects the geometry of the cutting process. Before the frictional conditions in metal cutting are considered, it is necessary to discuss the nature of friction between dry sliding surfaces.

Amontons' laws of friction, formulated in 1699 [29], state that friction is independent of the apparent area of contact and proportional to the normal load between the two surfaces. In 1785, Coulomb [30] verified these laws and made a further observation, that the coefficient of friction is substantially independent of the speed of sliding. The work of Bowden and Tabor [31] has contributed much to the explanation of these empirical laws.

Microscopic examination shows that even the most carefully prepared "flat" metallic surfaces consist of numerous hills and valleys. When two surfaces are placed together, contact is established at the summits of only a few irregularities (asperities) in each surface (Figure 2.24a). If a normal load is applied, yielding occurs at the tips of the contacting asperities, and the real area of contact A_r, increases until it is capable of supporting the applied load. For the vast majority of engineering applications, this real area of contact A_r is only a small fraction of the apparent contact area A_a and is given by

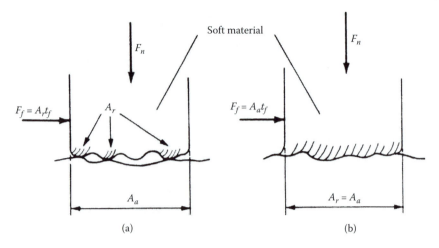

FIGURE 2.24 Suggested frictional behavior for a "soft" slider, where F_f = frictional force, F_n = normal force, A_r = real area of contact, A_a = apparent area of contact, and τ_f = shear strength of softer metal. (a) Sliding friction; (b) sticking friction.

$$A_r = \frac{F_n}{\sigma_y} \qquad (2.28)$$

where F_n is the normal force, and $_y$ is the yield pressure of the softer metal.

The adhesion resulting from the intimate metallic contact of these asperities has been termed *welding*, and when sliding takes place, a force is required for continual shearing of the welded junctions at the tips of these asperities. The total frictional force F_f is therefore given by

$$F_f = \tau_f A_r \qquad (2.29)$$

where τ_f is the shear strength of the softer metal.

Thus, from Equation 2.28 and Equation 2.29 the equivalent coefficient of friction is given by

$$\mu = \frac{F_f}{F_n} = \frac{\tau_f}{\sigma_y} \qquad (2.30)$$

Equation 2.30 shows that the coefficient of friction is independent of the apparent contact area, and since the ratio τ_f/σ_y would be expected to be substantially constant for a given metal, the frictional force is proportional to the normal load (i.e., μ is constant). These results are consistent with the laws of dry sliding friction.

During metal cutting, it has generally been observed that the mean coefficient of friction between the chip and the tool can vary considerably and is affected by changes in cutting speed, rake angle, and so on. This variance of the mean coefficient of friction results from the very high normal pressures that exist at the chip-tool interface. For example, when steel is machined, these normal pressures can be as high as 3.5 GN/m² and can cause the real area of contact to approach, or become equal to, the apparent contact area over a portion of the chip-tool interface (i.e., A_r/A_a equals unity). Thus, under these circumstances A_r has reached its maximum value and is constant. The frictional force F_f is still given by Equation 2.29 but is now independent of the normal force F_n, and the ordinary laws of friction no longer apply. Under these conditions the shearing action is no longer confined to surface asperities but takes place within the body of the softer metal (Figure 2.24b).

Consideration of frictional behavior in metal cutting has led [32,33] to the model of orthogonal cutting with a continuous chip and no built-up edge shown in Figure 2.25. Here the normal stresses between the chip and the tool are sufficiently high to cause A_r/A_a to approach unity over the region of length l_{st}

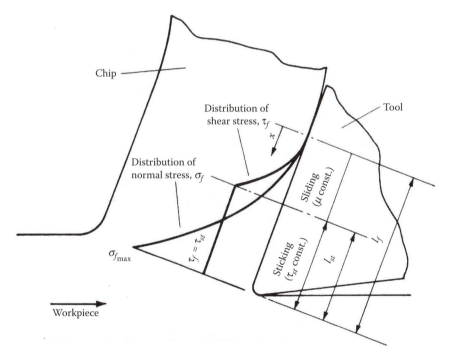

FIGURE 2.25 Model of chip-tool friction in orthogonal cutting, where $\sigma_{f\max}$ = maximum normal stress, τ_f = normal stress, τ_f = shear stress, τ_{st} = shear strength of chip material in the sticking region, l_f = chip-tool contact length, and l_{st} = length of sticking contact length. (After Zorev, N.N., *Int. Res. Prod. Eng.*, ASME, Ney York, 42–49, 965, 1963.)

adjacent to the tool cutting edge, termed the *sticking region*. In the length $l_f - l_{st}$, extending from the end of the sticking region to the point where the chip loses contact with the tool, the ratio A_r/A_a is less than unity, and therefore the coefficient of friction is constant; this region has been termed the *sliding region*.

In previous work [20], evidence of the sticking mode of frictional contact was produced by examination of the undersurface of the chip on specimens where the cutting action had suddenly been stopped. It was observed that in a region adjacent to the tool cutting edge, the grinding marks on the tool face were imprinted on the undersurface of the chip, indicating that no relative motion between the chip and the tool had occurred and that the real and apparent areas of contact are equal in this region. These observations have been confirmed optically using transparent sapphire tools and high-speed photography [33,34].

Under conditions of sticking friction, the mean angle of friction on the tool face will depend on the form of the normal stress distribution, the chip-tool contact length l_f, the mean shear strength of the chip material in the sticking region, and the coefficient of friction in the sliding region. Clearly, a single value of the mean angle of friction is insufficient to describe completely the frictional conditions on the tool face.

An analysis of the stress distribution on the tool face shown in Figure 2.25 has been presented by Zorev [32,33]. It was shown that the mean angle of friction is mainly dependent on the mean normal stress on the tool face, and this result may be used to explain the effect of changes in working normal rake γ_{n_e} on the mean friction angle β. As γ_{n_e} increases, the component of the resultant tool force normal to the tool face will decrease, and therefore the mean normal stress will decrease. However, the mean shear stress remains roughly constant and therefore an increase in γ_{n_e} would be expected to increase the mean angle of friction β. This result is in accordance with the findings of experimental work [23] where an increase in γ_{n_e} has been shown to result in an increase in β for a wide variety of work materials. The general form of the rake face stress distributions shown in Figure 2.25 have been observed experimentally from photoelastic measurements [36] and using split-tool dynamometers [37]. These methods do not allow the stresses to be determined very close to cutting edge, but there is evidence to indicate that the normal stress may be constant in the sticking region close to the cutting edge [37].

2.10 ANALYTICAL MODELING OF MACHINING OPERATIONS

In the period since Ernst and Merchant presented what was effectively the first analytical modeling approach to orthogonal cutting, a vast body of work has been devoted to investigation and analysis of the machining process. It is not intended in a basic text such as this one to present an exhaustive description of this work, but a number of comprehensive reviews have been made at various times [38–41]. However it is appropriate to outline some of the basic approaches to the analytical modeling of machining and to illustrate some of the more recent results that have been obtained.

2.10.1 MECHANISTIC MODELING OF MACHINING

Mechanistic models of machining processes are aimed at the accurate prediction of dynamic cutting forces for use in estimating other aspects of the cutting process including product quality, tool life, and process stability. The underlying assumption behind mechanistic methods is that the cutting forces are proportional to the uncut chip area and the constant of proportionality depends on the cutting conditions, process geometry, and work material properties. Early work using this basic approach was carried out by Koenigsberger and Sabberwal [42], but since then many models of three-dimensional cutting processes such as turning, boring, milling, and drilling have been developed [43].

An underlying assumption is that the force system at any point on the cutting edge can be represented by an oblique machining process as shown in Figure 2.26. The tool orientation is defined by the inclination angle λ_s and the rake face orientation by the normal rake angle γ_{n_e}. The magnitudes of the normal and frictional forces are assumed proportional to the area of the uncut chip A_c and are given by:

$$F_n = K_n A_c \qquad (2.31)$$

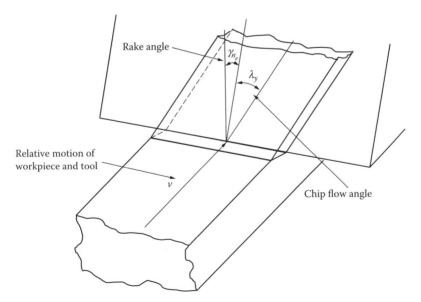

Rake angle

γ_{n_e}

λ_y

Relative motion of workpiece and tool

v

Chip flow angle

FIGURE 2.26 Oblique machining process, where v = cutting velocity, γ_{n_e} = working normal rake, and λ_γ = chip flow angle.

$$F_f = K_f A_c \qquad\qquad (2.32)$$

where

K_n = specific normal pressure
K_f = specific friction pressure

The specific pressure coefficients depend on the tool-workpiece combination, cutting conditions and cutting geometry. These coefficients have the same physical dimensions as the specific cutting energy (Section 2.5) and are determined from representative experimental data in a process called *calibration*. The resulting normal force is perpendicular to the rake face and the friction force lies in the plane of the rake face in the direction of chip flow defined by the chip flow angle λ_γ.

For specific cutting processes, the instantaneous uncut chip thickness and chip area must be determined. The more complex cutting edge geometries can be divided into a number of elemental oblique cutting edges each behaving as described above (Figure 2.27). The total force acting is obtained as the sum of these elemental oblique cutting process forces. The chip flow angle for each elemental edge in milling and drilling can be assumed to follow Stabler's law [44] (see Chapter 7 also), from which the chip flow angle is shown to be equal to the edge inclination angle. For the curved edges in turning and face milling, the effective chip flow direction is determined by breaking the edge into a large number of small straight edges of varying orientation, each of which is assumed to be a simple oblique cutting edge. Stabler's law can be assumed for each elemental cutting edge and the unified chip flow direction determined as the weighted average of all of the individual chip flow directions, with the weighting for each element determined, for example, from the magnitude of the friction force at the elemental edge.

This overall modeling approach has been adapted to a wide range of process geometries [38–40,43]. Instantaneous forces can be determined including the effect of such cutting edge errors as run out, chipped cutting edges and so on. These models can be incorporated into a feed back loop involving the dynamics of the machine structure for the analysis of machine tool stability.

2.10.2 SLIP LINE FIELD ANALYSIS

An approach to analyzing plastic deformation is slip line theory. The analysis by Lee and Shafer [23] described in Section 2.8.2 is a simplified slip line field solution. In general slip line theory applies to plane strain conditions only and the material properties are simplified to rigid perfectly plastic, i.e., no elastic strains occur and plastic flow occurs when the applied shear stress reaches a critical value k that is assumed to remain constant for varying strains, strain rates, and temperatures. This means that the approach is restricted to steady state (continuous chip), orthogonal cutting conditions.

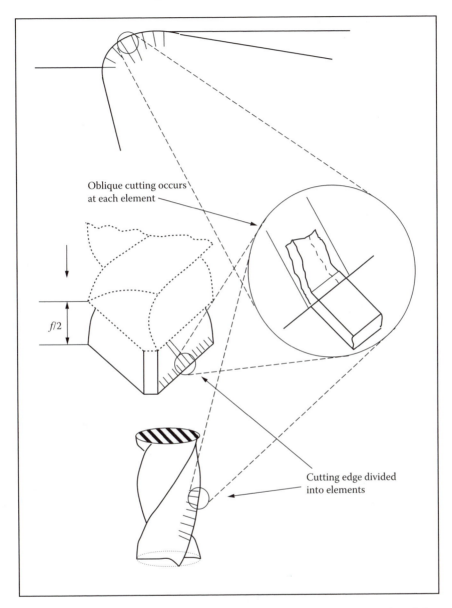

FIGURE 2.27 Approximate representations of complex cutting edges as a series of elemental oblique cutting edges.

Analysis of the stress in a plane strain loaded material shows that there are two orthogonal directions in which the shear stresses are a maximum, but these directions generally vary from point to point. A line tangential to the maximum shear stress is called a *slip line*. Consequently a complete set of orthogonal curvilinear slip lines in a plastic region is known as a *slip line field*. A slip line field must be compatible with boundary conditions and velocity fields, together with meeting some basic conditions. Firstly the boundary between material in a plastic state and material that has not yielded is a slip line. Thus in machining the boundaries between the primary deformation zone and the work and the primary deformation zone and the chip are slip lines. Similarly the boundary between the secondary deformation zone and the chip is also a slip line. Other basic conditions are that slip lines must meet a free surface at an angle of 45 degrees and at a friction surface, such as the rake face contact zone, where the friction stress is given by *mk*, slip lines must intersect the surface at an angle $\frac{1}{2}\cos^{-1}m$. Once a suitable slip line field has been constructed then equilibrium conditions enable forces and contact stresses to be determined.

A significant finding from slip line modeling is that the machining process is not uniquely defined. This means that specification of the rake angle and rake face friction factor *m* does not determine the resulting chip thickness. In practice multiple slip line fields can be constructed with different tool-chip contact lengths and chip thicknesses. Figure 2.28 shows three examples introduced by various workers [12], for the same rake angle and friction factor *m*. The first (Figure 2.28a) is the field proposed early on by Lee and Shaffer [23] and described in Section 2.8.2. Figure 2.28b shows one of a family of fields proposed by Kudo [44]. In this case the chip is formed straight, but the lower chip velocities near the tool face cause the chip to become curved. Figure 2.28c is one of a family of fields developed by Dewhurst [19] that model chip curvature more effectively. It can be readily seen that different chip shapes and thicknesses result from the same specified input conditions. Other forms of slip line field can also be derived, including a general form [45] that can be decomposed into the fields introduced by others, including the single shear plane model.

For steady state orthogonal machining, slip line field solutions provide a good insight into possible chip flows. In addition rake face contact stress distributions similar to those observed experimentally can be determined. However, the non-uniqueness of the possible solutions is a significant limitation and this mainly results from the use of a rigid plastic material model. This latter assumption further means that the effects of variable flow stress properties with strain, strain rate, etc., that are known to have large effects in machining, cannot be taken into account.

2.10.3 PREDICTIVE MODELS FOR ORTHOGONAL CUTTING

The single shear plane formulation developed by Ernst and Merchant was effectively the first predictive model for the machining process. While this has been shown to be consistent with many observations of the effect of various parameters on the chip formation, the inadequacies resulting from the simplification of a single shear plane have been outlined above. The shear plane is a velocity discontinuity and consequently the effects of strain hardening and strain rate, which is very high in actual cutting, cannot be taken into account. Consequently,

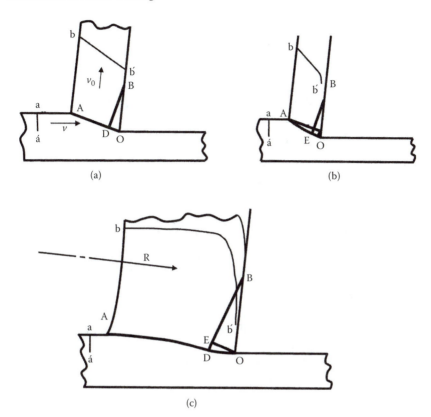

(a)

(b)

(c)

FIGURE 2.28 Three possible slip line field configurations for orthogonal cutting, with rake angle $\gamma_{n_e} = 5^0$ and friction factor $m = 0.9$: (a) Lee and Shafer model (From Lee, E.H. and Shaffer, B.W., *J. Appl. Physics*, Vol. 18(4), 405, 1951.); Kudo model (From Kudo, H., *Int. J. Mech. Sci.*, Vol. 7, 43–55, 1965.); Dewhurst model (From Dewhurst, P., *Proc. Royal Society*, London, Vol. A360, 587–610, 1978.). The line bb′ represents the distortion of the line aa′ after chip formation.

a number of analyses of the metal-cutting process using a thick shear-zone model have been developed and Oxley and his co-workers have carried out pioneering work in this area [12,26,28,46,47].

The starting point has been visioplasticity studies, in which observed flow patterns are used to deduce strain rates, strains, and temperatures. From these experimentally derived slip line fields that take into account flow stress variations can be obtained. Based on such experimental observations, an early model utilized a parallel-sided, thick, shear zone [28]. This model retained many of the features of the single shear-plane model, but allowed the effects of strain hardening, strain rate, and temperature to be included in the analysis. Over the years this approach has been refined into a predictive model for orthogonal cutting. The theory is based on a model of chip formation shown in Figure 2.29 that is derived from

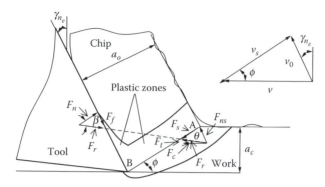

FIGURE 2.29 Predictive model for orthogonal cutting, where F_r = resultant tool force, F_c = cutting force, F_t = thrust force, F_s = shear force on shear plane, F_f = frictional force on tool face, F_n = normal force on tool face, ϕ = shear angle, γ_{n_e} = working normal rake, β = mean friction angle on tool face, a_c = undeformed chip thickness, a_o = chip thickness, v = cutting speed, v_s = shearing velocity on the shear plane, and v_0 = chip velocity.

the plasticity analyses of experimental flow fields. Plane-strain, steady-state conditions are assumed.

The shear plane, AB, near the center of the zone in which the chip is formed, (found from the same geometric construction as used in defining the shear plane in the shear plane model [21]) and the tool/chip interface are both assumed to be directions of maximum shear stress and maximum shear strain-rate. The basis of the theory is to analyze the stress distributions along AB and the tool/chip interface in terms of the shear angle ϕ (angle made by AB with cutting velocity v), work material properties, etc., and then to select ϕ so that the resultant forces transmitted by AB and the interface are in equilibrium. In this model, the tool is assumed to be perfectly sharp, meaning that no plowing forces exist. Once ϕ is known then the chip thickness a_0 and the various components of the cutting force can be determined from the following geometric relationships:

$$a_0 = a_c \cos\left(\phi - \gamma_{n_e}\right)\big/\sin\phi \tag{2.33}$$

$$F_c = F_r \cos\left(\beta - \gamma_{n_e}\right) \tag{2.34}$$

$$F_t = F_r \sin\left(\beta - \gamma_{n_e}\right) \tag{2.35}$$

$$F_f = F_r \sin\beta \tag{2.36}$$

$$F_n = F_r \cos\beta \tag{2.37}$$

$$F_n = \frac{F_s}{\cos \lambda} = \frac{\tau_s a_c a_w}{\sin \phi \cos \lambda} \qquad (2.38)$$

where α_c is the undeformed chip thickness, a_w is the width of cut, τ_s is the shear flow stress along AB, and the forces and angles described are as shown in Figure 2.29.

By starting at the free surface just ahead of A and applying the appropriate stress equilibrium equation along AB, it can be shown that, for $0 < \phi \leq \pi/4$, the angle made by the resultant F_r with AB is given by

$$\tan \lambda = 1 + 2\left(\frac{1}{4}\pi - \phi\right) - Cb \qquad (2.39)$$

where C is the constant in an empirical strain-rate relationship

$$\dot{\gamma}_{AB} = Cv_s/l_s \qquad (2.40)$$

in which $\dot{\gamma}_{AB}$ is the maximum shear strain-rate at AB, v_s is the shear velocity, l_s is the length of the shear plane, and b is the strain-hardening index of the material in

$$\sigma = \sigma_1 \varepsilon^b \qquad (2.41)$$

in which σ and ε are the uniaxial (effective) flow stress and strain, and σ_1 and b are constants that define the stress/strain curve for given values of strain rate and temperature. The geometry in Figure 2.29 also defines the angle λ as

$$\lambda = \phi + \beta - \gamma_{n_e} \qquad (2.42)$$

The temperature at AB is found from

$$\theta_s = \theta_0 + \eta\left[\frac{1 - \Gamma_t F_s \cos \gamma_{n_e}}{\rho c a_w \cos\left(\phi - \gamma_{n_e}\right)}\right] \qquad (2.43)$$

where θ_0 is the initial work temperature, F_s is the shear force along AB, $\eta(0 < \eta \leq 1)$ is a factor which allows for the fact that not all of the plastic work of chip formation has occurred at AB, ρ and c are the density and specific heat of the work material, respectively, and Γ is the proportion of heat conducted into the work, restricted to values between zero and one, that is estimated from empirical equations developed from experimental data given by Boothroyd [48]

$$\Gamma = 0.5 - 0.35 \lg\left(R \tan\phi\right) \text{ for } 0.04 < R \tan\phi \le 10.0 \qquad (2.44)$$

$$\Gamma = 0.3 - 0.15 \lg\left(R \tan\phi\right) \text{ for } R \tan\phi > 10 \qquad (2.45)$$

with R a nondimensional thermal number given by

$$R = \rho c v a_c / k \qquad (2.46)$$

where k is the thermal conductivity of the work material.

The strain at AB is given by

$$\gamma_{AB} = \frac{1}{2} \cos\gamma_{n_e} \Big/ \sin\phi\cos\left(\phi - \gamma_{n_e}\right) \qquad (2.47)$$

The average temperature at the tool/chip interface from which the average shear flow stress at the interface is determined is taken as

$$\theta_{int} = \theta_0 + \frac{1 - \beta_t F_s \cos\gamma_{n_e}}{\rho c a_c a_w \cos\left(\phi - \gamma_{n_e}\right)} + \Psi\Delta\theta_m \qquad (2.48)$$

where $\Delta\theta_m$ is the maximum temperature rise in the chip and the factor $\Psi(0 < \Psi \le 1)$ allows for θ_m being an average value. The temperature rise $\Delta\theta_m$ can be estimated using the methods described in Chapter 3, with the tool-chip contact length l_f calculated as

$$l_f = \frac{a_c \sin\lambda}{\cos\beta\sin\phi}\left\{1 + \frac{Cb}{3\left[1 + 2\left(\pi/4 - \phi\right) - Cb\right]}\right\} \qquad (2.49)$$

which is derived by taking moments about B of the normal stresses on AB to find the position of F_r and then assuming that the normal stress distribution at the tool face is uniform so that F_r intersects the tool a distance $l_f/2$ from B.

The above equations are sufficient to calculate cutting forces, temperatures and so on, for given cutting conditions so long as the appropriate work material properties are known [46,47]. With appropriate assumptions these predictions have been shown to be in good agreement with experimental results (Figure 2.30a). The results shown in Figure 2.30b demonstrate the predicted effect of increased cutting speed on cutting forces and the shear angle. Experimental results show that, when machining materials such as steel in the usual cutting range speed range (up to 400 m/min), an increase in cutting speed results in significant

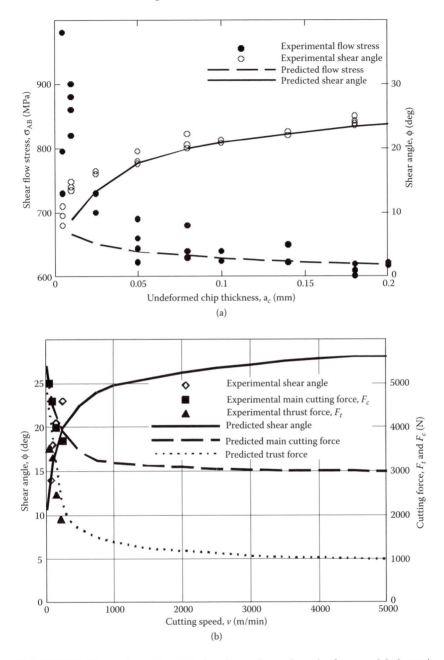

FIGURE 2.30 Comparison of predicted and experimental results from model shown in Figure 2.29. (a) Shear flow stress and shear angle against undeformed chip thickness; (b) cutting forces and shear angle variation with cutting speed. (Adapted from Oxley, P.L.B., Mechanics of Machining: An Analytical Approach to Assessing Machinability, Ellis Horwood, Chichester, England, 1989.)

reduction in the cutting forces, and this is associated with an increase in the shear angle. The predicted results can be extended to ultra high-speed cutting ranges and would indicate that this trend should continue. However, it is likely that the chip formation process may change significantly from that assumed, as indicated in Section 2.3.4.

This basic approach has also been extended to the analysis of oblique machining with nose radius tools [46,47]. A comprehensive description of this modeling approach and its applications can be found elsewhere [46].

2.10.4 FINITE ELEMENT ANALYSIS

The analytical approaches described in the preceding sections have severe limitations if they are to be applied to practical machining operations, in particular those involving three-dimensional effects and nonsteady cutting conditions. Considerable attention has been given to the use of finite element analysis to machining [12,39–41]. This approach offers the best hope for the comprehensive modeling of three-dimensional and nonsteady machining, although at considerable computational effort.

The basis of all finite element methods is the approximation of a material continuum by an assembly of small finite elements for which the relevant variables and quantities are determined only at the nodes of the elements. In between the nodes, the values of the variables, etc., are determined by interpolation. The accuracy of the solution can be improved by increasing the number of elements, although with associated increases in the computing power and time required for the simulation. In order to completely model machining the flow of work material from the workpiece into the chip must be considered, but analysis of the deformation and/or wear of the tool and machine elements may also be necessary.

In modeling the plastic material flow there are two basic approaches for assigning elements, both of which have advantages and disadvantages (Figure 2.31):

1. Fixing the elements in space and allowing material to flow through them (Eulerian technique)
2. Dividing the material into elements that move (and distort) with the flow (Lagrangian technique)

The main advantage of the Eulerian technique is that the shapes of the elements do not alter with time, so associated coefficients require computing only once and this considerably reduces the computational complexity. All machining operations involve a free surface of the chip and determining its position is one of the intended outcomes of the analysis. This presents a problem in the Eulerian technique since placement of the fixed elements is not obvious and an iterative approach to locating the element mesh boundaries is therefore necessary. For this reason the Eulerian technique is inherently more suitable for analysis of steady flow conditions. Thus, in machining, the modeling of intermittent cutting, transition conditions at tool entry and exits and cyclic chip formation mechanisms

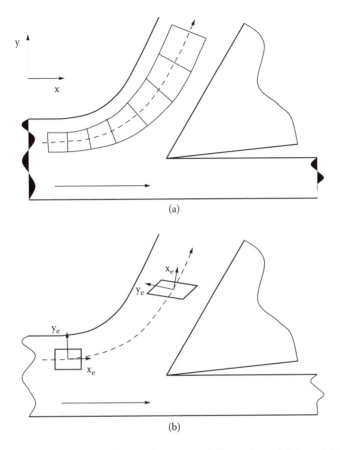

FIGURE 2.31 Approaches to finite element modeling of machining. (a) Eulerian approach, material flows through a fixed element mesh; (b) Lagrangian approach, elements move with material and distort and rotate with position. (Adapted from Childs, T.H.C., et al., *Metal Machining: Theory and Applications*, Arnold, London, 2000.)

are difficult or not possible. Another problem to be faced is the description of material property changes with strain, strain rate, and temperature as the material flows through the element mesh.

In the Lagrangian technique the elements change shape (for machining severely) and orientation as they flow through the deformation zone. Thus the coefficients associated with each element must be continually updated, increasing the computational complexity considerably and this can also lead to problems of geometric nonlinearity. The elements can become so distorted that periodic remeshing is necessary. However, although the computational complexity is greater, the Lagrangian technique is more suitable for the modeling of nonsteady conditions.

A finite element analysis can be made more accurate by using smaller elements, but computation times are increased dramatically as the number of elements is increased. A common feature of the finite element analysis of machining is that the element mesh can be made finer (smaller elements) in regions where variables change rapidly with position. For modeling of the machining process, a finer mesh is generally needed in the primary and secondary deformation zones. Using the Eulerian technique, this is straightforward as the elements in the fixed array can be made smaller in appropriate regions. For the Lagrangian technique, in order to improve the computational efficiency, the mesh should be made course to start with and then be refined as the material passes through the deformation zones, before being made course again. This progressive mesh refinement is an additional complication and is particularly a problem near to the cutting edge of the tool, where the work material divides and the chip is formed. Various approaches for dealing with this separation have been used. For the Eulerian technique the separation of material at the tool edge presents few problems.

Considerable progress has been made in the modeling of machining using the finite element method [39–41]. Flow patterns, cutting forces, temperature distributions, and chip formation mechanisms can be effectively simulated, producing results that are similar to experimental observations. This is particularly the case for orthogonal machining, but progress is also being made with more complex three-dimensional cutting operations. However, as with other predictive analyses of machining, the accuracy of the predictions is very dependent on how effectively the modeling of the work material properties for the conditions existing in the deformation zones, where strains, strain rates, and temperatures are high, is performed.

2.10.5 MODELING OF MATERIAL PROPERTIES

The successful application of any of the predictive modeling approaches to machining is highly dependent on how well the properties of the work material are represented at the conditions existing during chip formation. In particular the flow stress or instantaneous yield stress is influenced by temperature, strain, and strain rate. Thus suitable models of the constitutive behavior of the material are needed. A comprehensive discussion of the material property needs for modeling of the machining process has been given by Childs [48].

In general, the strains, strain rates, and temperatures in practical machining operations are much higher than for other deformation processes. These parameters are typically several orders of magnitude higher than can be handled by normally available materials testing equipment. Thus derivation of the appropriate material properties experimentally is difficult as the conditions existing during cutting are generally outside the range of available material testing equipment. This has led to the suggestion that the cutting process itself should be used as a material test for determining material properties [49,50]. However, this relies on the availability of an effective model of the machining process so that the required properties can be deduced from the experimental data.

Several material property relationships that take into consideration the effects of strain, strain rate, and temperature have been developed, and these have been utilized in the finite element and other analytical approaches to modeling of the machining process. One approach has been to express the flow stress of the work material as a function of strain and a parameter termed velocity-modified temperature, which combines the effects of temperature and strain rate [46]. Other semi-empirical constitutive models have been proposed and a comprehensive discussion of material property models can be found in [51]. Two of the more widely used constitutive models will now be outlined.

The material constitutive model developed by Johnson and Cook [52] relates flow stress to strain, strain rate, and temperature as follows:

$$\sigma = \left(\sigma_0 + \sigma_1 \varepsilon^b\right)\left(1 + C_1 \dot{\varepsilon}_n\right)\left(1 - \theta_n^m\right) \tag{2.50}$$

where

σ = flow stress

ε = equivalent plastic strain

$\dot{\varepsilon}_n$ = normalized strain rate, $\dot{\varepsilon}/\dot{\varepsilon}_0$, in which $\dot{\varepsilon}_0 = 1s^{-1}$

$\sigma_0, \sigma_1, C_1, b, m$ are material dependent constants

and

$$\theta_n = \frac{\left(\theta - \theta_0\right)}{\left(\theta_{mt} - \theta_0\right)} \tag{2.51}$$

where

θ = absolute temperature

θ_0 = room temperature

θ_{mt} = material melting temperature

The latter term reduces the flow stress to zero at the melting temperature. The constants $\sigma_0, \sigma_1, C_1, b, m$ of this model must be fitted to data obtained from several representative material tests at various strains and strain rates. This model has been shown to be a good representation of the strain hardening effect, and it can be readily incorporated into finite element simulations of the machining process.

An alternative material constitutive model based on the dislocation dynamics of plastic flow has been proposed by Zerelli and Armstrong [53] and for a body centered cubic (BCC) material has the form:

$$\sigma = C_0 + C_2 \exp\left(-C_3\theta + C_4\theta \ln(\dot{\varepsilon}_n)\right) + C_5 \varepsilon^b \tag{2.52}$$

where C_0, C_2, C_3, C_4, C_5, and b are material dependent constants. Again these constants must be determined by fitting to a range of experimental test data.

Material property models such as these have been utilized in finite element modeling of machining with some success. The limitation of their application is that the material dependent constants must be determined from tests carried out on available material test equipment. These test conditions are generally at lower strains and strain rates than exist during machining, particularly at high cutting speeds, and the validity of the derived constants for these more extreme conditions may be uncertain.

PROBLEMS

1. In an orthogonal cutting test on mild steel the following results were obtained:

   a_c = 0.25 mm F_c = 900 N
   a_0 = 0.75 mm F_t = 450N
   a_w = 2.5mm γ_{n_e} = 0
   l_f = 0.5 mm

   Calculate:
   a. The mean angle of friction on the tool face β
   b. The mean shear strength of the work material τ_s
   c. The mean frictional stress on the tool face τ_f

2. In an orthogonal cutting test the following conditions were noted:
   Width of chip (width of cut) = 2.5 mm
   Undeformed chip thickness = 0.25 mm
   Chip thickness = 1.0 mm
   Working normal rake = –5 degrees
   Cutting force = 900 N
   Thrust force = 900 N

   Calculate:
   a. The shear angle
   b. The mean shear strength of the work material

3. In an experimental turning operation where low-carbon steel was being machined using a carbide tool, the following data were taken:
   Cutting force F_c = 1 kN
   Thrust force F_t = 0.5 kN
   Working normal rake γ_{n_e} = 20 degrees
   Feed f = 0.141 mm
   Working major cutting-edge angle κ_r = 45 degrees
   Depth of cut (back engagement) a_p = 5 mm

Cutting speed $v = 2$ m/s
Cutting ratio $r_c = 0.2$
Workpiece diameter $d_w = 100$ mm
Workpiece machined length $l_w = 300$ mm

Estimate from the data given:
a. The specific cutting energy of the work material p_s
b. The power required for machining P_m
c. The undeformed chip thickness a_c
d. The width of cut a_w
e. The shear angle ϕ
f. The mean angle of friction on the tool face β
g. The time taken to complete the machining operation

4. Show that in metal cutting when the working normal rake is zero, the ratio of the shear strength of the work material τ_s to the specific cutting energy p_s is given by

$$\frac{\tau_s}{p_s} = \frac{1 - \mu r_c}{1 + r_c^2}$$

where μ is the coefficient of chip-tool friction, and r_c is the cutting ratio.

5. In an orthogonal machining operation with a rake angle of 5 degrees and an undeformed chip thickness of 1 mm, the chip thickness is found to be 3 mm. What will the chip thickness be if the rake angle is increased to 15 degrees, assuming Ernst and Merchant's first theory holds, i.e.,

$$2\phi + \beta - \gamma_{n_e} = \frac{\pi}{2}$$

Assume the friction coefficient is not influenced by changes in rake angle.

6. Derive an expression for the specific cutting energy p_s in terms of the shear angle ϕ and the mean shear strength of the work material τ_s in orthogonal cutting. Assume that the shear-angle relationship of Ernst and Merchant applies, that is,

$$2\phi + \beta - \gamma_{n_e} = \frac{\pi}{2}$$

7. Assuming that the shear angle theory of Lee and Shaffer applies, namely,

$$\phi + \beta - \gamma_{n_e} = \frac{\pi}{4}$$

Show that the specific cutting energy p_s will be given by

$$p_s = \tau_s \left(1 + \cot \phi\right)$$

where τ_s is the mean shear strength of the work material.

8. For the orthogonal cutting of a particular work material, it is found that the length of chip-tool contact is always equal to the chip thickness a_0, and that the mean shear stress at the chip-tool interface is equal to the mean shear stress on the shear plane. Show that, under these circumstances, the mean coefficient of friction on the tool face μ must be equal to or less than 4/3 and that when it is equal to unity, the shear angle ϕ is equal to the working normal rake γ_{n_e}.

9. In cutting experiments using a sharp tool with a rake angle of zero degrees, it was found that the chip-tool contact length was equal to the chip thickness a_0. It was also found that the shear stress along the tool face was constant and was a constant proportion R of the shear strength of the work material on the shear plane.
   a. Derive an expression for the cutting force F_c in terms of a_c, a_w, τ_s, R, and ϕ.
   b. Differentiate the expression to obtain the value of ϕ to give minimum F_c.

10. Assume that in an orthogonal cutting operation, the frictional force F_f on the tool face is given by $K \tau_s A_0$ where K is a constant, τ_s is the apparent shear strength of the work material, and A_0 is the cross-sectional area of the chip. Show that the following relationship exists between the mean coefficient of friction μ on the tool face, the shear angle ϕ and the working normal rake γ_{n_e}.

$$\mu = \frac{K \cos^2\left(\phi - \gamma_{n_e}\right)}{K \sin\left(\phi - \gamma_{n_e}\right)\cos\left(\phi - \gamma_{n_e}\right) + 1}$$

11. For the machining of a particular work material, it is found that the shear angle ϕ is always equal to the working normal rake γ_{n_e}. Assuming that the shear strength of the work material in the sticking-friction region on the tool face is the same as the shear strength τ_s of the

material on the shear plane and that the length of the sticking region is equal to the chip thickness a_0, derive expressions for the cutting force F_c and the thrust force F_t in terms of τ_s, γ_{n_e} and the cross-sectional area of the uncut chip A_c. Also, calculate the value of γ_{n_e} for which F_t would be zero. Neglect the forces in the sliding-friction region on the tool face.

12. In machining tests it was found that for a particular material, the following relation applied:

$$2\phi + \beta - \gamma_{n_e} = \frac{\pi}{2}$$

Also, when the rake angle γ_{n_e} was 20 degrees, the coefficient of friction was 1.2.

a. What would the chip thickness be if in a turning operation the feed was 0.15 mm and the major cutting-edge angle was 60 degrees?

b. If the cutting force is limited to 3 kN and the specific cutting energy of the work material is 2 GJ/m³ what maximum depth of cut can be taken?

13. Assuming that in orthogonal machining the distribution of normal stress along the tool rake face is linear, becoming maximum at the cutting edge and reducing to zero at the end of the chip-tool contact length, derive an expression for the mean coefficient of friction μ between the chip and tool in terms of: the maximum normal stress σ_m, the normal stress σ_0 at which real and apparent areas of contact become equal, and the coefficient of friction in sliding region μ_s.

REFERENCES

1. Finnie, I., Review of the Metal Cutting Analyses of the Past Hundred Years, *Mech. Eng.*, Vol. 78, no. 8, 715–721, 1956.

2. Cocquilhat, M., Experiences sur la Resistance Utile Produite dans le Forage, *Annales Travaux Publics en Belgique*, Vol. 10, 199–215, 1851.

3. Hartig, E., *Versuche uber Leistung und Arbeitsverbrauch der Werkzeugmaschine*, 1873.

4. Time, I., Soprotiveniye Metallov i Dereva Rezaniju (Resistance of Metals and Wood to Cutting), St. Petersburg, 1870.

5. Tresca, H., Memoires sur le Rabotage des Metaux, Bull. de la Societe d'Encouragement pour l'Industrie Nationale, 585–607, 1873.

6. Mallock, A., The Action of Cutting Tools, Proc. Royal Society, London, Vol. 33, 127–139, 1881–1882.

7. Reuleaux, F., Uber den Taylor Whiteschen Werkzengstahl, in Verein zur Beforderung des Gewerbefleisses in Preussen, Sitzungsberichte, Vol. 79, 179, 1900.

8. Taylor. F.W., On the Art of Cutting Metals, Trans. ASME, Vol. 28, 31, 1906.

9. Ernst. H. and Merchant, M.E., Chip Formation, Friction, and High Quality Machined Surfaces, in Surface Treatment of Metals, Vol. 29, American Society of Metals, New York, 1941, 299.

10. Wallace, P.W., An Investigation on the Friction Between Chip and Tool in Metal Cutting, Fellowship Thesis, Department of Mechanical Engineering, University of Salford, England, 1962.

11. Stevenson, M.G. and Oxley, P.L.B., An Experimental Investigation of the Influence of Speed and Scale on the Strain-Rate in the Zone of Intense Plastic Deformation, Proc. Inst. Mech. Eng., London, Vol. 184, 561, 1970.

12. Childs, T.H.C., Maekawa, K., Obikawa, T., and Yamane, Y., Metal Machining: Theory and Applications, Arnold, London, 2000.

13. Komandhuri, R., Schoeder, T., Hazra, J., Von Turkovich, B.F., and Flom, D.G., On Catastrophic Shear Instability in High Speed Machining of AISI 4340 Steel, Trans. ASME, J. Eng. Ind., Vol. 104, 121–131, 1982.

14. Shaw, M.C. and Vyas, A., Chip Formation in the Machining of Hardened Steel, Ann. CIRP, Vol. 41/1, 71–75, 1993.

15. Komandhuri, R. and Brown, R.H., On the Mechanics of Chip Segmentation in Machining, Trans. ASME, J. Eng. Ind., Vol. 103, 33–51, 1981.

16. Davies, M.A. and Burns, T.J., Thermomechanical Oscillations in Material Flow During High-Speed Machining, Phyl. Trans. Royal Society, London, Vol. 359, 821–846, 2001.

17. Toups, R.M., An Analysis of the Natural Sharpness Radius of Cutting Tools, MS Thesis, Mechanical Engineering, Georgia Institute of Technology, Atlanta, September, 1961.

18. Hill, R., The Mechanics of Machining: A New Approach, J. Mech. Physics Solids, Vol. 3, 47–53, 1954.

19. Dewhurst, P., On the Non-Uniqueness of the Machining Process, Proc. Royal Society, London, Vol. A360, 587–610, 1978.

20. Wallace, P.W. and Boothroyd, G., Tool Forces and Tool-Chip Friction in Orthogonal Machining, J. Mech. Eng. Sci., Vol. 6(1), 74–87, 1964.

21. Merchant, M.E., Mechanics of the Metal Cutting Process, J. Appl. Physics, Vol. 16, No. 5, 267 and No.6, 318, 1945.

22. Bridgeman, P.W., Physics Review, Vol. 48, 825, 1935; Proc. American Academy of Arts and Science, Vol. 71, 386, 1937; J. Appl. Physics, Vol. 8(5), 328, 1937; J. Appl. Physics, Vol. 14(6), 273, 1943; The Physics of High Pressure, G. Bell & Sons, Ltd., London, 1949.

23. Lee, E.H. and Shaffer, B.W., The Theory of Plasticity Applied to a Problem of Machining, J. Appl. Physics, Vol. 18(4), 405, 1951.

24. Pugh., H.D., Mechanics of the Cutting Process, Proc. Inst. Mech. Eng. Conf. Technol. Eng. Manuf., London, 1958, 235.

25. Kobayashi, S., Herzog, R.P., Eggleston, D.M., and Thomsen, E.C., A Critical Comparison of Metal Cutting Theories with New Experimental Data, Trans. ASME, J. Eng. Industry, Vol.82B/4, 333–347, 1960.

26. Palmer, W.B. and Oxley, P.L.B., Mechanics of Orthogonal Machining, Proc. Institution Mech. Eng., London, Vol. 173, 623–634, 1959.

27. Nakayama, K., Studies on the Mechanism of Metal Cutting, Bull. Faculty Eng., Yokohama National University, Japan, Vol.7, 1, 1958.

28. Fenton, R.G. and Oxley, P.L.B., Mechanics of Orthogonal Machining: Predicting Chip Geometry and Cutting Forces from Work-Material Properties and Cutting Conditions, Proc. Institution of Mech. Eng., London, Vol. 184, 927, 1970.

29. Amontons, Histoire de l'Academie Royale des Sciences avec les Memoires de Mathematique et de Physique, Paris, 1969.

30. Coulomb, C.A., Memoires de Mathematique et de Physique de l'Academie Royale des Sciences, Paris, 1785.

31. Bowden, F.P. and Tabor, P., Friction and Lubrication of Solids, Oxford University Press, London, 1954.

32. Zorev, N.N., Interrelation between Shear Process Occurring Along Tool Face and on Shear Plane in Metal Cutting, Int. Res. Prod. Eng., ASME, New York, 42–49, 965, 1963.

33. Zorev, N.N., Metal Cutting Mechanics, Pergamon Press, Oxford, 1966.

34. Doyle, E.D., et al., Frictional Interactions in Continuous Chip Formation, Proc. Royal Society, London, Vol. A 366, 173, 1979.

35. Wright, P.K., Frictional Interactions in Machining: Comparison between Transparent Sapphire and Steel Cutting, J. Metals Technol, Vol. 8, 150, 1981.

36. Amini, E., Photo Elastic Analysis of Stresses and Forces in Steady Cutting, J. Strain Anal., Vol. 3, 206, 1968.

37. Barrow, G., et al., Determination of Rake Face Stress Distribution in Orthogonal Machining, Int. J. Machine Tool Design and Res., Vol. 22, 25, 1982.

38. Ehmann, K.F., Kapoor, S.G., DeVor, R.E., and Lazoglu, I., Machining Process Modeling: A Review, Trans. of ASME, J. Manuf. Sci. Eng., Vol. 119, 655–663, 1967.

39. Lutterfelt, C.A. van, Childs, T.H.C., Jawahir, I.S, Klocke, F., and Venuvinod, P.K., Present Situation and Future Trends in Modelling of Machining Operations, Ann. CIRP, Vol 47 (2), 587–626, 1998.

40. Jawahir, I.S. and Lutterfelt, C.A. van, Recent Developments in Chip Control Research and Applications, Ann. CIRP, Vol. 42(2), 659–694, 1993.

41. Byne, G., Dornfeld, D., and Denkena, B., Advancing Cutting Technology, Ann. CIRP, Vol. 53(2), 1–25, 2003.

42. Koenigsberger, F. and Sabberwal, A.J.P., An Investigation into the Cutting Force Pulsations During Milling Operations, Int. J. Machine Tool Design and Res., Vol.1, 15–33, 1961.

43. Kapoor, S.G., DeVor, R.E., Zhu, R., Gajjela, R., Parakkal, G., and Smithey, D., Development of Mechanistic Models for the Prediction of Machining Performance: Model Building Methodology, Proc. CIRP Int. Workshop on Modeling of Machining Operations, Atlanta, GA, 109–120, 1998.

44. Kudo, H., Some New Slip-line Solutions for Two-dimensional Steady-state Machining, Int. J. Mech. Sci., Vol. 7, 43–55, 1965.

45. Fang, N., Jawahir, I.S., and Oxley, P.L.B., A Universal Slip-line Model with Non-unique Solutions for Machining with Curled Chip Formation and a Restricted Contact Tool, Int. J. Mech. Sci., Vol. 43, 557–580, 2001.

46. Oxley, P.L.B., Mechanics of Machining: An Analytical Approach to Assessing Machinability, Ellis Horwood, Chichester, England, 1989.

47. Oxley, P.L.B., Development and Application of a Predictive Machining Theory, Proc. CIRP Int. Workshop on Modeling of Machining Operations, Atlanta, GA, 35–52, 1998.

48. Childs, T.H.C., Material Property Needs in Modeling Machining Operations, Proc. CIRP Int. Workshop on Modeling of Machining Operations, Atlanta, Ga., pp 193–202, 1998.

49. Altan, T., Shatla, M., and Kerk, C., Process Modeling in Machining. Part 1: Determination of Flow Stress Data, Int. J. Machine Tools and Manuf., Vol. 41, 1511–1534, 2001.

50. Ozel, T. and Zeren, E., A Methodology to Determine Work Material Flow Stress and Tool-Chip Interfacial Friction Properties by Using Analysis of Machining, Proc. ASME Int. Mechanical Engineering Congress, Anaheim, Nov. 13–19, 2004 (Paper IMECE2004-59176).

51. Bariani, P.F., Dal Negro, T., and Bruschi, S., Testing and Modelling of Material Response to Deformation in Bulk Metal Forming, Ann. CIRP, Vol. 53/2, 573–595, 2004.

52. Johnson, G.R. and Cook, W.H., A Constitutive Model and Data for Metals Subjected to Large Strains, High Strain Rates, and High Temperatures, Proc, 7th. Int. Symposium on Ballistics, The Hague, Netherlands, 541–547, 1983.

53. Zerelli, F.J. and Armstrong, R.W., Dislocation Mechanics Based Constitutive Relations for Material Dynamics Calculations, J. Appl. Physics, Vol. 67/5, 1816–1825, 1987.

3 Temperatures in Metal Cutting

During the cutting of metal, high temperatures are generated in the region of the tool cutting edge, and these temperatures have a controlling influence on the rate of wear of the cutting tool and on the friction between the chip and tool. Figure 3.1 shows the relationship between tool life, the determination of which is discussed in Chapter 4, and average cutting temperature. Because of the effect on tool wear, considerable attention has been paid in the past to the determination of the temperatures in the tool, chip, and workpiece in metal cutting. This chapter presents a basic analysis of the heat and temperature generation in metal cutting, together with some discussion on the measurement of temperatures during machining.

3.1 HEAT GENERATION IN METAL CUTTING

It has been stated earlier that the rate of energy consumption during machining P_m is given by

$$P_m = F_c v \tag{3.1}$$

where F_c is the cutting component of the resultant tool force, and v is the cutting speed.

When a material is deformed elastically, the energy required for the operation is stored in the material as strain energy, and no heat is generated. However, when a material is deformed plastically, most of the energy used is converted into heat. In metal cutting the material is subjected to extremely high strains, and the elastic deformation forms a very small proportion of the total deformation; therefore, it may be assumed that all the energy is converted into heat.

Conversion of energy into heat occurs in the two principal regions of plastic deformation (Figure 3.2): the shear zone, or primary deformation zone, AB and the secondary deformation zone BC. If, as in most practical circumstances, the cutting tool is not perfectly sharp, a third heat source BD would be present due to friction between the tool and the new workpiece surface. However, unless the tool is severely worn, this heat source would be small and is neglected in the present analysis. Thus,

$$P_m = P_s + P_f \tag{3.2}$$

121

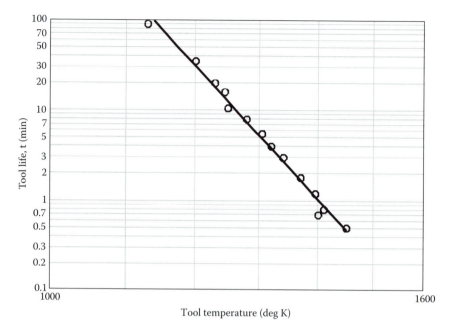

FIGURE 3.1 Experimental tool life and tool temperature results for machining heat-resistant steel with a P10 carbide tool. (Adapted from Oxley, P.L.B., *The Mechanics of Machining*, Ellis-Horwood Ltd., Chichester, U.K., 1989.)

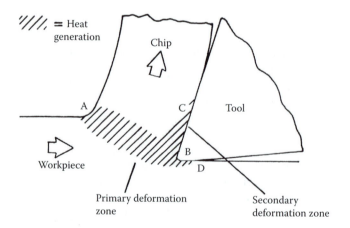

FIGURE 3.2 Generation of heat in orthogonal cutting.

where P_s is the rate of heat generation in the primary deformation zone (shear-zone heat rate), and P_f is the rate of heat generation in the secondary deformation zone (frictional heat rate).

P_f is given by $F_f v_o$, where F_f is the frictional force on the tool face, and v_o is the velocity of chip flow, which is given by vr_c. Thus, if P_f and P_m are known, P_s may be obtained from Equation 3.2.

To understand how heat is removed from these zones by the work-piece, chip, and tool materials, it is first necessary to consider the transfer of heat in a material moving relative to a heat source.

3.2 HEAT TRANSFER IN A MOVING MATERIAL

Consider the element ABCD (Figure 3.3), which has unit thickness and through which the material flows in the x direction. The point A has coordinates x, y, and the material at this point is assumed to have an instantaneous temperature θ. Coordinates and temperatures at points B, C, and D are shown in Figure 3.3.

Heat is transferred across the boundaries AB and CD by conduction because of the temperature gradients in the x direction and by transportation because of the flow of heated material across these boundaries. Across BC and AD, heat can only be transferred by conduction because there is no flow of material across these boundaries. It can be shown that if the net heat flow into the element is zero (no heating within the element), then

$$\frac{\partial^2 \theta}{\partial y^2} + \frac{\partial^2 \theta}{\partial x^2} - \frac{R}{a}\frac{\partial \theta}{\partial \theta} = 0 \tag{3.3}$$

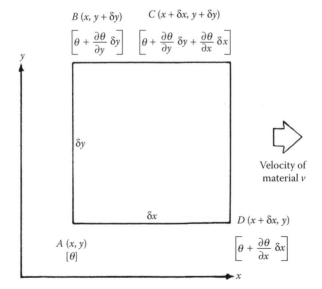

FIGURE 3.3 Element through which heated material flows.

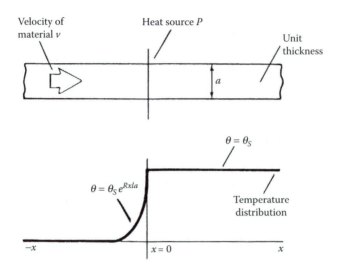

FIGURE 3.4 Temperature distribution in a fast-moving material for a one-dimensional case, where $\Theta_s = P/cva$.

where R is given by $\rho cva/k$ and is known as the thermal number and where

k = thermal conductivity
c = specific heat capacity
ρ = density
v = velocity of the material relative to the heat source

In metal cutting the thermal number is taken to be $\rho cva_c/k$, where v is the cutting speed, and a_c is the undeformed chip thickness.

The solution of Equation 3.3 is possible for only the simplest boundary conditions. It is useful to consider the solution of this equation for a one-dimensional case, and this one-dimensional case is depicted in Figure 3.4 for a metal moving at high speed. Here it is seen that a point in the material approaching the heat source is heated very rapidly, reaches its maximum temperature at the heat source, and then remains at constant temperature. With this result in mind, temperature distribution in metal cutting will now be considered.

3.3 TEMPERATURE DISTRIBUTION IN METAL CUTTING

Figure 3.5 shows an experimentally determined temperature distribution in the workpiece and the chip during orthogonal metal cutting [2]. This is a typical temperature distribution for orthogonal chip formation. As a point X in the material, which is moving toward the cutting tool, approaches and passes through the primary deformation zone, it is heated until it leaves the zone and is carried away within the chip. Point Y, however, passes through both deformation zones, and it is heated until it has left the region of secondary deformation. It is then

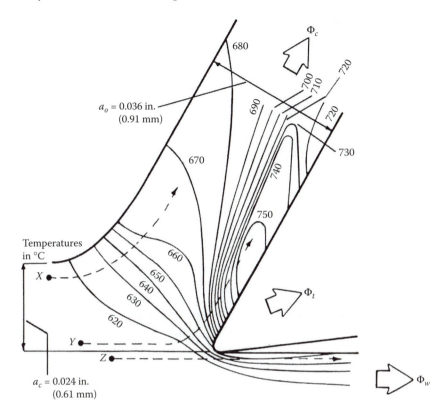

FIGURE 3.5 Temperature distribution in workpiece and chip during orthogonal cutting (obtained from an infrared photograph) for free-cutting mild steel where the cutting speed is 75 ft/min (0.38 m/s), the width of cut is 0.25 in. (6.35 mm), the working normal rake is 30 degrees, and the workpiece temperature is 611°C. (After Boothroyd, G., *Proc. Inst. Mech. Eng.*, Vol. 177, 789, 1963.)

cooled as the heat is conducted into the body of the chip, and eventually the chip achieves a uniform temperature throughout. Thus the maximum temperature occurs along the tool face some distance from the cutting edge. Point Z, which remains in the workpiece, is heated by the conduction of heat from the primary deformation zone. Some heat is conducted from the secondary deformation zone into the body of the tool. Thus,

$$P_m = \Phi_c + \Phi_w + \Phi_t \tag{3.4}$$

where

P_m = total rate of heat generation

Φ_c = rate of heat transportation by the chip

Φ_w = rate of heat conduction into the workpiece

Φ_t = rate of heat conduction into the tool

Because the chip material near the tool face is flowing rapidly, it has a much greater capacity for the removal of heat than the tool. For this reason Φ_f usually forms a very small proportion of P_m and may be neglected.

3.3.1 TEMPERATURES IN THE PRIMARY DEFORMATION ZONE

The rate of heat generation in the primary deformation zone is P_s, and a fraction of this heat, Γ, is conducted into the workpiece; the remainder $[1 - \Gamma]P_s$, is transported with the chip. Thus, the average temperature rise θ_s of the material passing through the primary deformation zone is given by

$$\theta_s = \frac{(1-\Gamma)P_s}{\rho c v a_c a_w} \tag{3.5}$$

where a_w is the width of the chip.

Equation 3.5 shows that if Γ is known for a given set of cutting conditions, θ_s may be determined.

Several theoretical analyses of the temperatures in the workpiece and shear zone have been carried out, two of the more successful being those of Weiner [3] and Rapier [4]. Figure 3.6 shows the idealized model of the cutting process employed; it was assumed that the primary deformation zone could be regarded as a plane heat source of uniform strength, that no heat was lost from the free

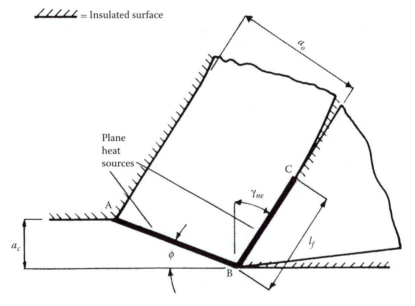

FIGURE 3.6 Idealized model of cutting process employed in theoretical work on cutting temperatures.

surfaces of the workpiece and chip, and that the thermal properties of the work material were constant and independent of temperature. The problem was to solve Equation 3.3, the basic heat transfer equation, within the boundaries of the workpiece. An exact solution to this problem was not found possible and further assumptions were necessary. The most useful suggestion was made by Weiner [3] who assumed that no heat was conducted in the material in the direction of its motion. This assumption was shown to be justified in metal cutting because at high speeds the transfer of heat in the direction of motion is mainly by transportation, and the conduction term can be neglected. This means that Equation 3.3 can be rewritten as

$$\frac{\partial^2 \theta}{\partial y^2} - \frac{R}{a}\frac{\partial \theta}{\partial \theta} = 0 \qquad (3.6)$$

Weiner was able to solve Equation 3.6 within the stipulated boundary conditions for the workpiece, and he produced an equation expressing Γ (the proportion of P_s conducted into the workpiece) as a unique function of $R\ tan\ \phi$ (where ϕ is the shear angle). This theoretical relationship between Γ and $R\ tan\ \phi$ is compared with experimental data in Figure 3.6, where it is seen that the theory has slightly underestimated results at high values of $R\ tan\ \phi$ (i.e., at high speeds and feeds). In the theory where a plane heat source was assumed, heat can only flow into the workpiece by conduction, whereas in reality heat is generated over a wide zone, part of which extends into the workpiece. The effect of this wide heat generation zone becomes increasingly important at high speeds and feeds and explains the slight disagreement between theory and experiment at high values of $R\ tan\ \phi$.

3.3.2 TEMPERATURES IN THE SECONDARY DEFORMATION ZONE

The maximum temperature in the chip occurs where the material leaves the secondary deformation zone (point C, Figure 3.1) and is given by

$$\theta_{max} = \theta_m + \theta_s + \theta_0 \qquad (3.7)$$

where
  θ_m = temperature rise of the material passing through the secondary deformation zone
  θ_s = temperature rise of the material passing through the primary deformation zone (given by Equation 3.5)
  θ_o = initial workpiece temperature

In an analysis of the chip temperatures, Rapier [4] assumed that the heat source resulting from friction between the chip and tool was a plane heat source of uniform strength (Figure 3.5) and was able to solve Equation 3.6 within the boundary conditions shown. The following expression was obtained:

$$\frac{\theta_m}{\theta_f} = 1.13\sqrt{\frac{R}{l_o}} \tag{3.8}$$

where

θ_m = maximum temperature rise in the chip owing to the frictional heat source in the secondary deformation zone

l_o = length of the heat source divided by the chip thickness (l_f/a_o)

R = thermal number

The average temperature rise of the chip resulting from the secondary deformation θ_f (frictional heat source) is given by

$$\theta_f = \frac{P_f}{\rho c v a_c a_w} \tag{3.9}$$

A later comparison of Equation 3.8 with experimental results [2] showed that Rapier's theory considerably overestimated θ_m, and it was suggested that this overestimation occurred because during cutting, the friction between the chip and tool causes severe deformation of the chip material and the resulting heat source extends some distance into the chip. The boundary conditions shown in Figure 3.8 are thought to approximate more closely the real conditions, and an analysis based on this revised model [2] yielded results that agreed with experimental data. These results, shown in Figure 3.9, indicate the effect of variations in the width of the uniformly distributed heat source. When these curves are used, l_o can be estimated from the wear on the tool face, and the width of the heat source can be estimated from a photomicrograph of the chip cross section. A typical chip cross section is shown in Figure 3.10; where the lines of maximum grain elongation are curved, it may be assumed that the material has passed through the secondary deformation zone.

3.3.3 EXAMPLE

To illustrate the application of the theories and equations derived in this chapter, a worked example is now presented. In this example the maximum temperature along the tool face is estimated for the following conditions during the orthogonal cutting of mild steel:

Working normal rake $\gamma_{n_e} = 0$
Cutting force F_c = 890 N (200 lbf)
Thrust force F_t = 667 N (150 lbf)

Cutting speed $v = 2$ m/s (394 ft/min)
Undeformed chip thickness $a_c = 0.25$ mm (0.009 8 in.)
Width of cut $a_w = 2.5$ mm (0.098 in.)
Cutting ratio $r_c = 0.3$
Length of contact between chip and tool $l_f = 0.75$ mm (0.03 in.)

The total heat generation rate is given by Equation 3.1. Thus,

$$P_m = F_c v = 890(2) = 1780 \text{ J/s } (100 \text{ Btu/min})$$

The rate of heat generated by friction between the chip and tool is given by:

$$P_f = F_f v_0 = F_f v r_c$$

In the present example γ_{n_e} equals 0, and therefore F_f equals F_t. Thus,

$$P_f = F_t v r_c = 667(2)0.3 = 400 \text{ J/s } (22.7 \text{ Btu/min})$$

The rate of heat generation from shearing is given by Equation 3.2. Thus,

$$P_s = P_m - P_f = 1380 \text{ J/s } (77.3 \text{ Btu/min})$$

To estimate the temperature rise using Equation 3.5, it is first necessary to obtain from Figure 3.6; to obtain the value of $R \tan \phi$ is required.

The thermal number R is given by cva_c/k and assuming that for mild steel ρ is 7200 kg/m³ (0.26 lb/in.³), k is 43.6 J/smK [302.6 Btu (in.)/ (h) (ft²) (°F)], and c is 502 J/kgK [0.12 Btu/(lb) (°F)]:

$$R = \frac{7200(502)(2)(0.00025)}{43.6}$$

When the rake angle γ_{n_e} equals zero, $\tan\phi$ equals r_c. Thus,

$$R \tan \phi = 41.5(0.3) = 12.45$$

The proportion of the shearing heat conducted into the workpiece is now obtained from Figure 3.7. Thus,

$$\Gamma = 0.1$$

If the appropriate values are substituted into Equation 3.5,

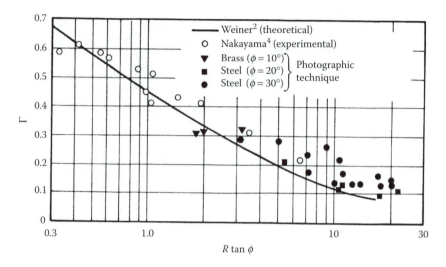

FIGURE 3.7 Effect of $R \tan \phi$ on division of shear-zone heat between chip and workpiece, where Γ = the proportion of shear zone heat conducted into the workpiece, R = thermal number, and ϕ = the shear angle. (After Boothroyd, G., *Proc. Inst. Mech. Eng.*, Vol. 177, 789, 1963.)

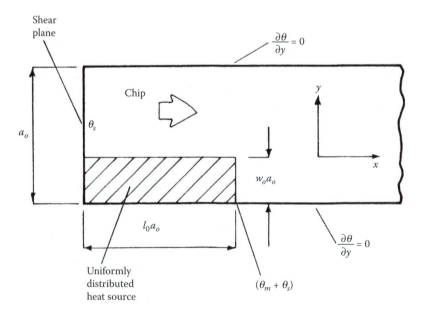

FIGURE 3.8 Revised boundary condition for chip.

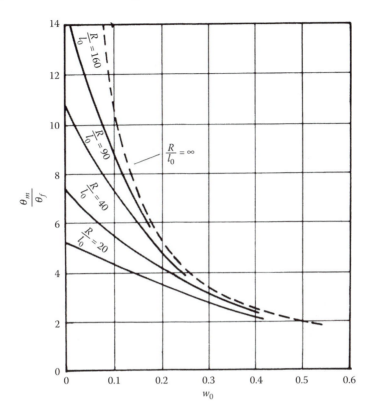

FIGURE 3.9 Effect of width of secondary deformation zone on chip temperatures, where R = thermal number, $l_0 a_0$ = chip-tool contact length, $w_0 a_0$ = width of secondary deformation zone, Θ_m = maximum temperature rise in the chip, and Θ_f = mean temperature rise in chip. (After Boothroyd, G., *Proc. Inst. Mech. Eng.*, Vol. 177, 789, 1963.)

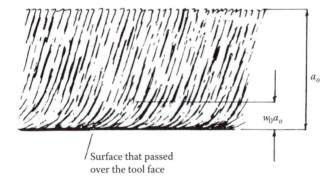

FIGURE 3.10 Grain deformation in chip cross section.

$$\theta_f = \frac{(1-\Gamma)P_s}{\rho c v a_c a_w}$$

$$= (1-0.1)(1380)/(7200)(502)(2)(0.00025)(0.0025)$$

$$= 275°C \ (495 \ °F)$$

This value (275°C or 495°F) is the mean temperature rise in the primary deformation zone, and since it is a temperature difference, it can be expressed in either degrees Centigrade (°C) or Kelvin (K).

The mean temperature rise of the chip from the secondary deformation zone (frictional heat source) is given by Equation 3.9.

Thus, neglecting the heat conducted into the cutting tool,

$$\theta_s = \frac{400}{4.518} = 88.5°C(159.3°F)$$

To obtain the ratio θ_m/θ_f and hence θ_m from Figure 3.9, it is necessary to estimate the values of w_0 and R/l_0.

The width of the secondary deformation zone divided by the chip thickness w_0 is assumed to be 0.2 for mild steel under unlubricated cutting conditions. Also,

$$l_0 = \frac{l_f r_c}{a_c} = \frac{0.75(0.3)}{0.25} = 0.9$$

Therefore,

$$\frac{R}{l_0} = \frac{41.5}{0.9} = 46.1$$

The ratio θ_m/θ_f may now be obtained from Figure 3.8, thus,

$$\theta_m = 4.2(88.5) = 372°C(670°F)$$

If it is assumed the workpiece is at a room temperature θ_0 of 22°C (72°F), the maximum temperature along the tool rake face is given by Equation 3.7. Thus,

$$\theta_{max} = \theta_m + \theta_s + \theta_o = 372 + 275 + 22 = 669°C(1234°F)$$

It should be noted that in these calculations, the thermal properties of the material are assumed to be constant and independent of temperature. With many engineering materials, however, the specific heat capacity and thermal conductivity

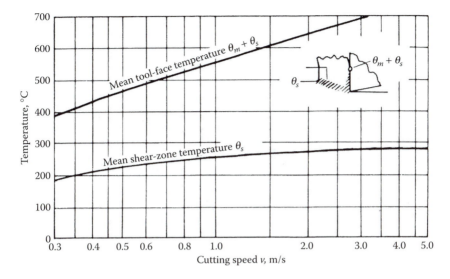

FIGURE 3.11 Effect of cutting speed on cutting temperatures (theoretical).

vary considerably with changes in temperature. If more accurate predictions of tool temperatures are required, the relationships between the thermal properties of the material and temperature must be known and care taken to use the appropriate values in the calculations.

3.3.4 EFFECT OF CUTTING SPEED ON TEMPERATURES

If the tool forces and the cutting ratio do not vary with changes in cutting speed, for the conditions used in the preceding example the relationships between temperature and cutting speed shown in Figure 3.11 are obtained. Here it is seen that the mean shear-zone temperature θ_s, increases slightly with increasing cutting speed and then tends to become constant, whereas the maximum tool face temperature $(\theta_m + \theta_s)$ increases rapidly with increasing cutting speed.

3.3.5 PREDICTION OF TEMPERATURE DISTRIBUTIONS IN MACHINING

Slip line field modeling [5,6] and finite element modeling [5,7,8] of machining enable temperature distributions to be determined and these are shown to be similar in form to the experimentally observed distributions such as that shown in Figure 3.5. Quantitative prediction of the peak temperatures is very dependent on appropriate assumptions being made for material property models and boundary conditions for the simulations [7]. This is illustrated by Figure 3.12 that shows a comparison between predicted peak temperatures and those obtained experimentally for some different material property models assumed. Also included is the prediction obtained using the analytical method described by Tlusty [8].

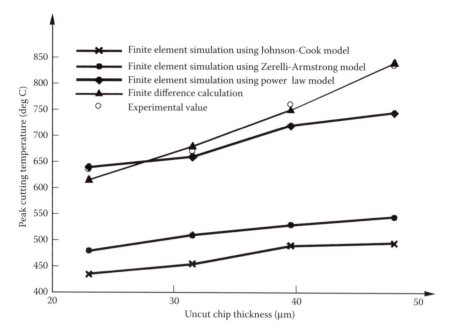

FIGURE 3.12 Comparison of experimental peak cutting temperatures with those pre-dicted by finite element modeling using different material property models. (Adapted from Davies, M.A., Cao, Q., Cooke, A.L., and Ivester, R, *Ann. CIRP*, Vol. 52/1, 77–80, 2003.)

3.4 THE MEASUREMENT OF CUTTING TEMPERATURES

A number of methods have been developed for the measurements of temperatures in metal cutting. Some of these methods only make it possible for the average cutting temperatures to be determined, but effective methods are available for determining temperature distributions in the workpiece, chip, and tool near the cutting edge.

3.4.1 WORK-TOOL THERMOCOUPLE

A technique widely used to study cutting temperatures is the work-tool thermo-couple technique. In this technique the electromotive force (emf) generated at the junction between the workpiece and tool is taken as a measure of the tem-peratures in this region. A typical work-tool thermocouple arrangement on a lathe is shown in Figure 3.13. It is important when using this technique to insulate the thermocouple circuit from the machine and to use the same circuit when cali-brating the thermocouple. It may be assumed that the reading given by this method is an indication of the mean temperature along the chip-tool interface. This technique has been used extensively in the past to investigate the effects of changes in cutting conditions on cutting temperatures and to obtain empirical

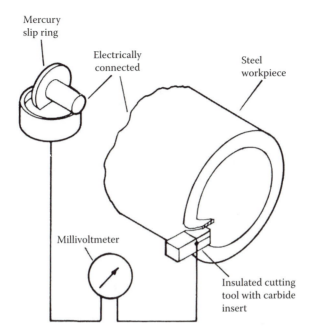

Mercury
slip ring

Electrically
connected

Steel
workpiece

Millivoltmeter

Insulated cutting
tool with carbide
insert

FIGURE 3.13 Work-tool thermocouple circuit.

relationships between temperature and cutting-tool wear rate. However, the work-tool thermocouple method is limited because it gives no indication of the distribution of temperature along the tool rake face.

There are a number of sources of error in using the work-tool thermocouple. In particular, the tool and work materials are not ideal elements for a thermocouple. Consequently, the emf tends to be low and the emf/temperature calibration nonlinear. The work-tool thermocouple must be calibrated against a standard thermocouple. Each tool and workpiece material combination must be calibrated separately. In addition, it is unlikely that the emf determined with a stationary tool, used for calibration, is the same as that obtained for an equivalent temperature during cutting when the work material is being severely strained. Further details of the calibration of work-tool thermocouple setups are given in [5].

3.4.2 DIRECT THERMOCOUPLE MEASUREMENTS

Direct thermocouple measurements can be made during cutting. The results shown in Figure 3.7 were obtained [9] using the thermocouple technique illustrated in Figure 3.12. In these experiments the rig was first run without cutting, and the reading on the millivoltmeter resulting from the rubbing action of the constantan wire on the workpiece was noted.

This reading was subsequently subtracted from the readings taken while cutting was in progress. With this method, the temperatures at selected points

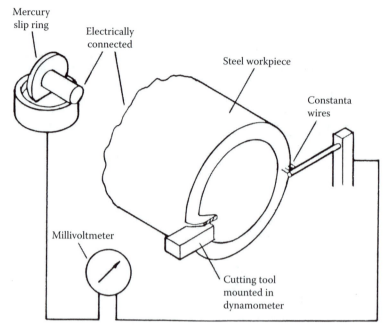

FIGURE 3.14 Arrangement for measurement of workpiece temperatures using the thermocouple technique. (After Nakayama, K., *Bull. Faculty Eng., Yokohama National University (Japan)*, Vol. 5, no. 1, 1956.)

around the end face of the tubular workpiece were measured and then used to calculate the proportion of the shear-zone heat conducted into the workpiece.

Direct measurement of temperatures can be made by making a hole in the tool close to the cutting edge and inserting a thermocouple to measure the temperature at a particular position [5]. This can then be repeated with holes in various positions to give an estimate of the temperature distributions. Significant errors may occur where the temperature gradients are steep, as the holes for the thermocouples may cover a considerable range of temperature. In addition the presence of the holes may distort the heat flow and temperature fields in the tool.

3.4.3 RADIATION METHODS

When the tool-workpiece area can be observed directly, cameras and film sensitive to infrared radiation can be used to determine temperature distributions. The result shown in Figure 3.14 was obtained from an infrared photograph of the cutting operation [2]. In the technique used to produce this result a furnace of known temperature distribution was photographed simultaneously with the cutting operation using an infrared-sensitive plate, enabling the optical density of the plate to be calibrated against temperature. Some of the experimental results are shown in Figure 3.7 and experimental confirmations of the curves in Figure 3.9 were obtained by this method.

For the result shown in Figure 3.5 the workpiece was preheated because of the relatively low sensitivity of the infrared photographic plates available at that time. Improvements in infrared-sensitive films and development of thermal imaging video cameras now make it possible to determine temperature distributions for workpieces at room temperature. Modern miniature electronic photo detectors arranged in a focal plane array system enable temperature distributions to be determined with resolutions as low as 5 μm [7].

3.4.4 Hardness and Microstructure Changes in Steel Tools

The room-temperature hardness of hardened steel decreases after reheating, and the loss of hardness is related to the temperature and time of heating. The hardness decrease is the result of changes in the microstructure or the steel. These structural changes can be observed using optical and electron microscopes. These changes provide an effective means of determining temperature distributions in the tool during cutting. Microhardness measurements on tools after cutting can be used to determine constant-temperature contours in the tool [10], but the technique is time-consuming and requires very accurate hardness measurements.

The structural changes in the material take place gradually, but it has been observed that for some high-speed steels distinct modifications occur at approximately 50°C intervals between 600 and 900°C [11]. This permits temperature measurements with an accuracy of ±25°C within the heat-affected region. Metallographic examination of the tool after cutting makes it possible for temperature distributions in the tool to be determined, but requires experienced interpretation of the observed structural changes. This method has been used to study temperature distributions in high-speed steel lathe tools and drills [11–13]. The main limitation of this method of temperature estimation is that it can be used only within the range of cutting conditions suitable for high-speed steel and when relatively high temperatures are generated.

PROBLEMS

1. If, when machining at high cutting speeds, the heat conducted into the cutting tool becomes negligible and the heat conducted into the workpiece is 4% of the total heat generated, derive an expression for the final chip temperature θ in terms of the:
   a. Specific cutting energy p_s
   b. Specific heat capacity of the work material c
   c. Density of the work material ρ

2a. Show that for orthogonal cutting metal with a tool of zero rake angle, the rate of heat generation P_s in the shear zone is given by

$$P_s = F_c v(1 - \mu r_c)$$

where

F_c = cutting force
v = cutting speed
μ = mean coefficient of friction on the tool face
r_c = cutting ratio

2b. For the same conditions, calculate the mean shear-zone temperature rise θ_s, when the metal has a specific cutting energy of 2.8 GN/m², $\mu = 1.0$, $r_c = 0.2$, and 10% of the shear-zone heat is conducted into the workpiece. Assume for the work material that $\rho = 7200$ kg/m³ and $c = 500$ J/kgK.

2c. What would be the value of θ_s, if the cutting speed were doubled and the proportion of shear-zone heat conducted into the workpiece remained the same?

3. During some machining experiments, it was found that for the range of conditions studied, the following assumptions could be made:
   a. The heat conducted into the cutting tool was negligible.
   b. The proportion of the heat generated in the shear zone conducted into the workpiece was 0.2.
   c. The maximum temperature rise in the chip due to the frictional heat source θ_m followed the relation

$$\theta_m = \theta_f \sqrt{R}$$

   where θ_f is the mean temperature rise of the chip due to the frictional heat source, and R is the thermal number.
   d. The heat generated due to friction was equal to 20% of the heat generated in the shear zone.
   Derive an expression for the maximum temperature in the chip above the initial workpiece temperature in terms of the specific cutting energy of the workpiece p_s, the specific heat c, and density of the workpiece and the thermal number R.

4. It was found from experiments where the chips produced during shaping were caught in a calorimeter that the mean chip-temperature rise was 500°C. The cutting conditions were as follows:

   Cutting speed = 1 m/s
   Undeformed chip thickness = 0.5 mm
   Width of chip = 5 mm
   Working normal rake = 45 degrees
   Cutting force = 6000 N
   Thrust force = 0

Chip thickness = 1.2 mm

Length of workpiece = 300 mm

After 100 strokes of the shaper, the insulated workpiece was immersed in a calorimeter and its additional heat content found to be 30 kJ. Neglecting the heat conducted into the cutting tool, calculate the proportion of the shear-zone heat conducted into the workpiece.

REFERENCES

1. Oxley, P.L.B., *The Mechanics of Machining*, Ellis-Horwood Ltd., Chichester, U.K., 1989.
2. Boothroyd, G., Temperatures in Orthogonal Metal Cutting, *Proc. Inst. Mech. Eng.*, Vol. 177, 789, 1963.
3. Weiner, J.H., Shear Plane Temperature Distribution in Orthogonal Cutting, *Trans. ASME*, Vol. 77, no. 8, 1331, 1955.
4. Rapier, A.C., A Theoretical Investigation of the Temperature Distribution in the Metal Cutting Process, *British J. Appl. Phys.*, Vol. 5, no. 11, 400, 1954.
5. Childs, T.H.C., Makamura, K., Obikawa, T., and Yamara, Y., *Metal Machining — Theory and Applications*, Arnold, London, 2000.
6. Chandrasekaran, H. and Thuvander, A., Tool Stresses and Temperature in Machining Proc. CIRP Int. Workshop on Modeling of Machining Operations, Atlanta, May 19 , 1998, 247–256.
7. Davies, M.A., Cao, Q., Cooke, A.L., and Ivester, R., On the Measurement and Prediction of Temperature Fields in Machining AISI 1045 Steel, *Ann. CIRP*, Vol. 52/1, 77–80, 2003.
8. Tlusty, J., *Manufacturing Processes and Equipment*, Prentice Hall, New Jersey, 2000.
9. Nakayama, K., Temperature Rise of Workpiece During Metal Cutting, *Bull. Faculty Eng., Yokohama National University (Japan)*, Vol. 5, no. 1, 1956.
10. Trent, E.M., *Metal Cutting*, 3rd ed., Butterworths-Heinemann, London, 1991.
11. Wright, P.K. and Trent, E.M., Metallographic Methods of Determining Temperature Gradients in Cutting Tools, *JISI*, Vol. 211, 364, 1973.
12. Mills, B., Wakeman, D.W., and Aboukhushaba, A., A New Technique for Determining the Temperature Distribution in High-speed Steel Cutting Tools Using Scanning Electron Microscopy, *Ann. CIRP*, Vol. 29, 73–76, 1980.
13. Mills, B. and Mattishaw, T.D., The Application of Scanning Electron Microscopy to the Study of Temperatures and Temperature Distributions in M2 High-speed-steel Twist Drills, *Ann. CIRP*, Vol. 30, 15–18, 1981.

4 Tool Life and Tool Materials

4.1 INTRODUCTION

Cutting-tool life is one of the most important economic considerations in metal cutting. In roughing operations the various tool angles, cutting speeds, and feed rates are usually chosen to give an economical tool life. Conditions giving a very short tool life are uneconomical because tool-grinding and tool replacement costs are high. On the other hand, the use of very low speeds and feeds to give long tool life is uneconomical because of the low production rate. Clearly, any tool or work material improvements that increase tool life will be beneficial. Considerable efforts have been made over the years to develop new and improved materials with better life. To form a basis for such improvements much effort has been made to understand the nature of tool wear and other forms of tool failure.

The life of a cutting tool can be brought to an end in various ways, but these ways may be separated into two main groups:

1. The gradual or progressive wearing away of certain regions of the face and flank of the cutting tool
2. Failures bringing the life of the tool to a premature end

4.2 PROGRESSIVE TOOL WEAR

The fundamental nature of the mechanism of wear can be very different under different conditions. In metal cutting, three main forms of wear are known to occur: adhesion, abrasion, and diffusion wear.

In adhesion wear [1], the wear is caused by the fracture of welded asperity junctions between the two metals. In metal cutting, junctions between the chip and tool materials are formed as part of the friction mechanism; when these junctions are fractured, small fragments of tool material can be torn out and carried away on the underside of the chip or on the new workpiece surface. The conditions that exist in metal cutting are well suited to adhesive wear as new material surfaces uncontaminated with oxide films are continually produced, and this facilitates the formation of welded asperity junctions.

The form of wear known as abrasion wear occurs when hard particles on the underside of the chip pass over the tool face and remove tool material by mechanical action. These hard particles may be highly strain-hardened fragments of an unstable built-up edge, fragments of the hard tool material removed by adhesion wear, or hard constituents in the work material, including oxide scales on the work surface.

Solid-state diffusion occurs when atoms in a metallic crystal lattice move from a region of high atomic concentration to one of low concentration [2,3]. This process is dependent on the existing temperature, and the rate of diffusion increases exponentially with increases in temperature. In metal cutting, where intimate contact between the work and tool materials occurs and high temperatures exist, diffusion can occur where atoms move from the tool material to the work material. This process, which takes place within a very narrow reaction zone at the interface between the two materials and causes a weakening of the surface structure of the tool, is known as diffusion wear.

4.3 FORMS OF WEAR IN METAL CUTTING

The progressive wear of a tool takes place mainly in two distinct ways (Figure 4.1):

1. Wear on the tool face characterized by the formation of a crater and resulting from the action of the chip flowing along the face
2. Wear on the flank where a wear land is formed from the rubbing action of the newly generated workpiece surface

4.3.1 CRATER WEAR

The crater formed on the tool face conforms to the shape of the chip underside and is restricted to the chip-tool contact area (Figure 4.1). In addition, the region adjacent to the cutting edge where sticking friction or a built-up edge occurs is subjected to relatively slight wear.

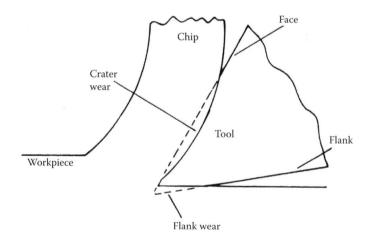

FIGURE 4.1 Regions of tool wear in metal cutting.

It was seen in Chapter 3 that in metal cutting, the highest temperatures occur some distance along the tool face; at high cutting speeds these temperatures can easily reach the order of 1000°C or more. Under these high-temperature conditions, high-speed steel tools wear very rapidly because of thermal softening of the tool material. Although carbide-tool materials retain their hardness at these high temperatures, solid-state diffusion can cause rapid wear.

In experimental work the maximum depth of the crater is usually a measure of the amount of crater wear and can be determined by a surface-measuring instrument.

Under very high-speed cutting conditions, crater wear is often the factor that determines the life of the cutting tool: the cratering becomes so severe that the tool edge is weakened and eventually fractures. However, when tools are used under economical conditions, the wear of the tool on its flank, known as flank wear, is usually the controlling factor.

4.3.2 FLANK WEAR

Wear on the flank of a cutting tool is caused by friction between the newly machined workpiece surface and the contact area on the tool flank. Because of the rigidity of the workpiece, the worn area, referred to as the flank wear land, must be parallel to the resultant cutting direction. The width of the wear land is usually taken as a measure of the amount of wear and can be readily determined by means of a toolmaker's microscope.

Figure 4.2 shows a typical graph of the progress of flank wear land width, VB, with time or distance cut for a given cutting speed. The curve can be divided into three regions:

1. The region AB where the sharp cutting edge is quickly broken down and a finite wear land is established
2. The region BC where wear progresses at a uniform rate
3. The region CD where wear occurs at a gradually increasing rate

Region CD is thought to indicate the region where the wear of the cutting tool has become sensitive to the increased tool temperatures caused by the presence of a wear land of such large proportions. Clearly in practice, it is advisable to regrind or dispose of the tool before wear enters the last region (region CD in Figure 4.2) where rapid breakdown occurs.

4.3.3 TOOL-LIFE CRITERIA

A tool-life criterion is defined as a predetermined threshold value of a tool-wear measure or the occurrence of a phenomenon. In practical machining operations the wear of the face and flank of the cutting tool is not uniform along the cutting edge; therefore, it is necessary to specify the locations and degree of the wear when deciding on the amount of wear allowable before regrinding or disposal of

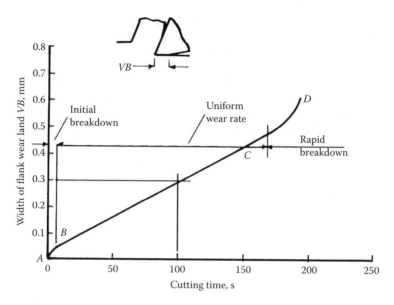

FIGURE 4.2 Development of flank wear with time for a carbide tool at a cutting speed of 1 m/s.

the tool. Figure 4.3 shows a typical worn single point tool. As shown in the figure, the amount of cratering varies along the active cutting edge, and the crater depth KT is measured at the deepest point of the crater (section A – A). It can be seen that flank wear is usually greatest at the extremities of the active cutting edge. Contributions at the corner tend to be more severe than those in the central part of the active cutting edge because of the complicated flow of the chip material in that region. The width of the flank wear land at the tool corner (zone C) is designated VC. At the opposite end of the active cutting edge (zone N) a groove or wear notch often forms, because in this region the work material tends to work harden from the previous processing operations. The width of the wear land at the notch is designated VN.

In the central portion of the active cutting edge (zone B), the wear land is usually fairly uniform. However, to allow for variations that may occur, the average wear-land width in this region is designated VB, and the maximum wear-land width is designated VB_{max}.

The criteria recommended in the ISO standard dealing with tool-life testing [4] follow.

4.3.4 COMMON CRITERIA FOR HIGH-SPEED STEEL OR CERAMIC TOOLS

The criteria recommended by the ISO to define the effective tool life for high-speed steel tools or ceramic tools are:

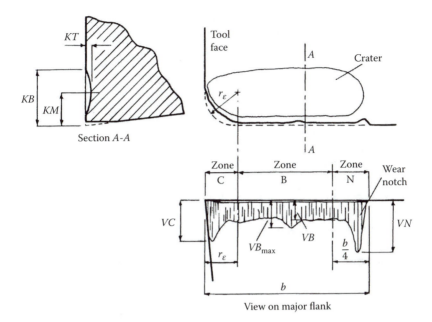

FIGURE 4.3 Some features of single-point-tool wear in turning operations. (After International Standards Organisation, ISO 3635, ISO Geneva, 1993.)

1. Catastrophic failure, or
2. $VB = 0.3$ mm if the flank is regularly worn in zone B, or
3. $VB_{max} = 0.6$ mm if the flank is irregularly worn, scratched, chipped, or badly grooved in zone B.

4.3.5 COMMON CRITERIA FOR SINTERED-CARBIDE TOOLS

For sintered-carbide tools, one of the following criteria is recommended:

1. $VB = 0.3$ mm, or
2. $VB_{max} = 0.6$ mm if the flank is irregularly worn, or
3. $KT = 0.06 + 0.3f$, where f is the feed.

4.3.6 TOOL LIFE

Tool life is defined as the cutting time required to reach a tool-life criterion. The most significant factor affecting tool life when the work material, tool material, and tool shape are chosen for a particular machining operation is the cutting speed, but other machining parameters have been found to have a secondary effect on tool life.

In studying the optimization of machining processes, it is necessary therefore to know the relationship between tool life and cutting speed for the conditions under examination. Early work on this subject was carried out by Taylor [5], who produced an empirical equation that can be written as follows:

$$\frac{v}{v_r} = \left(\frac{t_r}{t}\right)^n \tag{4.1}$$

where

n = constant
v = cutting speed
t = tool life
v_r = reference cutting speed given a known tool life of t_r

Referring again to Figure 4.2, which shows the results of a tool-wear test, it can be seen that the tool life for a criterion of VB equal to 0.3 mm is 100 s. If this test were repeated for various cutting speeds, each test would give a different value of tool life for a mean wear-land width of 0.3 mm. If all the results were then plotted on logarithmic scales, a graph similar to that shown in Figure 4.4 would result. It can be seen that the points fall on a straight line and therefore

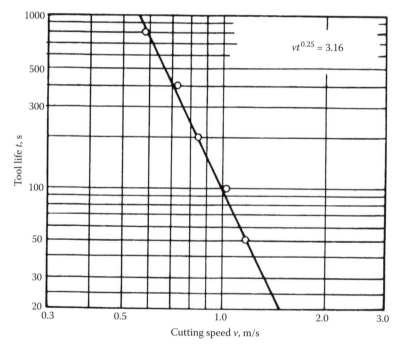

FIGURE 4.4 Typical relationship between tool life and cutting speed.

can be represented by Equation 4.1, where $(1/n)$ is the slope of the line, and (v_r, t_r) is one point on the line.

Thus from the graph n is 0.25 and if the reference speed v_r is chosen arbitrarily as 1 m/s, t_r is 100 s. Equation 4.1 now becomes

$$vt^{0.25} = 100^{0.25} = 3.16$$

It will be shown later, in Chapter 6, how such information can be used in the optimization of machining processes.

This basic relationship between cutting speed and tool life has been found to apply well to a wide range of materials and cutting conditions. However at very high speeds it appears that this relationship may change or at least different values of n, etc. may apply [3]. In practice the tool life shows a significant variation for repeated tests with the same set of cutting conditions. Thus a considerable number of tests must usually be conducted to establish the constants in Taylor's equation for a particular tool/work material combination with an appropriate degree of confidence. In addition, other machining parameters, in particular feed and tool engagement (depth of cut), are found to have an effect on tool life, but to a much smaller extent than cutting speed. This has led to the development of other tool life relationships, including extended versions of Taylor's equation. Some of these other tool life relationships are summarized in Table 4.1. Similar to the basic Taylor equation, the various constants for a particular tool/work material combination must be established empirically through a well-designed set of cutting tests. The last tool life equation in Table 4.1 represents the dependence on cutting temperatures as identified in Figure 3.1.

TABLE 4.1
Selected Empirical Tool Life Relationships

Tool Life Equations	Equation[a]
Taylor's basic equation	$vt^n = C$
Taylor's extended equation	$t = \dfrac{C}{v^{1/n} f^{1/n_1} a_e^{1/n_2}}$
Taylor's extended equation	$v = \dfrac{C}{t^n f^{n_1} d^{n_2} (BHN/200)^{n_3}}$
Temperature-based equation	$\theta t^{n_4} = C$

[a] n, n_1, n_2, n_3, n_4, C are constants.

4.3.7 PREMATURE TOOL FAILURE

Many cutting tools, particularly single-point cutting tools, are provided with carbide inserts, either brazed to the shank or attached mechanically (in the latter case they are known as "throwaway" or "disposable" inserts). These carbide inserts are extremely brittle, and the tools should be handled carefully since sudden loads caused by dropping the tool or engaging a large cut suddenly may fracture the insert. These tools are also susceptible to fracture from transient thermal stresses, such as those occurring during intermittent cutting (as in the milling operation) or from localized cooling when coolant is inadequately applied during cutting. Damage may also be caused by careless grinding or incorrect brazing techniques such that cracks may be formed by thermal stresses, thus weakening the cutting edge.

Although most research work on tool wear has dealt, in the past, with the progressive flank and crater wear of cutting tools, the ways in which the life of a tool may be prematurely brought to an end can be extremely important in practice.

4.3.8 THE EFFECT OF A BUILT-UP EDGE

The presence of a built-up edge on the tool face during cutting can affect the tool-wear rate in various ways, sometimes decreasing the life of a cutting tool and sometimes increasing it. With an unstable built-up edge the highly strain-hardened fragments, which adhere to the chip undersurface and the new work-piece surface, can increase the tool-wear rate by abrading the tool faces. However, when very hard materials (such as cast iron) are cut, the presence of a stable built-up edge can be beneficial [6]. A stable built-up edge protects the tool surface from wear and performs the cutting action itself (Figure 4.5).

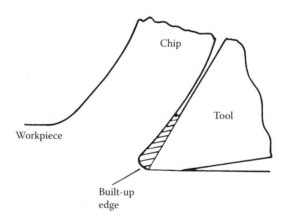

FIGURE 4.5 Built-up edge protecting tool face. (After Trent, E.M., *Production Engineer* (*London*), Vol. 38, no.3, 105, 1959.)

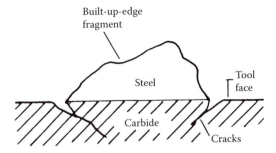

FIGURE 4.6 Crack forms in carbide insert during cooling. (After Trent, E.M., *Production Engineer* (*London*), Vol. 38, no. 3, 105, 1959.)

A built-up edge can also contribute to sudden tool failures when tools with carbide inserts are used. For example, when a tool is suddenly disengaged, a portion of built-up edge (welded during cutting to the tool face) may be torn off, taking with it a fragment of tool material. A built-up edge can also be detrimental to tool life during the cooling of the tool at the end of a cut. Because the coefficient of linear expansion of steel is approximately twice that of a carbide tool material, cracks may be introduced in the insert where the attached portion of built-up edge has contracted, during cooling, by an amount greater than the contraction of the carbide [6] (Figure 4.6).

4.3.9 THE EFFECT OF TOOL ANGLES

In general, it can be said that poor cutting conditions giving increased specific cutting energies and increased tool temperatures result in higher tool-wear rates. Because an increase in rake usually leads to an improvement in cutting conditions, a longer tool life would be expected. However, when the tool rake is large, the cutting edge is mechanically weak, resulting in higher wear rates and shorter tool life. Therefore, for an otherwise constant set of cutting conditions, an optimum rake exists giving a maximum tool life.

A typical relationship between rake angle and tool life is shown in Figure 4.7, where the optimum rake is approximately 14 degrees when cutting high-strength steel with a high-speed steel tool [7]. Experience has shown that the optimum rake is roughly constant for given work and tool materials, and in practice the values given in Table 4.2 are used.

In certain circumstances, for example, when milling with carbide cutters, where the impact loading at each revolution of the cutter might tend to fracture the insert, a negative rake is used to achieve greater strength.

Experience has shown that the width of the flank wear land is usually the limiting factor determining the life of the cutting tool. However, it has been shown [8] that the physical conditions of stress, temperature, and speed that determine the wear rate on the tool flank are constant along the wear land and for reasonably small wear lands these physical conditions are not greatly affected by changes

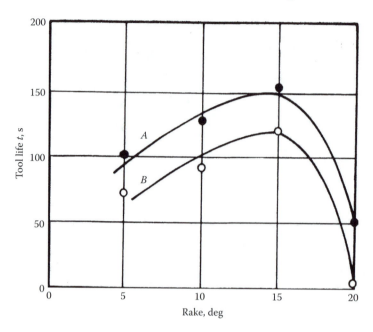

FIGURE 4.7 Effect of tool rake on tool life, where work material is high-strength steel and the tool material is high-speed steel. In curve *A* the feed is 0.127 mm (0.005 in.) and the cutting speed is 0.66 m/s (130 ft/min); in curve *B* the feed is 0.508 mm (0.02 in.) and the cutting speed is 0.41 m/s (80 ft/min). (After Cherry, J., Practical Investigation in Metal Cutting, *Production Engineer* (*London*), Vol. 41, no. 2, 90, 1962.)

TABLE 4.2
Recommended Normal Rake Angle (Degrees)
for Roughing Operations

Work Material	High-Speed Steel	Carbide
Cast iron, cast brass	0	0
Brass and bronze	8	3.5
Soft brass and high-speed steel	14	3.5
Mild steel	27	3.5
Light alloys	40	13

in the wear-land width. For these two reasons it would be expected that the wear rate of the tool material normal to the resultant cutting direction would be constant and independent of the normal clearance. Figure 4.8 shows that the rate of increase of flank wear-land width, on the other hand, is dependent on the flank clearance. From the figure, with zero tool rake,

$$\left(VB\right) = \left(NB\right)\cot\gamma_{n_e} \tag{4.2}$$

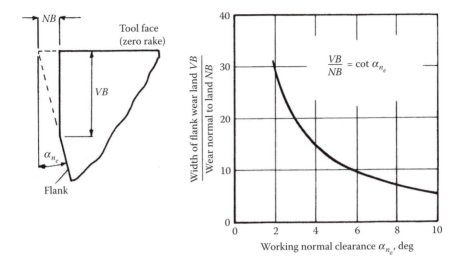

FIGURE 4.8 Effect of normal clearance on flank wear.

$$\frac{VB}{NB} = \frac{\dot{VB}}{\dot{NB}} = \cot\gamma_{n_e} \qquad (4.3)$$

where

$\dot{VB}$ = rate of increase of flank wear-land length

$\dot{NB}$ = rate of removal of tool material normal to the cutting direction

γ_{n_e} = working normal clearance

Hence, for an assumed constant wear rate normal to the cutting direction, the flank wear rate $\dot{VB}$, which determines the tool life, is proportional to the clearance, $\cot\gamma_{n_e}$. The relationship is illustrated in Figure 4.8, where it is seen that for small values of clearance an increase in clearance will give a significantly reduced wear rate. It is clear, however, that in practice the normal clearance cannot be made too large without running the risk of weakening the tool edge. Experience has shown that with most work materials, clearance angles of 8 degrees with high-speed steel tools and 5 degrees with carbides give the best compromise between these conflicting requirements.

4.3.10 THE EFFECT OF SPEED AND FEED ON CRATERING AND BUILT-UP-EDGE FORMATION

In metal cutting, increases of speed or feed result in an increase in the temperature on the tool face. At low speeds increases in tool-face temperatures tend to reduce friction at the chip-tool interface and hence tend to prevent the formation of a built-up edge; at high speeds increases in tool-face temperatures tend to increase the rate of crater wear. The relative effects of changes in speed and feed are shown clearly by graphs prepared by Trent [5]; two examples of the effects of changes in feed and speed are presented in Figure 4.9. These graphs show, for a given work and tool material combination, the ranges of speed and feed where a built-up edge occurs, where rapid cratering occurs, and where, under very severe conditions, extremely rapid breakdown of the tool occurs because of deformation of the tool at the cutting edge (a result of high temperatures and stresses). Such graphs are extremely useful when comparing the performance of various tool materials. The results in Figure 4.9 show, for example, that when titanium is added to a tungsten-carbide tool material, crater wear occurs at a significantly higher range of speeds and feeds when carbon steel is machined. Thus the useful range of speeds and feeds is increased by the use of tungsten-titanium-carbide tool materials for machining carbon steels.

4.3.11 Tool Damage Models

Studies of the different mechanisms of wear in cutting tools have led to the development of tool wear rate (rate of volume loss) or tool damage models. For cemented carbide tools the wear rate is dominated by the temperature related diffusion process, particularly for higher cutting speeds [9–11]. A fundamental wear rate equation has been derived by considering abrasive and diffusion wear [12], as follows:

$$\frac{dV_w}{dt} = v(\theta, f)G + D\exp\left(-E/R\theta\right) \tag{4.4}$$

where

V_w = the volume lost by wear
v = the cutting speed
f = feed
E = process activation energy
R = universal gas constant
θ = cutting temperature
G, D = constants dependent on tool and work material

The first term is for abrasive wear and is related to the cutting distance. The second term represents the diffusion wear. This equation can be used to relate wear rate to average cutting temperature of the tool. As cutting temperature rises to above 800°C the abrasive wear term is found to be negligible [9]. The constants G and D must be determined by a few cutting tests for the specific tool/workpiece combination.

A wear rate model based on adhesive wear [11,13,14] has been developed of the form:

$$\frac{dV_w}{dt} = A\sigma_n v_s \exp\left(-B/\theta\right) \tag{4.5}$$

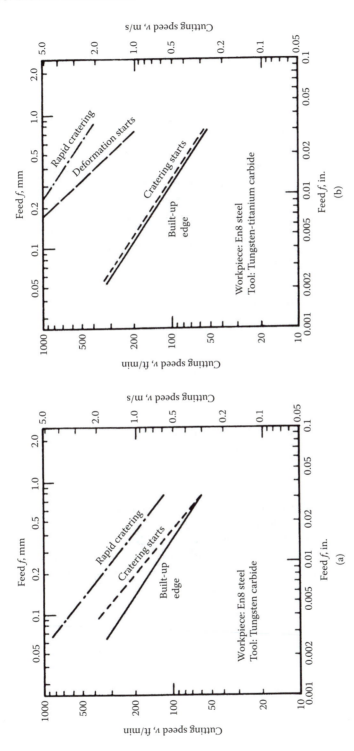

FIGURE 4.9 (a) Tool-wear chart for a tungsten-carbide tool; (b) Tool-wear chart for a tungsten-titanium-carbide tool. (After Trent, E.M., *Production Engineer* (*London*), Vol. 38, no. 3, 105, 1959.)

where

σ_n = normal stress

v_s = sliding velocity

A, B = constants dependent on the tool/workpiece materials

In this study it was shown that flank wear and crater wear obey similar mechanisms. Again the constants A and B must be determined by selected cutting tests. Equation 4.5 has been used in conjunction with finite element simulation to model wear of carbide tools in orthogonal cutting [15].

It should be noted that, at higher temperatures when the abrasive wear term in Equation 4.4 could be neglected, that the two equations are essentially of similar form, with unknown constants dependent on the materials involved, even though these equations are based on different wear mechanisms. In consequence both could be fitted, by suitable choice of constants, to the same experimental results. These wear rate models confirm the dependency of tool wear and hence tool life on the average cutting temperatures.

4.4 THE TOOL MATERIAL

4.4.1 Basic Requirements of Tool Materials

Cutting tools are required to operate under high loads at elevated temperatures, often well in excess of 1000°C. In addition, severe frictional conditions occur between the tool and chip and between the tool and the newly machined workpiece surface. The main requirements for cutting tool materials are:

1. High-temperature physical and chemical stability; high hot-hardness is of particular importance
2. High wear resistance
3. High resistance to brittle fracture

High performance in all of these attributes simultaneously is generally not possible. For example, materials with increased high-temperature resistance and high wear resistance will generally have reduced resistance to brittle fracture. The development of tool materials is generally a compromise between these separate requirements and their relative importance depends on the specific applications.

Other factors that influence the performance of cutting tools are:

1. Relative hardness of the tool and work material
2. Abrasive particles, such as scale, on the surface of the workpiece
3. Chemical compatibility of the tool and work material
4. Cutting temperatures
5. Condition of the machine tool; rigidity, for example
6. Type of machining operation, in particular whether continuous or interrupted cuts occur, which is important for tool materials with a low resistance to brittle fracture.

TABLE 4.3
Major Classes of Tool Material

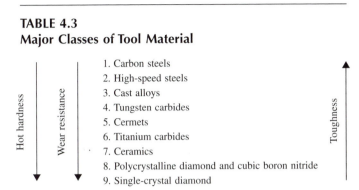

4.4.2 MAJOR CLASSES OF TOOL MATERIALS

The major types of cutting-tool materials are shown in Table 4.3. The earliest materials used were hardened carbon steels, but these are restricted to low cutting speeds and temperatures, as the martensite present, which gives the material its hardness, softens at 250°C and above. Consequently, carbon steel tools are now only used for machining soft materials, as in woodworking.

One of the greatest steps forward in the machining of metals was made when Taylor [5], in collaboration with White, discovered the heat-treatment process used in producing high-speed steel-cutting tools. Use of these cutting tools made higher metal removal rates possible owing to the improved tool-wear behavior. Since Taylor's heat-treatment discovery, developments in metallurgical science and technology have led to other new tool materials, such as cast alloys, cemented carbides, and, more recently, sintered oxides or ceramics.

Practical experience has shown which tool materials are most suited to particular operations. With single-point tools, where tool manufacturing problems are not serious, the desirable property is a high hardness value that is maintained at high temperatures. Unfortunately an increase in hardness is usually accompanied by a decrease in the impact strength of the material; an increase in hardness also increases tool manufacturing problems. Consequently, the materials found most suitable for continuous cutting with single-point cutting tools often may not be used for multipoint cutting tools. Since single-point cutting tools form only a small proportion of the total number of cutting tools used in practice, a significant proportion of metal cutting is still performed with tools manufactured from high-speed steel, but other materials have taken over many of the applications that some years ago would have been carried out with high-speed steel. At the present time more than 50% of the cutting tools used world wide are based upon cemented carbide materials, with 40% being made from high-speed steels and the remaining 10% from all of the other material classes [16]. However, in terms of the percent of material removed, cemented carbides represent 2½ to 3½ times the amount removed by high speed-steel tools [17].

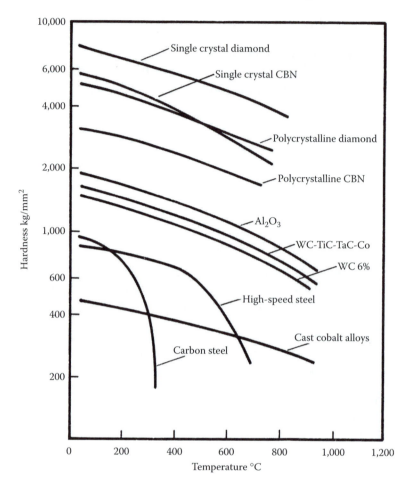

FIGURE 4.10 Variation of hardness with temperature for some tool materials. (Adapted from Almond, E.A., *Towards Improved Tool Performance*, National Physical Laboratory Conference, Metals Society, London, 1982.)

The materials shown in Table 4.3 are arranged in order of increasing temperature resistance and hot hardness. However, within each material class properties vary significantly such that there is a considerable amount of overlap between adjacent classes. Figure 4.10 shows the effect of temperature on the hardness of some of these materials [18]. All of these materials, with the exception of single-crystal diamond, consist of a softer phase separating very hard particles. Increasing the proportion of the harder-phase material increases the hardness but reduces the toughness of the material. Recent developments in tool materials have included the coating of these basic materials with thin, high-wear-resistant layers to improve performance, particularly for high-speed steel and tungsten carbide tools. The main classes of tool materials will now be discussed in greater detail.

4.4.3 High-Speed Steel

High-speed steels (HSS) are alloy steels with alloying elements consisting mainly of tungsten (about 18%) and chromium (about 4%); they may also contain cobalt, vanadium, or molybdenum. During the heat-treatment process, carbides of the alloying elements are formed and dispersed throughout the material as very hard particles. Cobalt, however, dissolves to substitute for iron atoms in the crystal matrix. Over 20 grades of HSS tool materials are in common use, each with its own advantages and limitations [19]. These can be divided into three main categories: tungsten, molybdenum, and molybdenum-cobalt based grades. Table 4.4 shows the composition of some of the standard grades of HSS. High-speed steel-cutting tools are relatively inexpensive and tough, but have limited hot hardness and can only be used for cutting temperatures up to 550°C. High-speed steel can be hot-rolled and forged to rough shapes, then machined, prior to heat treatment and finish grinding. It is still a common tool material, particularly for monolithic tools such as drills, taps, milling cutters, and so on, but for turning operations other materials are now used almost exclusively.

In recent years high-speed steel cutting tools have been available with a thin (1 to 2 μm) coating of titanium nitride (TiN), which has an equivalent hardness of Rockwell C 80-85. The high-speed steel provides a relatively ductile, shock-resistant core to the tool, while the coating has exceptional wear resistance and lower friction than the uncoated high-speed steel.

TABLE 4.4
Composition and Properties of Standard High Speed Steels

Grade	C %	W %	Mo %	Cr %	V %	Co %	Wear Resistance	Toughness	Hardness	Cost
1. Conventional steels										
T1	0.75	18.00	—	4.00	1.00	—	4	8	5	5
M1	0.80	1.75	8.50	3.75	1.15	—	4	10	5	3
M2	0.85	6.00	5.00	4.00	2.00	—	5	10	5	3
2. Cobalt steels										
M33	0.88	1.75	9.50	3.75	1.15	8.25	5	5	8	5
T5	0.80	18.00	—	4.25	2.00	8.00	5	4	8	6
T6	0.80	20.00	—	4.50	1.75	12.00	5	2	9	8
3. High vanadium steels										
M3	1.05	6.00	5.00	4.00	2.40	—	6	6	6	4
M4	1.30	5.50	4.50	4.00	4.00	—	9	6	6	4
T15	1.50	12.00	—	4.50	5.00	5.00	10	9	9	6
4. High-hardness Co steels										
M42	1.10	1.50	9.50	3.75	1.15	8.25	6	9	9	5
M44	1.15	5.25	6.50	4.25	2.00	12.00	6	3	10	6

Rating 1 (low) to 10 (high)

Another recent development is HSS tools produced by powder metallurgy. This form of processing results in a more uniform structure, with very little carbide segregation, that results in improved life and toughness.

4.4.4 Cast Alloy Tools

Cast alloy tools contain no iron and are cast into their final shape. They consist of cobalt, chromium, tungsten, and carbon, the carbide phase being about 25 to 30%, by volume, of the matrix. The tools are usually cast in graphite chilled molds to produce a fine grained hard surface made up of complex carbides. The core of the tool is a relatively tough carbide enriched material. These tools can usually be used at slightly higher speeds than high-speed steels and have less tendency to form a built-up edge during machining. They find the widest applications in the machining of cast iron, malleable iron, and the hard bronzes.

4.4.5 Cemented Carbide Tools

Cemented carbides are the most commonly used cutting tool materials at present. Cemented tungsten carbides are made by mixing tungsten powder and carbon at high temperatures in the ratio of 94 and 6% respectively, by weight. This compound is then combined with cobalt, and the resulting mixture is compacted and sintered in a furnace at about 1400°C. The cobalt acts as a binder phase, and the percentage of hard carbide particles varies from 60 to 95%. Adjustment of the type, size, and concentration of the hard particles allows a range of properties to be obtained.

The tungsten-carbide-cobalt material can maintain high hardness values at temperatures as high as 1200°C and can therefore be used at much higher cutting speeds than high-speed steel or cast-alloy tool materials. However, as mentioned earlier, cemented carbides are not as tough and cannot be shaped after sintering. For this reason they most often take the form of inserts, either brazed on or clamped on (Figure 4.11) and there are a number of different clamping methods from different manufacturers in use. Solid round tools can also be produced. The clamped-on inserts are thrown away after all the cutting edges have been worn and consequently are often referred to as disposable or throw-away inserts. Straight tungsten carbides are the strongest and most wear resistant but are subject to rapid cratering when machining steels. To improve resistance to cratering, tantalum carbide and titanium carbide are added to the basic composition, resulting in tungsten-tantalum carbide or tungsten-titanium carbide.

Cemented titanium carbide is also available in the form of disposable inserts. This material is more wear resistant than straight tungsten carbide but has the disadvantage of lower strength. However, developments in the manufacture of this material have led to the production of superior grades that allow cutting speeds approaching those of ceramics; these superior grades can be used in the semi-rough cutting of steels and certain cast irons.

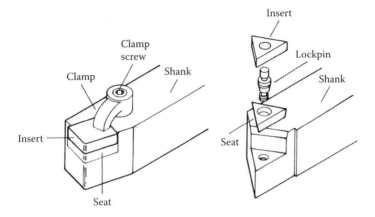

FIGURE 4.11 Typical styles of clamped-on (throwaway) insert tools.

Cemented carbides are available in many different grades [16] which differ in hardness and wear resistance. In general, increasing the cobalt content of the material increases the toughness but reduces the hardness. All cemented carbides can be used at elevated temperatures compared to high-speed steel, but these materials are relatively brittle and can fracture easily when interrupted cuts are used. Selection of the most appropriate grade for a given application is aided by standard classifications, such as that developed by ISO [20] and the American C system. In the ISO classification, cemented carbides are divided into three main categories: P, for materials that produce long chips, including mainly steels; M, for more demanding work materials such as stainless and heat resistant alloys; and K, for short chipping materials such as cast irons, hardened steels, and nonferrous alloys. The letter designations are combined with numbers indicating the application suitability, such as (01) for light finishing and (50) for heavy rough machining. The C Classification follows a similar approach and Table 4.5 compares the two classifications. In practice these classifications are a rough guide to selection as tool manufacturers offer a wide range of grades and proprietary compositions. In addition the classifications have not kept up with tool developments, in particular the widespread use of tool coatings. More specific grade recommendations can be obtained from data published by tool manufacturers.

New grades of cemented carbide are continually being developed. One major trend has been to use smaller grain size carbide powders (down from 10 μm to 1 μm). This has resulted in increased densities, with associated improvements in hardness and toughness. Research is also taking place into "nano-phase" carbides, with particle sizes of the order of 0.1–0.2 μm. Developments in processing technology include the cobalt enrichment of the surface layers, such that the cobalt binder phase is increased locally in these layers. This leads to improved toughness without loss of overall hardness and wear resistance of the bulk of the material. The cobalt-enriched layer is approximately 0.013 to 0.025 mm thick

TABLE 4.5
Standard Classification of Cemented Carbide Tool Materials

| Designation | | Work | Composition % | | | | Machining |
ISO	American	Material	WC	Co	TiC	TaC	Applications
P01	C8		64	3	25	8	Finishing
P10	C7		76	6	12	6	
P20		Steels					Roughing
P30	C6		82	8	8	2	
P40	C5		70	12	6	12	Interrupted
P50							cuts
M10		Steels, cast iron,					Finishing
M20		nonferrous metals					
M30							
M40							Roughing
K01	C4	Cast iron,	97	3			Finishing
K10	C3	nonferrous metals,	96	4			
K20	C2	nonmetallic	94	6			
K40	C1	materials	94	6			Roughing

and may contain as much as three times the cobalt concentration as the bulk material. The enrichment is achieved through diffusion of nitrogen, titanium, and cobalt during sintering.

4.4.6 CERMET TOOLS

Cermets are cemented materials that use hard particles other than tungsten carbide, including titanium carbide, titanium nitride, titanium carbonitride, and possibly molybdenum carbide. The metal binder phase is often a mixture of cobalt and nickel. Cemented carbides and cermets are produced using similar processes. Relative to tungsten carbide, cermets are more wear resistant and of similar hardness. Strength, toughness, and thermal shock resistance are generally lower. Current applications of cermets are in interrupted cutting, such as semi-finish and finish milling. Cermet cutting tools can handle high cutting speeds with moderate feeds.

4.4.7 CERAMIC TOOLS

Ceramic tool materials are made from sintered aluminum oxide (Al_2O_3) and various other boron and silicon nitride powders; these powders are mixed together and sintered at about 1700°C. Sintered oxides can be used at cutting speeds two to three times those employed with tungsten carbides, are very hard, and have a high compressive strength and low thermal conductivity. They are, however, extremely brittle and may not be used where shock and vibrations occur. They are only available in the form of throwaway inserts.

Three types of ceramic cutting tools are mainly used: aluminum oxide (Al_2O_3), SiAlON (a combination of silica, aluminum, oxygen, and nitrogen), and silicon nitride (Si_3N_4). Aluminum oxide is the most common and performs best for finishing and semi finishing, with noninterrupted cuts. SiAlON offers greater resistance to mechanical shock and possesses the ability to withstand the higher thermal shocks that arise from the use of coolants. SiAlON cannot be used with carbon steels because of chemical incompatibility [19]. Silicon nitride tools have attributes similar to SiAlON, but with improved toughness and the ability to withstand high temperatures and loads.

Ceramic cutting tools have found application principally in turning and milling cast irons and superalloys and in the finishing of hardened steels. The particular advantage over cemented carbides is the ability of these materials to withstand high temperatures, with softening occurring at around 2200°C. Thus very hard materials can be made machinable by cutting at speeds that develop enough heat to raise the cutting temperatures to around 1800°C.

The most recent developments involving ceramic tools are ceramic-ceramic composites, which consist of an Al_2O_3 matrix reinforced by tiny silicon carbide whiskers. These whiskers distribute the cutting forces more evenly, which results in a considerable increase in fracture toughness compared to cermets for example. These composite materials are suitable for roughing and finishing cuts and appear particularly useful for machining nickel-based alloys, cast iron, and steels.

4.4.8 POLYCRYSTALLINE TOOLS

Polycrystalline cutting tools are fabricated by compacting small particles of diamond (synthetic and natural) or cubic boron nitride (CBN) under high pressure and temperature conditions. Single-crystal diamonds are used for fine-finishing operations. Although natural diamonds are the hardest materials known, their crystalline structure results in nonuniform strength. Polycrystalline tool materials are more homogeneous, with improved strength and durability, but with slightly reduced hardness compared to natural diamond. Cubic boron nitride is less expensive than diamond, but over 20 times more costly than carbide. The high cost is counterbalanced by a considerably increased tool life (up to 50 times longer than tungsten carbide).

The tool inserts are made by bonding a thin layer of the polycrystalline material to a cemented carbide substrate, which provides shock resistance, while the polycrystalline material provides very high wear resistance and cutting edge strength (Figure 4.12). At high temperatures CBN is chemically nonreactive to iron and nickel, and it has high resistance to oxidation. These materials are thus well suited to cutting of hardened ferrous and high-temperature alloys. Development of these super hard materials has enabled cutting speeds to be increased dramatically for machining many materials. In addition, the machining of steels in their hardened state has become relatively common place [21].

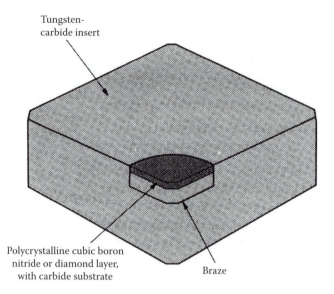

Tungsten-
carbide insert

Polycrystalline cubic boron
nitride or diamond layer,
with carbide substrate

Braze

FIGURE 4.12 Typical polycrystalline tool insert.

4.4.9 TOOL COATINGS

The basic requirement for efficient rough machining is a tool material that exhibits
the good toughness while having superior wear resistance. While the development
of different tools materials can lead to improvements in these basic properties
another approach is to provide a hard wear resistant coating onto a relatively
tough basic tool material. In recent years this has lead to the development of a
wide variety of different tool coatings, such that it is estimated that 70–75% of
all carbide tools now sold are coated [22]. This trend is not restricted to carbide
tools but applies equally to the other tool materials including HSS. These coatings
are between 5 and 8 μm thick and are found to practically eliminate interdiffusion
between the chip and the tool.

Coatings improve wear resistance, increase tool life, and enable higher cutting
speeds to be used. This enables a broader application range for a given tool
material. The improved wear resistance results from high hardness of the coating
material, together with its chemical inertness and lower coefficient of friction.
Eventually when the coating has been worn away by abrasion, the wear rate
becomes similar to the uncoated tool.

A wide variety of hard-coating materials is used and often these are applied
in multiple layers of more than one material. The most common coating used is
titanium nitride (TiN) and is often seen as the gold-colored coating on many
carbide and HSS tools. Titanium nitride has high hardness and a low coefficient
of friction that reduces erosion, abrasion, and wear. Other coating materials
include titanium carbonitride (TiCN), titanium aluminum nitride (TiAlN), aluminum oxide (Al_2O_3), and titanium diboride (TiB_2). Titanium carbonitride is a blue

gray coating that is harder than TiN and has a low thermal conductivity. As a coating aluminum oxide possesses excellent oxidation resistance, together with high wear resistance. It becomes less conductive as temperature increases and hence acts as an effective heat barrier. It is particularly useful as a coating for high speed and dry machining applications.

The newest coatings consist of multiple layers of different materials to the extent that single layers of one material are now unusual. For example, layers of Al_2O_3 and TiN are often combined. The outer layers of TiN provide a low friction and chemically inert surface for a new tool. As the TiN wears through, the additional wear-resistant layer of Al_2O_3 is exposed. This combination is particularly effective for high-speed machining of cast irons and steels. Other combinations of coatings are used and the number of layers may go up to 13 or more in some cases.

Two basic methods of applying hard coatings on tool materials are used. These are chemical vapor deposition (CVD) and physical vapor deposition (PVD). The process selected depends on the tool material composition, tool geometry, and application. However, CVD is more commonly used for tool coatings than PVD.

Chemical vapor deposition is a thermochemical process. In CVD the tools are heated in a sealed reaction chamber to around 1000°C, at atmospheric pressure in an inert atmosphere. Gaseous hydrogen and volatile compounds supply the metallic and nonmetallic constituents of the coating materials. For example, for TiN coatings, titanium tetrachloride, hydrogen, and nitrogen are introduced into the chamber. The chemical reactions form TiN on the tool surfaces. The process can be used for coatings of TiN, TiC, TiCN, and Al_2O_3. Diamond thin films can also be produced by CVD. The thickness of CVD coatings ranges from 5 to 20 μm. Coating materials usually have a higher coefficient of thermal expansion than cemented carbides, so the coatings are usually in residual tension at room temperature. The high processing temperatures result in good bonding of the coatings and the substrate materials.

Physical vapor deposition (PVD) processes are carried out in high vacuum conditions at temperatures in the range of 200 to 500°C. Particles of the coating material are deposited physically onto the surface of the workpiece. Three variants of the process exist: arc evaporation, sputtering, and ion plating. In arc evaporation the coating material is evaporated by a number of arc evaporators to produce highly reactive plasma of ionized vapor of the coating material. The vapor condenses on the substrate material coats. Uniform coatings on complex shapes can be obtained. In sputtering, an electric field ionizes an inert gas (usually argon). The positive ions bombard the coating material and cause ejection of atoms (sputtering). The atoms then condense onto a heated workpiece to produce a good bond. Ion plating is a generic term that describes the combination of sputtering and vacuum deposition. All PVD processes are line-of-sight processes and thus require moving fixtures to coat complex shapes. A big difference with PVD is the relatively low processing temperature that results in a finer grain structure in the coating and a smooth bright coating with a low coefficient of friction is produced. The coating is also free of the thermal cracks that are common for CVD. The process is particularly useful for coating sharp edges and complex chip control geometries, as PVD coatings are generally thinner than those produced by CVD.

4.5 TOOL GEOMETRIES

The general configuration of the cutting tools is dependent on the specific machining operation being used: turning, milling, drilling, etc. With the development of modern hard tool materials these basic tool forms are now produced with replaceable inserts for the cutting edges. This has resulted in a wide variety of insert shapes [23]. Obviously some of these shapes are dictated by the specific operations performed, such as for thread cutting and inserts for part-off tools. Details of the insert geometry, such edge preparations, can have a significant influence tool life and tool failure.

Most turning operations are carried out with coated, indexable carbide inserts. A small number of basic insert shapes are used. Details of these insert geometries, such as tool point angles, edge radii, and the edge preparation, have an influence on tool life. Most turning inserts also contain features in the rake face in the form of grooves, dimples, etc. These are for chip control purposes and are discussed in more detail in Chapter 8.

Large tool point angles are stronger and are preferred for rough machining operations. Smaller angles (35–55 degrees) are useful for more intricate finishing work but reduce the insert strength for roughing passes. In general the largest possible angle should be used dependent on the workpiece geometry to be machined. The tool nose radius affects insert strength and surface finish. A large tool nose radius provides better heat dissipation and improved tool life. The nose radius also influences chip formation.

Sharp edges are relatively weak and break easily. Thus some form of edge preparation is an important part of tool insert geometry. These preparations include a honed radius, chamfer, land or some combination of all three (Figure 4.13). A

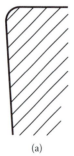
(a)
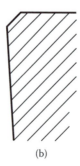
(b)
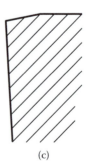
(c)

FIGURE 4.13 Tool edge preparations. (a) Honed radius; (b) chamfer; (c) land.

radius is applied to most corners to round off the sharp edges. For carbide inserts radii of 5–25 μm (0.0002–0.003 in.) are used. Chamfers break the tool corner. A land is a relief back from the cutting edge on the insert face and effectively strengthens the cutting edge by redirecting the tool nose forces into the body of the insert.

Tool rake angles can be positive or negative, with the latter being preferred for rough machining due to increased insert strength that results. Negative rake angles may also increase in the number of usable cutting edges on the insert by allowing the insert to be inverted to utilize the edges on the lower insert face. However, negative rake angles produce thicker chips and in general result in higher cutting forces.

4.6 THE WORK MATERIAL

The term *machinability* is often applied to work materials to describe their machining properties; it can have several meanings depending on the cutting process under consideration. When it is stated that material A is more machinable than material B, this can mean that a lower tool-wear rate is obtained with material A, or a better surface finish can be achieved with material A, or that less power is required to machine material A. Clearly, with finishing processes, tool wear and surface finish are the most important considerations; with roughing operations, tool wear and power consumption are important.

It should be noted that any statement regarding machinability may only apply under the particular set of circumstances existing when the observation was made. For example, under a given set of conditions a better surface finish may be obtained with material A than material B; however, under another set of conditions, say with a different tool material, the situation may be reversed. Similar behavior can occur with the other criteria of machinability, tool-wear rate, and power consumption. To complicate the situation further, a certain group of materials may be placed in one order of machinability on a tool-wear basis, but in a different order if surface-finish or power-consumption criteria are applied. Clearly, the term machinability can have little meaning except in a loose qualitative sense.

Even though the term is meaningful only in a loose qualitative sense, many attempts have been made to obtain a quantitative measure of machinability — a machinability index or number. A method for producing such an index, if the results were meaningful, would be most helpful, particularly to steel manufacturers who must check the machining properties of their work materials and therefore would welcome a quick and reliable checking method. Many ingenious methods for producing machinability data have been proposed, and some are still used today. Although these methods are of doubtful meaning, they can be used to measure the variation in some machining property of materials having the same specification. It would be most difficult to prove that the results of these tests yield quantitative information on machining properties of practical interest, although experience has shown that these results do give some guide.

4.6.1 Tool Wear and Machinability Testing

The production of such tool-wear data as those shown in Figure 4.2, Figure 4.4, and Figure 4.9 is extremely tedious and involves a series of carefully taken measurements over long periods of time. These tests, of necessity, consume a considerable amount of work material and often several experimental tools. The additives and inclusions in commercial work materials can vary due to segregation in such a manner that considerable variations in tool-wear rate occur from the inside of the specimen to the outside; variations can also occur along the length of a specimen. Considerable variations between tool materials of nominally the same specification can also occur. With carbide tools, specimen inserts placed together in the furnace during the sintering process are known to exhibit different wear performances. With high-speed steel tools, the problem is even greater: the tempering temperatures are so critical that the slightest variation in heat-treatment conditions results in large variations in performance.

For the reasons given above, the results of extended tool-wear tests can often be misleading because of the danger that the properties of both the workpiece and specimen tools may vary during a series of tests. The ISO-recommended tool-life test for turning [4] attempts to minimize these problems by specifying the tool and work materials and the test conditions very closely. However, some method of rapid wear testing, consuming a small quantity of work material and requiring only one specimen tool is clearly desirable and several testing methods have been proposed.

One method of rapid testing is known as the accelerated wear test, and one example of this type of test involves a facing operation in a lathe. In this test the workpiece is rotated at high speed, and the tool is traversed from the workpiece centerline radially outward. Conditions are chosen such that complete failure of the tool occurs before the tool reaches the outside diameter of the workpiece. The radial distance machined is taken as a measure of the machining characteristics of the particular tool/workpiece combination. Although this method is useful as a comparative test for various work or tool materials, the results yield little information regarding the fundamental nature of wear in machining and do not give an absolute measure of the performance of the tool, a measure which could be related to its behavior in practical circumstances.

Another type of rapid wear test involves a special technique for measuring extremely small amounts of wear. In this method the working surfaces of the cutting tool are carefully prepared and polished, the tool is then used to cut the test workpiece for a short time, and the small amount of wear occurring on the tool flank is measured. The disadvantage of this method is that the wear characteristics measured are those occurring in the initial tool breakdown period and may not always be representative of the behavior of a tool in the later stages of wear. A further technique for measuring small amounts of tool wear is the radioactive-tracer technique. In this method the test-cutting tool is irradiated by a radioactive source before the cutting test. The chips produced during the test are collected and their activity measured with a radioactivity counter; the reading is taken as a measure of the wear rate of the cutting tool. This method is also suitable only as a comparative test.

In a more recent proposal for rapid wear testing, the tool is initially prepared by grinding an artificial wear land on its flank. Increases in the magnitude of this wear land from a small amount of machining are measured, and thus the wear rate of the tool in the central linear portion of the wear curve (Figure 4.2) is obtained. In this test, care must be taken to grind the initial wear land with a slightly negative "clearance" to establish the required frictional contact between the tool and workpiece at the beginning of the test.

In all the theories of progressive tool wear, the rate of wear would be expected to increase both with increased metallic contact at the interface and with increased temperature; thus it might be expected that some relationship exists between tool wear and chip-tool friction and between tool wear and temperature. That a relationship of this type does exist was found true in the past and explains why, under many conditions, measurements of chip-tool friction or cutting temperatures can give a useful guide to the wear behavior of a given work-tool combination. Such measurements are used as machinability tests and have the great advantage that they require very little work material.

4.6.2 Factors Affecting the Machinability of Metals

In general it can be said that high hardness will give poor machinability because power consumption and temperatures and hence tool-wear rates will be high. However, many other factors affect machinability, and it can also be said in general that pure metals tend to adhere to the working surfaces of the cutting tool and give high friction and high tool-wear rates. For example, when pure iron is machined, cutting forces and tool-wear rates are extremely high, almost as high as those obtained when some of the very tough alloy steels are machined.

The deliberate addition of sulfur, lead, and tellurium to nonferrous metals, as well as to steels, makes possible increased production rates and improved surface finish, but how these additives function is not completely understood. It appears certain that these additives reduce the metallic contact between the tool and work material; hence they reduce friction and tool-wear rates. These constituents do not greatly affect the room-temperature mechanical properties of the parent metal, but the high-temperature properties can be seriously affected. The effect of sulfur, lead, or tellurium on the high-temperature properties of metals of course limits the use of these so-called free-machining metals, but nevertheless they find wide application and demand for them is high. The lack of theory on the effects of the workpiece material on tool wear creates difficulties for engineers and metallurgists involved in production. For example, with increasing automation in machining processes, it becomes more important to maintain consistent tool-wear characteristics, but it is well known that very large differences in tool wear occur when cutting metals of nominally the same specification. These differences are usually not related to the material properties normally measured.

Particular difficulties are encountered with metals developed for high-temperature applications such as those metals used in gas turbines. Addition of free-machining elements is not usually possible with these metals, and therefore it is left to the tool engineer and production engineer to solve the machining problems. Some of these problems are so difficult that other machining techniques, such as electrochemical machining, may be more economical to use; these techniques are discussed in Chapter 14. However, the development of harder and more temperature resistant tool materials has enabled many of the more difficult to machine materials to be readily machined. Additionally, other easier to machine materials can be economically machined at much higher speeds and material removal rates. This has led to the term high performance machining (HPM) [24], meaning either machining at much higher speeds (high speed machining (HSM)) [25] or the machining of materials in their hardened state (hard machining).

4.7 HIGH SPEED MACHINING

Developments in ultra-hard cutting tool materials, together with improved machine tool spindle and drive systems, have led to the ability to machine at much higher cutting speeds than previously possible. High speed machining (HSM) allows high material removal rates while maintaining relatively low feeds. Thus chip loads and cutting forces can be moderately low. This enables complex shapes to be economically machined from the solid, including items with very thin sections. This is particularly useful if the advantages of HSM can be used to replace assembly operations to yield significant economic benefits. Figure 4.14 shows a comparison between the manufacture of a sheet metal fabricated component for a military helicopter and a replacement part produced by HSM from a solid blank of material [26]. It can be seen that when the total cost of assembly and fabrication of the part is taken into account that the HSM part is significantly less expensive to produce, together with an associated reduction in weight of the part.

A number of benefits from HSM can be identified [25]:

- Increased machining accuracy, particularly when machining thin webs, etc., due to reduced chip loads
- Improved surface finish and reduction in surface damage layers
- Reduced burr formation
- Improved chip disposal due to better chip breaking
- Possibilities of improved stability against chatter vibrations as a result of wider stability lobe spacing at higher speeds (see Chapter 9)
- Simplified tooling

A limitation of using HSM for the more difficult machine materials can be reduced tool life and hence the overall costs of machining must be considered.

Most high speed machining is carried out on light alloy materials for which tool life is not a limitation, but increasingly HSM is being utilized for other

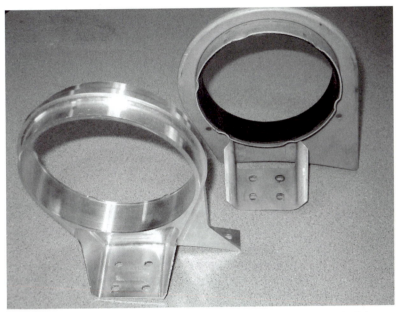

High speed machined part (left)	Sheet metal fabrication (right)
1 machined part	5 sheet metal parts
2 hours manufacturing time	32 hours manufacturing time
45% less cost	20 tools needed
10% weight reduction	19 rivets required
Tooling costs virtually eliminated	

FIGURE 4.14 Cost reduction from replacing a sheet metal fabrication with a single high speed machined part. (Adapted from Herrera, A., Proc. 12th Int. Forum on Design for Manufacture and Assembly, Newport, RI, June, 1997, Published by Boothroyd Dewhurst, Inc., Wakefield, RI.)

materials. The range of cutting speeds that are regarded as HSM varies with the work material [25]. Figure 4.15 shows the ranges of cutting speeds for conventional (low speed) machining and for HSM, together with an intermediate transition range, for various generic classes of workpiece materials. As can be seen cutting conditions regarded as HSM for titanium alloys and some alloys steels are conventional for many aluminum alloys.

For most materials chip formation in the high speed range results in shear localized chips as discussed in section 2.3.4. This is particularly the case for titanium alloys, alloy steels, and nickel-based alloys. In general complete separation of chip segments occurs at very high speeds. This transition is illustrated for AISI 4340 alloy steel in Figure 4.16 [27]. For the easier to machine materials such as aluminum alloys, then continuous chip formation can be observed for all cutting speeds.

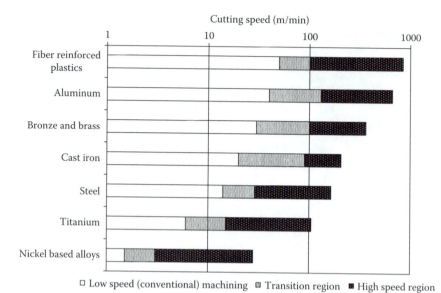

FIGURE 4.15 Speed ranges for conventional and high speed cutting for different material classes. (Adapted from Schulz, H. and Moriwaki, T., *Ann. CIRP*, Vol. 41/2, 637–643, 1992.)

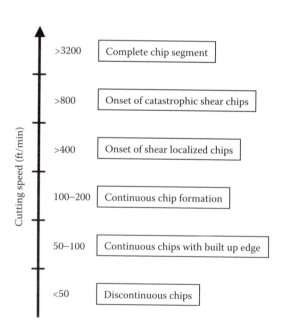

FIGURE 4.16 Speed ranges for different types of chip formation for AISI 4340 steel. (Adapted from Komandhuri, R., et al., *ASME, J. Eng. Ind.*, Vol. 104, 121–131, 1982.)

High cutting speeds present some problems with coolant application, in particular getting coolant to the tool tip is very difficult. For this reason mist application or dry cutting is most frequently used. Success has also been achieved with air cooling in some cases.

4.8 HARD MACHINING

Many machine components are produced from hardened steel and their functional behavior is influenced significantly by the fine finishing processes used during the final stages of manufacture. The usual manufacturing sequence for a hardened steel shaft for example would be rough turning, finish turning, and hardening followed by finish grinding. With the development of modern hard tool materials (ceramics and CBN), an alternative manufacturing process is turning of the material in its hardened state (hard turning or hard machining). Hard machining has increased flexibility relative to grinding and has the ability to manufacture a complex geometry in one setup. Thus for suitable applications hard machining is a competitive alternative to finish grinding. There are a number of potential advantages from hard turning. Material removal rates are much higher than grinding and machining times can be substantially reduced [21]. There is a reduction in the number of separate machines required with associated reductions in lost production time from setups and transfer between workstations. A comprehensive review of developments in the cutting of hardened steel can be found in [21].

For hard turning to replace grinding operations it must be possible to produce equivalent finished surfaces, with similar tolerances and surface roughness. Part geometry and surface finish comparable to grinding can be obtained by hard turning if machines with high rigidity are used with sharp cutting tools. However, worn cutting tools have been found to cause surface damage in hard turned components.

The chip formation processes in hard machining generally result in shear localized and segmental chips. Local thermal softening and crack initiation at the free surface influence considerably the localized shearing [28]. The machining of hardened steel generally results in a thin rehardened surface layer of material with an over tempered region just beneath. This layer is usually referred to as a white layer, because as a result of the micro-structural changes in the rehardened layer, it appears white after chemical etching. White layers consist of over 60% austenite [21] and the formation is caused through being heated above the austenizing temperature followed by rapid self-quenching due to heat loss into the bulk of the material. Residual stresses are also caused in the surfaces that may affect the service performance of the component.

White layers and residual stress formation are influenced considerably by the sharpness of the cutting tool. With sharp tools the surface micro-structure remains much the same as the bulk of material and the surface residual stresses tend to be compressive. As the tool flank wear increases, friction increases and the thermal load on the surface layer grows. This leads to the development of white layers and a greater tendency to induce tensile residual stresses in the surface layers of the part [24] (Figure 4.17).

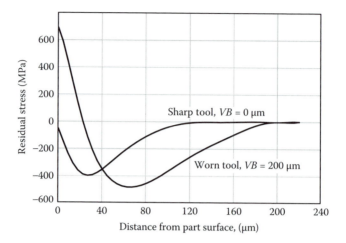

FIGURE 4.17 Surface layer residual stresses for the machining of hardened steel with sharp and worn tools. (Adapted from Byne, G., Dornfeld, D., and Denkena, B., *Ann. CIRP,* Vol. 53/2, 1–25, 2003.)

The main advantages of hard machining over grinding are for the manufacture of the more complex parts as different shapes and surfaces can be achieved with a single tool. In addition internal and external machining can often be done on the same machine tool. Consequently significant potential for reducing machining times exists for suitable applications.

PROBLEMS

1. A single-point cutting tool has a zero rake angle and a clearance angle of 2 degrees. By what percentage would the life of the tool between regrinds be increased if a clearance angle of 8 degrees were provided? (Assume the tool is reground after a specified amount of flank wear has taken place and that the rate of wear of the tool normal to the wear land is constant.)

2. From the results of tests for a tool clearance of 8 degrees, it was found that the tool life was 240 s for a cutting speed of 1 m/s, and 12 s for a cutting speed of 5 m/s. Assuming that Taylor's tool-life relationship holds, that is,

$$vt^n = v_r t_r^n$$

   a. The constant n
   b. The percentage change in tool life if the tool clearance angle is decreased from 8 to 3 degrees
   c. The new value of v_r when $t_r = 1$ (after the change in clearance angle)

3. In a multipass rough-turning operation, the tool life is observed to obey an equation

$$t = v^{-3} f^{-0.8} a_p^{-0.1}$$

where v, f, and a_p are the cutting speed, feed, and depth of cut, respectively, in SI units. The system normally operates with $v = 1$ m/s, $f = 0.5$ mm, and $a_p = 10$ mm. Estimate the effect (percent increase or decrease) on tool life, cutting force, and power of doubling the metal removal rate by increasing (a) the speed, (b) the feed, and (c) the depth of cut.

REFERENCES

1. Bowden, F.P. and Tabor, D., *Friction and Lubrication of Solids*, Oxford University Press, London, 1954.
2. Trent, E.M., *Metal Cutting*, 3rd. ed., Butterworths-Heinemann, London, 1991.
3. Childs, T.H.C., Maekawa, K., Obikawa, T., and Yamane, Y., *Metal Machining: Theory and Applications*, Arnold, London, 2000.
4. International Standards Organisation (ISO), Tool Life Testing with Single-Point Turning Tools, ISO 3635, ISO Geneva, 1993.
5. Taylor, F.W., On the Art of Cutting Metals, *Trans. ASME*, Vol. 28, 31, 1906.
6. Trent, E.M., Tool Wear and Machinability, *Production Engineer (London)*, Vol. 38, no. 3, 105, 1959.
7. Cherry, J., Practical Investigation in Metal Cutting, *Production Engineer (London)*, Vol. 41, no. 2, 90, 1962.
8. Boothroyd, G., Eagle, J.M., and Chisholm, A.W.J., Effect of Tool Flank Wear on the Temperatures Generated during Metal Cutting, *Adv. Machine Tool Design and Res.*, Vol. 7, 667, 1967.
9. Mathew, P., Use of Predicted Cutting Temperatures in Determining Tool Performance, *Int. J. Machine Tools and Manuf.*, Vol. 29, no. 4, 135–149, 1989.
10. Molnari, A. and Nouari, M., Modeling of Tool Wear by Diffusion in Metal Cutting, *Wear*, Vol. 252, 135–149, 2002.
11. Usui, E., Hirota, A., and Masuko, M., Analytical Prediction of Three Dimensional Cutting Process—Part 3: Cutting Temperature and Crater Wear of Carbide Tool, *Trans. ASME*, Vol. 100, 222–228, 1978.
12. Takeyama, H. and Murata, T., Basic Investigations on Tool Wear, *Trans. ASME, J. Eng. for Ind.*, Vol. 85, 33–38, 1963.
13. Kitagawa, T., Maekawa, K., Shirakashi, T., and Usui, E., Analytical Prediction of Flank Wear of Carbide Tools in Turning Plain Carbon Steel (Part 1) — Characteristic Equation of Flank Wear, *Bull. Japan Soc. Precision Eng.*, Vol. 22, no. 4, 263–269, 1988.
14. Kitagawa, T., Maekawa, K., Shirakashi, T., and Usui, E., Analytical Prediction of Flank Wear of Carbide Tools in Turning Plain Carbon Steel (Part 2) — Prediction of Flank Wear, *Bull. Japan Soc. Precision Eng.*, Vol. 23, no. 2, 126–134, 1989.

15. Yen, Y.-C., Sohner, J., Weule, H., Schmidt, J., and Altan, T., Estimation of Tool Wear of Carbide Tool in Orthogonal Cutting Using FEM Simulation, Proc. of 5th CIRP International Workshop on Modeling of Machining Operations, 149–160, Purdue University, May 21–22, 2002.

16. Destafani, J., Cutting Tools 101, *Manuf. Eng.*, September, 57–69, 2002.

17. Tlusty, J., *Manufacturing Processes and Equipment*, Prentice Hall, New Jersey, 2000.

18. Almond, E.A., Towards Improved Tests Based on Fundamental Properties, in *Towards Improved Tool Performance*, National Physical Laboratory Conference, Metals Society, London, 1982.

19. Drozda, T.J. and Wick, C. eds., *Tool and Manufacturing Engineers Handbook*, Vol. 1, Machining 4th ed., Society of Manufacturing Engineers, 1983.

20. International Standards Organization, Classification of Cemented Carbides, ISO 513, 1975 (E).

21. Tönsoff, H.K., Arendt, C., and Ben Amor, R., Cutting Hardened Steel, *Ann. CIRP*, Vol. 49/2, 547–566, 2000.

22. Destafani, J., Cutting Tools 101: Coatings, *Manuf. Eng.*, October, 47–55, 2002.

23. Destafani, J., Cutting Tools 101: Geometries, *Manuf. Eng.*, November, 41–49, 2002.

24. Byne, G., Dornfeld, D., and Denkena, B., Advancing Cutting Technology, *Ann. CIRP*, Vol. 53/2, 1–25, 2003.

25. Schulz, H. and Moriwaki, T., High Speed Machining, *Ann. CIRP*, Vol. 41/2, 637–643, 1992.

26. Herrera, A., Design for Manufacture and Assembly Applied to the Design of the AH64D Helicopter, Proc. 12th Int. Forum on Design for Manufacture and Assembly, Newport, RI, June, 1997 (Published by Boothroyd Dewhurst, Inc., Wakefield, RI).

27. Komandhuri, R., Schoeder, T., Hazra, J., Von Turkovich, B.F., and Flom, D.G., On Catastrophic Shear Instability in High Speed Machining of AISI 4340 Steel, *Trans. ASME, J. Eng. Ind.*, Vol. 104, 121–131, 1982.

28. Konig, W., Berthold, A., and Koch, K.F., Turning versus Grinding — A Comparison of Surface Integrity Aspects and Attainable Accuracy, *Ann. CIRP*, Vol. 42/1, 39–43, 1993.

5 Cutting Fluids and Surface Roughness

5.1 CUTTING FLUIDS

Cutting fluids are applied to the chip formation zone to improve the cutting conditions (compared to dry cutting conditions). These improvements can take several forms, depending on the tool and work materials, the cutting fluid, and to a large extent the cutting conditions. The two most important ways in which a cutting fluid can act are as a coolant and as a lubricant. The lubricating action is more important at low cutting speeds, whereas the cooling effect is more important at higher cutting speeds due to the large increases in heat generated by the chip removal process.

Many practical cutting fluids have a mineral or vegetable oil base, mineral oil being the more widely used [1–3]. However, there are available synthetic fluids that are oil-free and based upon polymers, together with organic and inorganic compounds. Some of these oils are made to be applied as an emulsion with water (water-miscible cutting fluids) — the remainder are used neat (undiluted with water), either plain or having various additives. In general, water-based emulsions are used when cooling action is the most important requirement because these emulsions have a much larger heat-conducting capacity than neat oils. Neat oils are used when lubricating action is the most important consideration; their use is usually confined to slower speed cutting operations such as screw cutting, broaching, and gear cutting.

There are four basic categories of cutting fluid:

Straight or neat oils — These are used undiluted and are usually mineral oils but often include other lubricants such as fats, vegetable oils, and esters, together with high-pressure compounds based on chlorine, sulfur, and phosphorus.

Soluble oils — These consist of oil with emulsifiers that allow the oil to disperse in water. These oils are used in diluted form with a usual concentration range of 3 to 15%. These fluids are widely used in industry and are generally the least expensive cutting fluids available.

Synthetic fluids — These are oil-free solutions formulated from alkaline inorganic and organic compounds, usually with the addition of corrosion inhibitors. They are used diluted with water, with a concentration range from 3 to 10%. These fluids often provide the best cooling performance.

Semi-synthetic fluids — These fluids are essentially a combination of synthetic and soluble oil fluids and exhibit characteristics of both types.

175

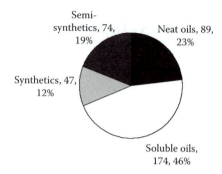

FIGURE 5.1 Quantities of the different types of cutting fluids used in the U.S. manufacturing industry.

They consist of natural and synthetic emulsifiers, oil droplets, and clear emulsions. They offer good corrosion resistance, lubrication, and tolerance to contamination.

The overall quantities of these different types of cutting fluids that are used in industry are shown in Figure 5.1. It can be seen that around 80% of cutting fluids in use are water-based emulsions of oils and/or synthetic fluids.

The overall cost of applying cutting fluids is significant. A study in the European automotive industry [4,5] indicates that the cost of cutting fluids makes up nearly 20% of the total manufacturing cost in machining, whereas the cost of the actual cutting tools is only around 7.5%. Thus there are significant benefits in reducing the quantities used or from extending the useful life of cutting fluids.

5.2 THE ACTION OF COOLANTS

Several advantages are to be gained by applying a coolant to the cutting process:

1. An increase in tool life through temperature reduction in the region of the tool cutting edge
2. Easier handling of the finished workpiece
3. A reduction in the thermal distortion caused by temperature gradients generated within the workpiece during machining

The last two factors are most important in the grinding process. A secondary and sometimes overlooked advantage of cutting-fluid application is the flushing action, which assists the removal of chips from the cutting area. Properly placed fluid nozzles can help prevent the packing of chips in flutes of drills or milling cutters.

Quantitative information on the effect of a coolant on cutting-tool life has been obtained mainly from drill life tests; these tests clearly demonstrated improvements in drill life from the application of a coolant. It seems most likely that these improvements are caused by some reduction in temperature in the region of the drill point, but work on cutting temperatures described in Chapter 3 suggests that the heat losses from the exposed surface of the tool and workpiece have quite a small effect on the temperatures in the region of the tool cutting edge. It can only be assumed, therefore, that the tool-wear rate is extremely sensitive to the small changes in temperature in the region of the wearing surfaces that result from the application of a coolant.

Holmes [1] has described the various general characteristics of water-miscible cutting fluids (Table 5.1) and has developed the guide to their selection presented in Table 5.2. Although developed sometime ago, these guide lines are still applicable to the modern day coolants available.

TABLE 5.1
General Characteristics of Water-Miscible Cutting Fluids

Type of Lubricant	Characteristics
Emulsifiable, Soluble Oils	
Generable-purpose	Used at dilutions between 1:10 and 1:40 and give a millky emulsion. Used for general-purpose machining.
Clear-type	Used at dilutions between 1:20 and 1:60. Their high emulsifier content results in emulsions that vary from translucent to clear. Used for grinding or general-purpose machining.
Fatty	Used at concentrations similar to those for general-purpose soluble oils and a similar (milky-emulsion) appearance. The fat content makes them particularly effective for general machining operations on nonferrous metals.
Extreme-pressure	Generally contain sulferized or chlorinated extreme-pressure additives. Used at dilutions between 1:10 and 1:20 where a higher performance than that given by general-purpose, clear-type, or fatty soluble oil is required.
Chemical Solutions	
Grinding fluids	Essentially solutions of chemical rust inhibitors in water. Used at dilutions between 1:50 and 1:100 for grinding operations on iron and steel.
General-purpose, synthetic cutting fluids	Contain mainly water-soluble rust inhibitors and surface-active load-carrying additives. Used at dilutions between 1:10 and 1:40 for cutting and higher dilutions for grinding. Most are suitable for both ferrous and nonferrous metals.
Extreme-pressure, synthetic cutting fluids	Similar in characteristics to general-purpose, synthetic cutting fluids, but containing extreme-pressure additives to give higher machining performance when used with ferrous metals.

Source: Adapted from Holmes, P.M., *Ind. Lubr. Tribology*, Vol. 23, 2, 47–55, 1971.

TABLE 5.2
Guide to the Selection of Water-Miscible Cutting Fluids

Machining operation	Workpiece material					
	Free-machining nonferrous alloys	Tough nonferrous alloys	Free-machining and low-carbon steels	Medium carbon steels	High-carbon and alloy steels	Stainless and heat-resistant alloys
Grinding	General-purpose or clear-type, soluble oils		Clear-type, soluble oil, or chemical grinding fluid			
Turning	General-purpose oil, soluble oil	General purpose or fatty, soluble oils	General-purpose, soluble oil, or synthetic fluid		Extreme pressure, soluble oil or synthetic fluid	
Milling	General-purpose or fatty, soluble oils			Extreme-pressure, soluble oil or synthetic fluid	Extreme-pressure, soluble oil, or synthetic fluid (neat cutting oils may be necessary)	
Drilling			Fatty or extreme-pressure, soluble oils; synthetic fluids			
Gear shaping	General-purpose or fatty, soluble oils			Extreme-pressure, soluble oil or synthetic fluid	Neat cutting oils preferable	
Hobbing				Extreme-pressure, soluble oil or synthetic fluid		Neat cutting oils preferable
Broaching				Extreme pressure, soluble oil or synthetid fluid		
Tapping	General-purpose or fatty, soluble oils (neat cutting oils may be preferable)			Extreme-pressure, soluble oil or synthetic fluid (neat cutting oils may be preferable)	Neat cutting oils preferable	

Source: After Holmes, P. M., Ind. Lubr. Tribology, Vol. 23, 2, 47–55, 1971.

5.3 THE ACTION OF LUBRICANTS

5.3.1 BOUNDARY LUBRICATION

Bowden and Tabor [6] have shown that under conditions of high temperature and pressure and low sliding velocity, a hydrodynamic film cannot be maintained, and thus direct contact occurs. Frictional resistance to motion under these conditions arises from a combination of asperity shearing and viscous shearing of the fluid; this is known as *boundary lubrication* and is greatly affected by the nature of the adsorbed fluid layer (boundary layer), a few molecules thick, on the surfaces of the sliding metals. A lubricant boundary layer functions by reducing the area of intimate metallic contact between the two surfaces. Some lubricants having "reactive ingredients" (extreme-pressure additives of sulfur, phosphorus, or chlorine) can offer considerably greater surface protection during sliding by formation of a solid lubricant layer. This layer of solid lubricant, often the product of a chemical reaction between the lubricant and the metal surface, remains effective at temperatures up to its melting point.

When two metallic surfaces, each having an absorbed solid lubricant layer, are placed together and a normal load applied, the resulting deformation of the surface asperities causes some penetration of the lubricant layer, and metallic contact is established (Figure 5.2).

If the real area of contact A_r is thought of as the area supporting the load and γ_m is the proportion of this area in which metallic contact occurs, the frictional force F_f required to continually shear the junctions between surface asperities can be written as

$$F_f = A_r \left[\gamma_m \tau_1 + \left(1 - \gamma_m\right) \tau_2 \right] \tag{5.1}$$

where τ_1 and τ_2 are the shear strengths of the softer metal and the softer lubricant layer, respectively. Thus, if γ_m remains essentially constant for given values τ_1

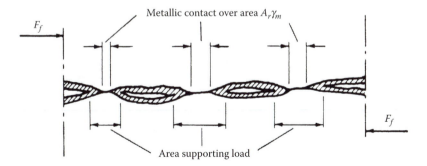

FIGURE 5.2 Sliding between surfaces with a solid lubricant layer, where F_f = friction force, τ_1 = shear strength of the softer metal, and τ_2 = shear strength of the lubricant layer.

and τ_2, the frictional force will be proportional to A_r, so that Amontons' law will hold as for unlubricated sliding surfaces. If the temperature at the interface rises until the lubricant layer melts, the layer loses its rigidity, and a large increase in metallic contact occurs; if τ_1 is greater than τ_2, F_f increases. As the interface temperature increases further, the frictional force and surface damage become characteristic of unlubricated sliding contact, although the lubricant is visibly present on the surface. The lubricant layer is now said to be desorbed or completely mobile. Thus in Equation 5.2, as the melting point of the lubricant boundary layer is exceeded, γ_m tends to unity and F_f tends to $A_r\tau_1$, thus approaching dry sliding-friction conditions.

5.3.2 LUBRICATION IN METAL CUTTING

Under certain conditions the application of a lubricant to the process can result in a reduction in friction on the tool face; this friction reduction on the tool face can cause a reduction in power consumption, an increase in tool life, and, most important, an improvement in the surface finish of the machined component by reducing the occurrence of a built-up edge.

The action of cutting lubricants has in the past provided a fascinating topic for fundamental research. It has been known for some time that a considerable reduction in the mean coefficient of friction on the tool face can be achieved, under certain conditions, by application of the correct lubricant. However, many accepted engineering lubricants can have little or no effect on the cutting operation. It has also been shown that other fluids and chemicals (e.g., carbon tetrachloride), never used as lubricants in normal engineering applications, can have startling effects on the cutting process, reducing power consumption by as much as 60% in certain cases.

The extremely high pressures existing in the region of the chip-tool interface during machining do not allow complete hydrodynamic lubrication, where the chip and tool would be separated by a thin film of fluid. Indeed, it has been known for some time that the lubricating action of cutting fluids is mainly of a chemical nature. For example, note the lubricating effect of carbon tetrachloride (CCl_4) when copper is machined: it can be seen from Figure 5.3 that at low cutting speeds friction at the chip-tool interface is reduced considerably by the application of the fluid. It has been shown that this effect results from formation of a low-shear-strength film of copper chloride at the chip-tool interface, which acts as a boundary lubricant, preventing, to a large extent, intimate metallic contact between the chip and tool. Under dry cutting conditions, the intimate contact between chip and tool results in an extensive secondary deformation zone that has the appearance of a stable built-up edge. The effect of the lubricant is to reduce the intimacy of chip-tool contact, thereby eliminating the built-up edge and secondary deformation. This elimination of the built-up edge and secondary deformation causes a reduction in friction at the chip-tool interface, reducing the forces required to form the chip.

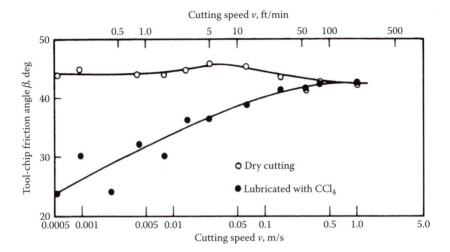

FIGURE 5.3 The effect of cutting speed on the lubricating action of carbon tetrachloride, where the work material is copper, the undeformed chip thickness is 0.25 mm (0.01 in.), the chip-tool contact length is 0.3 mm (0.012 in.), and the tool rake is 45 degrees. (After Cassin, C. and Boothroyd, G., *J. Mech. Eng. Sci.*, Vol. 7, 1, 67–81, 1965.)

5.3.3 CHARACTERISTICS OF AN EFFICIENT LUBRICANT IN METAL CUTTING

The work of Cassin and Boothroyd [7] has led to the suggestion that for a pure chemical compound to act as an efficient lubricant in metal cutting it should:

1. Have a small molecular size to allow rapid diffusion and penetration to the chip-tool interface
2. Contain a suitable reactive ingredient that, on reaction with the work material, forms a compound of lower shear strength that acts as a boundary lubricant
3. Be sufficiently unstable to be broken down under the temperatures and pressures existing at the chip-tool interface

Carbon tetrachloride (CCl_4), chloroform ($CHCl_3$), trichloroethane (CH_2CCl_3), and certain other chlorinated hydrocarbons have been found to be efficient lubricants when cutting many metals at low speeds. One important exception to this, however, is the machining of lead. The compound formed when machining lead in the presence of a chlorinated fluid is lead chloride, and this compound has a higher shear strength than lead itself. Thus under these circumstances an increase in the tool-chip friction occurs following application of the fluid.

The main disadvantage in using pure compounds such as those mentioned above is their toxicity. The use of carbon tetrachloride in any process involving

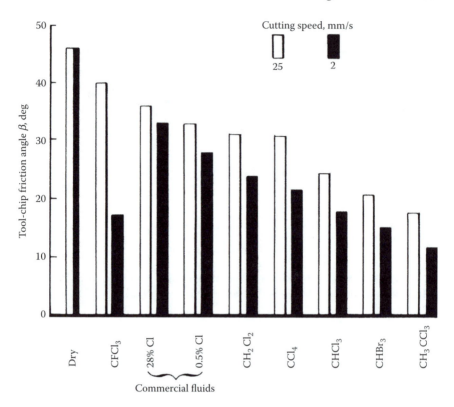

FIGURE 5.4 The lubricating action of various fluids, where the work material is copper, the undeformed chip thickness is 0.25 mm (0.01 in.), the chip-tool contact length is 0.3 mm (0.012 in.), and the tool rake is 45 degrees (all fluids are applied liberally). (After Cassin, C. and Boothroyd, G., *J. Mech. Eng. Sci.*, Vol. 7, 1, 67–81, 1965.)

high temperatures is out of the question since when heated, carbon tetrachloride gives off a very poisonous gas (phosgene). However, one form of methyl chloroform (inhibited trichloroethane), an extremely efficient lubricant under low-speed conditions, has been found to have relatively low toxicity, and successes with its use in practice have been reported in broaching and tapping.

Figure 5.4 indicates the magnitude of the lubricating effect when copper is machined with a variety of compounds and two commercial fluids containing chlorinated additives. It can be seen that the lubricating effect of the commercial fluids is relatively small, and this small effect probably results from insufficient penetration of these fluids to the chip-tool interface during machining. Clearly, greater benefits could be obtained if a neat compound could be found that is safe to use under a range of practical cutting conditions.

The effectiveness of all cutting lubricants diminishes as the cutting speed is increased. An example of this is illustrated in Figure 5.3 where the application of CCl_4 during the machining of copper had no effect on the chip-tool friction

TABLE 5.3
Characteristics of Neat Cutting Oils

Oil No.	Description	Viscosity (μm²/s at 38°C)
1	Very low-viscosity, inactive oil containing fatty and chlorinated additives	5.5
2	Inactive oil containing sulfurized fatty extreme-pressure additive	22.5
3	Inactive oil similar to No. 2 but with higher additive content	19.7
4	Multipurpose, chlorinated extreme-pressure oil with antistick-slip additives	37.8
5	Active oil containing free sulfur and sulfurized fat	25.0
6	Similar to No. 5 but with higher additive content	19.7
7	Inactive, extreme-pressure oil containing chlorinated and fatty additives; light in color	40.2
8	Low-viscosity, active oil containing free sulfur and sulfurized fatty additives	12.5
9	Special-purpose, highly chlorinated, active exteme-pressure oil	12.5

angle at cutting speeds greater than 0.5 m/s (100 ft/mm). This diminished effect may be explained partly by a loss of penetration of the fluid to the chip-tool interface at these higher speeds and partly by the increased temperatures, which, according to the discussion in section 5.3.1 would be expected to reduce the effectiveness of the solid boundary lubricant formed.

The preceding discussion has been concerned with the effect of a lubricant on tool-chip friction during machining, which in itself is of little interest to the practical machinist. The machinist is more interested in the resulting surface finish, particularly at low cutting speeds where the effect of a built-up edge would be most pronounced. Clearly, a reduction in chip-tool friction would imply a reduction in the size of a built-up edge or its complete elimination, and this would give an improved surface finish. It may be accepted, therefore, that measurements of the effects of lubricants on tool-chip friction give an adequate guide to their performance in practice.

Table 5.3 and Table 5.4 present the characteristics of neat cutting oils and a guide to their selection [13].

5.4 APPLICATION OF CUTTING FLUIDS

The cutting fluid cannot perform its function unless it is effectively delivered to the cutting zone. Fluids selected for lubricating properties must be applied in such a way that a film can form on the sliding surfaces. Fluids chosen for cooling must gain access to the region of the tool cutting edge. This is usually achieved by flooding the cutting area with fluid, which effectively removes the heat generated by deformation of the work material. Additional problems occur in the

TABLE 5.4
Guide to the Selection of Neat Cutting Oils

Machining operation	Workpiece material					
	Free-machinine nonferrous alloys	Tough nonferrous alloys	Free-machining and low-carbon steels	Medium carbon steels	High-carbon and alloy steels	Stainless and heat-resistant alloys
Thread or form grinding	2	3	5	6 or 7		7
Turning	2 or 4	3 or 4		3, 4, or 5	6 or 4	6 or 7
Milling				4 or 5	6 or 7	
Drilling				5 or 6	6 or 8	8 or 9
Gear shaping				5 or 6		7 or 8
Hobbing	3		3 or 5	6 or 8		7 or 9
Broaching	3 or 4		6		8 or 9	9
Tapping	2	2 or 3			6 or 8	8 or 9

The numbers refer to the oils listed in Table 5.3

Source: Holmes, P.M., *Ind. Lubr. Tribology*, Vol. 23, 2, 47–99, 1971.

application of cutting fluids in the grinding process owing to the high wheel speeds, and these are discussed in Chapter 9.

Continuous application of cutting fluids is generally preferred to intermittent applications, which cause temperature fluctuations. These temperature variations can lead to the formation of micro cracks in hard and brittle tool materials, which may ultimately reduce the life of the cutting tools. Four main fluid application methods are used in practice for general cutting operations: manual, flood, jet, and mist application. A secondary method in some specific applications, such as drilling, is to deliver the cutting fluid through the tool.

5.4.1 MANUAL APPLICATION

Cutting fluids and lubricants are often applied manually, particularly for small-batch and jobbing shop production. Manual application with a brush by the operator is the easiest and least expensive method of fluid delivery. The main disadvantages of manual application are: (1) intermittent fluid application, (2) poor chip removal, and (3) limited access of the fluid to the cutting zone.

5.4.2 FLOOD APPLICATION

The most common method of fluid application is to flood the tool, workpiece, and cutting zone with fluid delivered from a low-pressure nozzle directed appropriately. Flood application allows a continuous flow of fluid to the cutting zone and helps to remove chips from the cutting zone. The cutting fluid drains into a sump in the base of the machine, where it is filtered before being pumped back to the delivery nozzle.

The effectiveness of flood application of cutting fluids depends on the geometry of the cutting process being used. Careful orientation of the delivery nozzles is necessary to ensure that the cutting fluid is not deflected away from the cutting zone by the rotating workpiece or tool.

5.4.3 JET APPLICATION

A high pressure and hence higher velocity jet of fluid is directed at the cutting zone. Again fluid drains into the sump in the base of the machine for filtering and delivery back to the nozzle. This form of application can be very effective in removing chips from the cutting area and may be particularly useful for higher cutting speed operations, including grinding (see Chapter 9).

5.4.4 MIST APPLICATION

Cutting fluids can be applied as an air-carried mist. Mist application is usually applicable to processes for which cutting speeds are high and the uncut chip thickness is relatively small such as end-milling operations. This form of application can be used effectively when flooding is impractical, and it sometimes provides a means of applying cutting fluids to otherwise inaccessible areas. A disadvantage of mist application is the health risk due to inhalation of fluid droplets by the operator; consequently, good ventilation is required. Mist application is an important part of minimum quantity lubrication applications (see section 5.8).

5.4.5 THROUGH THE TOOL APPLICATION

In some processes where penetration of the cutting fluid into the cutting zone is particularly difficult such as drilling, application through suitable passages in the tool shank and cutting edge can be very effective. This type of application has long been the practice in deep-hole gun drilling, but many tool suppliers now offer ranges of drills for through the tool coolant application. This has been shown to improve tool life and more effectively flush chips from the cutting zone [8]. Developments have also occurred with the delivery of lubricants through the spindle and tools during end milling operations, in particular in association with near dry cutting conditions (see section 5.8).

5.5 CUTTING FLUID MAINTENANCE

The significant cost associated with cutting fluid application places particular importance on good maintenance of these fluids to extend useful life. The aging process of cutting fluids is gradual and because of this cutting fluid maintenance is often delayed or ignored. Such an approach is very much a false economy. A cutting fluid maintenance strategy will not only reduce replacement costs but also result in less frequent disposal costs.

Machining coolants do not wear out [2,9]. Over a period of time they become contaminated with a number of substances, including:

- Tramp oils from the machine hydraulic and lubrication systems
- Oxidation products as a result of the heat generated during machining
- Solid particles from the material being processed
- Microorganisms (bacteria, fungi, molds) that live and multiply on the fatty components in the fluids
- Separated emulsion oil

In addition evaporation alters the coolant concentration. If the concentration is regularly restored and the contamination removed, the life of coolants can be extended considerably.

Coolants fail when the contamination is not removed. The used fluid and contaminants flow into the sump or reservoir of the machine tool where they are mixed by hydraulic agitation. The large particles fall to the bottom of the sump and can be removed manually or automatically by a drag out conveyor system. It is important that the sump is of adequate size, as this will allow the oily contamination to rise to the top where it can be removed by skimmers. In addition barrier filters and centrifugal separators should be used to remove the residual substances.

Bacterial growth is a significant concern with coolants. The bacteria feed and multiply on the fatty acid and animal fat additives in the free oil contaminants. This process raises the acidity of the fluid, can result in bad odors, breakdown emulsions, and lead to operator health hazards in the form of skin rashes. The addition of toxic biocides can control this growth, but keeping the coolant clean is a more effective solution.

To ensure that the coolant system is working properly, periodic testing is essential. The pH levels should be regularly determined with dip strip or pH meters. The oil concentration values should be determined using refractometers. Allowing sample quantities of fluid to settle can assess solid particle concentrations. If these values are found to be outside acceptable limits, then remedial action can be taken. Table 5.5 [9] summarizes some potential problems with cutting fluids, together with some suggested means of control.

5.6 ENVIRONMENTAL CONSIDERATIONS

There are a number of potential environmental concerns with cutting fluids. As far as the direct working environment is concerned, there are two main considerations. Bacterial growth can lead to skin irritations and rashes. A second concern is the release of small fluid droplets (aerosols) into the atmosphere. The common methods of coolant application, together with agitation from fast moving machine elements and the chips produced, can result in significant cutting fluid aerosol formation. The U.S. Occupational Safety and Health Administration (OSHA) [10] recommended permissible exposure levels (PEL) for metalworking fluids concentrations is 5 mg/m^3. The National Institute of Occupational Safety and Health (NIOSH) has more stringent recommendations of 0.5 mg/m^3 [11]. It has been found that oil mist levels in U.S. automotive parts manufacturing facilities when traditional flood coolant application is used are generally of the order of 20 to 90 mg/m^3 [12]. Thus careful control of the aerosol concentrations produced from cutting fluid applications is necessary.

5.7 DISPOSAL OF CUTTING FLUIDS

When cutting fluids deteriorate to levels that result in the need for disposal care must be taken to adequately treat the fluids. Most companies will use an external hazardous material service or an internal wastewater treatment system. Several important factors [10] need to be considered:

1. Federal, state, and local regulations must be fully complied with
2. The method of treatment depends on the type of fluid

TABLE 5.5
Troubleshooting Guide for Cutting Fluids

Problem	Cause	Solution
Foaming	Concentration too high	Check pH, adjust concentration
	Machine cleaner in sump	Run machine; cleaner should dissipate
	Mechanical problem (crack in hose, sump level too low, crack in pump, pump pressure too high)	Check machinery and repair
	Soft water	Sample water and treat
	High tramp oil content	Skim off oil; check hydraulic lines for leaks
Rusting	Concentration too low	Adjust concentration
	Poor mixing (soluble oil)	Add concentrate to water
	High tramp oil content	Skim off oil; check hydraulic lines for leaks
Poor tool life	Concentration too low	Adjust concentration
	Wrong product being used	Consult cutting fluid suppliers
	Large amounts of biocide added to sump or system	Consult cutting fluid suppliers
	High tramp oil content	Skim off oil; check hydraulic lines for leaks
Odor	Low concentration	Adjust concentration
	Low pH	Check pH and consult cutting fluid suppliers
	High tramp oil content	Skim off oil; check hydraulic lines for leaks
Skin irritation	High concentration	Adjust concentration
	High pH	Check pH and consult cutting fluid suppliers
	High tramp oil content	Skim off oil; check hydraulic lines for leaks
	Dirty shop cloths	Use only clean cloths
	Allergies	Have operators checked for allergies;
	Out-of-shop influences	contact physician as necessary
Residue in machine	High concentration	Adjust concentration
	High tramp oil content	Skim off oil; check hydraulic lines for leaks
	Incorrect mixing	Consult suppliers
		Adjust coolant nozzles; check ventilation
	High misting operations	system

Source: Adapted from Kuchler, C., *CNC Mach.*, Vol. 6, Issue 20, Winter, 2002.

Straight oils can be recycled by cleaning and filtering in a similar manner to other lubricating oils. With water-diluted coolants, the water can be removed and the remaining solids disposed of separately. The primary stage involves removing the free oil (skimming) and suspended solids (filtration and settling). This is followed by ultra filtration and chemical treatment to separate fluids. The final stage uses carbon filtration or reverse osmosis for final purification of the water that can then safely be reused.

5.8 DRY CUTTING AND MINIMUM QUANTITY LUBRICATION

Concerns about coolant costs and environmental problems from large quantity cutting fluid application has led to recent developments in the use of dry or near dry cutting conditions. The latter has become known as minimum quantity lubrication (MQL) [13–17]. Comprehensive reviews of dry and near dry machining developments can be found in [4,17].

For dry cutting to be acceptable, steps must be taken to compensate for the primary functions of the cutting fluid. More friction and adhesion between the tool and workpiece will occur. The thermal loads on the tool and workpiece are greater (Figure 5.5) [4], which may result in increased levels of tool wear. The increased temperatures may result in thermal softening of the work material that can result in lower cutting forces. In addition chip formation may change and chip control may become more important. As a result, careful choice of tool materials and coatings is necessary. High temperature hardness is very important and modern cemented carbides, ceramics, cermets, and polycrystalline diamond (PCD) materials are highly suitable for dry cutting. The development of alternative ways of cooling the cutting zone has also been investigated [4,17].

Elimination of cutting fluids obviously removes the flushing action that is beneficial for the transport of chips and machining dust from the cutting zone. For this reason machine tool manufacturers have been developing vacuum suction

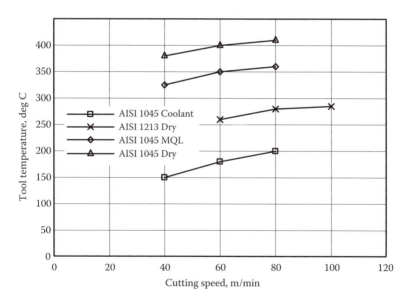

FIGURE 5.5 Effect of Dry and Minimum Quantity Lubrication on Tool Temperatures in Drilling Operations. (Adapted from Klocke, F. and Eisenblatter, G., *Ann. CIRP*, Vol. 46, No. 2, 519–526, 1997.)

systems for chip removal. For some processes such as boring, drilling, reaming, and tapping operations, the problems of chip removal may become critical and the tendency for chips to clog in drill flutes, etc., may be exacerbated by the increased cutting temperatures associated with dry machining.

The environment hazards of cutting fluids are removed by the use of dry cutting. However, the potential health problems from inhalation of machining dusts remain, together with associated flammability issues, but these are far less troublesome than the problems of fluid mist and aerosols.

Dry cutting has been found practical for many operations and materials. For others the application of MQL may be appropriate. Typical flow rates are 5 to 50 ml/hr, compared to typical flood cooling where around 10 l/min can typically be dispensed. The cutting fluids used are usually synthetic esters (chemically modified vegetable oils) and fatty alcohols. These have excellent lubrication and natural dissolving properties. Table 5.6 and Table 5.7 give the main characteristics of MQL fluids and corresponding application areas.

In order to apply these fluids, they are mixed with air to form an aerosol and delivered close to the cutting edge. This has led to the development of a number of MQL delivery systems that are commercially available. Systems are available in which the fluids are fed externally close to the cutting tool or internally through suitable passageways in the tools. Single or double channel delivery systems are available. With single channel delivery, the cutting fluid and air are mixed remote from the cutting edge. For double channel delivery, the cutting fluid and air are fed separately and mixed close to the cutting tool. A typical example for machining centers that employs internal two channel delivery and mixing through the

---

TABLE 5.6
Characteristics of Minimum Quantity Lubrication Fluids

Synthetic Esters	Fatty Alcohols
Chemically modified vegetable oils	Long-chained alcohols made from natural raw materials or from mineral oils

Good biodegradability
Low level of hazard to water
Toxicologically benign

Synthetic Esters	Fatty Alcohols
• High flash and boiling point with low viscosity	• Low flash and boiling point, with comparatively high viscosity
• Very good lubrication properties	• Poor lubrication properties
• Good corrosion resistance	• Better heat removal due to evaporation latent heat
• Inferior cooling properties	
• Vaporize with residuals	• Little residuals

Source: Adapted from Weinert, K., Inasaki, I., Sutherland, J.W., and Wakabayashi, T., *Ann. CIRP*, Vol. 52/2, 511–537, 2004.

TABLE 5.7
Main Areas of Application of MQL Fluids

Synthetic Esters	Fatty Alcohols
Machining Requirements	
• Primarily for reduction of friction	• Primarily for heat removal
• High quality surfaces demanded	• Examples are: sawing, turning, and milling of
• Workpiece materials with adhesion tendencies (prone to built up edge formation, etc.)	gray cast iron, machining of cast aluminum alloys
• Low cutting speeds and high specific cutting pressures	
• Lubrication of supporting and/or guiding rails	
• Examples are: drilling, fine boring, and thread forming operations in steel and aluminum	

Source: Adapted from Weinert, K., Inasaki, I., Sutherland, J.W., and Wakabayashi, T., *Ann. CIRP*, Vol. 52/2, 511–537, 2004.

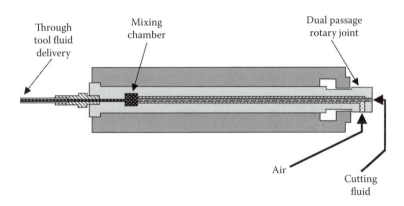

FIGURE 5.6 Schematic of minimum quantity lubrication system for delivering fluids through the center of a machine tool spindle. (Adapted from Makiyama, T., 4th NCMS Fall Workshop Series, October 2002 (also www.horcos.co.jp).)

spindle is shown in Figure 5.6 [15]. This can be used in conjunction with a vacuum suction system to remove chips and fluids from the cutting area. In general double channel delivery is more effective, giving better control of lubricant quantities and less escaping mist is generated [17]. Double channel delivery also makes it easy to switch from MQL to conventional fluid delivery.

Table 5.8 shows the extent to which dry and MQL machining has been successfully applied for different materials and processes [16]. Although these

TABLE 5.8
Utilization of Dry Cutting and Minimum Quantity Lubrication (MQL) in Various Machining Processes

Material:	Aluminum		Steel		Cast Iron
Processes	Cast Alloys	Wrought Alloys	High Alloy Steels, Bearing Steels	Free Cutting Steels, Quench and Tempering Steels	GG20–GGG70
Drilling	MQL	MQL	MQL	dry/MQL	dry/MQL
Reaming	MQL	MQL	MQL	MQL	MQL
Tapping	MQL	MQL	MQL	MQL	MQL
Thread forming	MQL	MQL	MQL	MQL	MQL
Deep hole drilling	MQL	MQL	—	MQL	MQL
Milling	dry/MQL	MQL	dry	dry	dry
Turning	dry/MQL	dry/MQL	dry	dry	dry
Gear milling	—	—	dry	dry	dry
Sawing	MQL	MQL	MQL	MQL	MQL
Broaching	—	—	MQL	dry/MQL	dry

Source: Adapted from Klocke, F. and Eisenblatter, G., *Ann. CIRP*, Vol. 46, No. 2, 519–526, 1997.

developments have taken place the majority of machining is still currently carried out with application of large quantities cutting fluids, but the trend towards dry and MQL machining is increasing [13].

5.9   SURFACE ROUGHNESS

The final surface roughness obtained during a practical machining operation may be considered as the sum of two independent effects:

1. The "ideal" surface roughness, which is a result of the geometry of the tool and the feed or feed speed
2. The "natural" surface roughness, which is a result of the irregularities in the cutting operation

5.9.1   IDEAL SURFACE ROUGHNESS

The ideal surface roughness represents the best possible finish that may be obtained for a given tool shape and feed and can be approached only if built-up edge, chatter, inaccuracies in machine-tool movement, and so on, are eliminated. The ideal surface finish (which would be obtained under these ideal conditions),

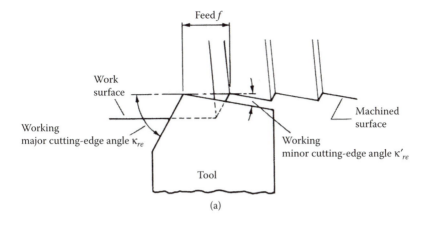

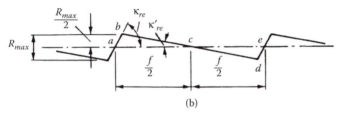

FIGURE 5.7 Idealized model of surface roughness for a cutting tool with a sharp corner, where κ_{r_e} = working major-cutting edge angle, κ'_{r_e} = working minor cutting edge angle, and f = feed. (a) Surface produced; (b) cross section through surface irregularities.

for a turning operation in which a sharp-cornered tool is used, is illustrated in Figure 5.7a.

For the purposes of quantitative comparisons and analysis, it is useful to be able to express the roughness of machined surfaces in terms of a single factor or index. The index most commonly used is known as the arithmetical mean value R_a and may be found as follows.

In the curve in Figure 5.7b, which shows a cross section through the surface under consideration, a mean line is first found that is parallel to the general surface direction and divides the surface in such a way that the sum of the areas formed above the line is equal to the sum of the areas formed below the line. The surface roughness R_a is now given by the sum of the absolute values of all the areas above and below the mean line divided by the sampling length.

Thus, for the example in Figure 5.7, the surface roughness value is given by

$$R_a = \frac{|area - abc| + |area - cde|}{f} \tag{5.2}$$

where f is the feed.

Since the areas *abc* and *cde* are equal,

$$R_a = \frac{2}{f}\left(area - abc\right) = \frac{R_{max}}{4} \tag{5.3}$$

where $R_{max}/2$ is the height of triangle *abc*.

It is interesting to note at this stage that the arithmetical mean value of surface roughness for a surface having uniform triangular irregularities is equal to one-quarter the maximum height of the irregularities.

Now by geometry

$$R_{max} = \frac{f}{\cot \kappa_{r_e} + \cot \kappa'_{r_e}} \tag{5.4}$$

where κ_{r_e} and κ'_{r_e} are the working major and minor cutting-edge angles, respectively. Substitution of Equation 5.4 in Equation 5.3 gives

$$R_a = \frac{f}{4\left(\cot \kappa_{r_e} + \cot \kappa'_{r_e}\right)} \tag{5.5}$$

Equation 5.5 shows that the arithmetical mean value for such a surface is directly proportional to the feed, and the curve in Figure 5.8 (where Equation 5.5

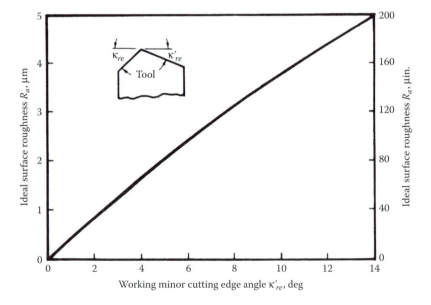

FIGURE 5.8 Effect of the minor-cutting-edge angle on an idealized model of surface roughness, where $\kappa_{r_e} = 45$ degrees, and $f = 0.1$ mm (0.004 in.).

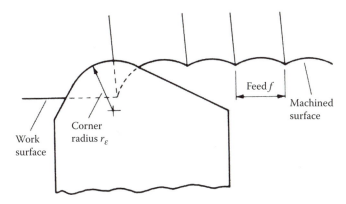

FIGURE 5.9 Idealized model of surface roughness for a tool with a rounded corner, where r_ε is the corner radius.

is plotted for a given feed and working major-cutting-edge angle) shows how the arithmetical mean value is affected by the working minor-cutting-edge angle.

Practical cutting tools are usually provided with a rounded corner, and Figure 5.9 shows the surface produced by such a tool under ideal conditions. Deriving a theoretical equation giving the arithmetical mean value for such a surface is rather more difficult than in the preceding example, but it can be shown that this roughness value is closely related to the feed and corner radius by the following expression:

$$R_a = \frac{0.0321f^2}{r_\varepsilon} \qquad (5.6)$$

where r_ε is the corner radius.

In Figure 5.10 the theoretical relationship between the surface-roughness value and the feed given by Equation 5.6 is compared with experimental results. In these experiments, work material (copper) and cutting conditions were carefully chosen such that the natural surface roughness was extremely low and no imperfections from the cutting action (chatter, built-up edge, etc.) were visible on the specimens. The operation was one of turning, and before each test the tools were carefully ground with the correct corner radius. Figure 5.10 shows that in these experimental results, the actual roughness of the specimens was close to the "ideal" for each feed used.

5.9.2 NATURAL SURFACE ROUGHNESS

In practice, it is not usually possible to achieve conditions such as those described above, and normally the natural surface roughness forms a large proportion of the actual roughness. One of the main factors contributing to natural surface roughness is the occurrence of a built-up edge. The built-up edge may be continually building up and breaking down, the fractured particles being carried away

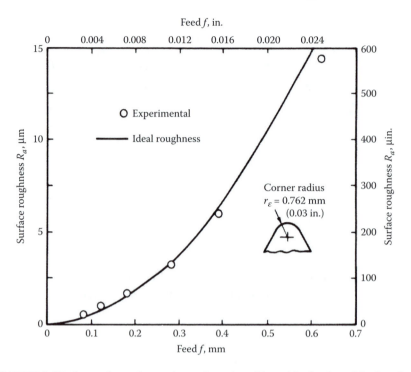

FIGURE 5.10 Comparison of experimental results with an idealized model of surface roughness.

on the undersurface of the chip and on the new workpiece surface. Thus, it would be expected that the larger the built-up edge, the rougher would be the surface produced, and factors tending to reduce the built up edge would give improved surface finish. Such factors would therefore be an increase in cutting speed, a change from, say, high-speed steel tool material to cemented carbide, the introduction of free machining materials such as leaded or resulfurized steels, the application of the correct cutting lubricant at low cutting speeds, and so on. An example of the effect of cutting speed is shown in Figure 5.11, where the actual surface roughness for a turned component is large at low cutting speeds and becomes reduced as the cutting speed is increased, until it approaches the ideal surface roughness at high cutting speeds.

Other factors that commonly contribute to natural surface roughness in practice are:

1. The occurrence of chatter or vibrations of the machine tool
2. Inaccuracies in machine tool movements such as the movement of the saddle on a lathe
3. Irregularities in the feed mechanism
4. Defects in the structure of the work material
5. Discontinuous chip formation when machining brittle materials

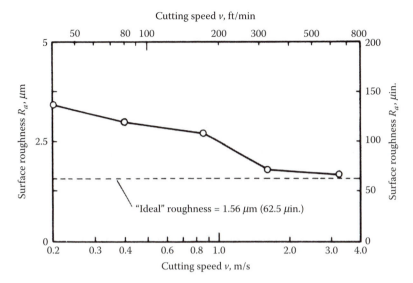

FIGURE 5.11 Effect of cutting speed on the surface roughness of turned specimens of mild steel.

6. Tearing of the work material when ductile metals are cut at low cutting speeds
7. Surface damage caused by such factors as chip flow

The preceding discussion has been confined to the surface roughness produced by single-point tool operations. When consideration is given to the surface produced by multipoint cutting tools, it must be realized that because of the slight inaccuracies in the manufacture of the cutting tools or because of the inaccuracies of the primary motion of the machine tool, one tooth on the cutter plays a dominant role in generating the machined surface. Thus, in a slab-milling operation, for example, the surface roughness obtained under ideal conditions can be obtained by assuming that the cutting tool has only one tooth. Making the appropriate substitutions in Equation 5.6 leads to the expression

$$R_a = \frac{0.0642}{d_t}\left(\frac{v_f}{n_t}\right)^2 \tag{5.7}$$

where

v_f = feed speed
d_t = diameter of the cutter
n_t = rotational frequency of the cutter

As a rough guide it can be stated that, in general, grinding processes will give surface roughness values from 0.05 to 2.5 μm (2 to 100 μin.), diamond turning will give surface roughness values from 0.1 to 1.0 μm (4 to 40 μin.), and milling and turning operations will give surface roughness values greater than 1 μm (40 μin.).

5.9.3 MEASUREMENT OF SURFACE ROUGHNESS

Before discussing the measurement of surface roughness, it is important to realize that other kinds of deviation from a perfectly smooth surface can occur. These deviations are called *surface flaws* and *waviness*. Surface flaws are widely separated irregularities that occur at random over the surface. They may be scratches, cracks, or similar flaws. Figure 5.12 illustrates some of the basic terminology associated with the assessment of surface texture [18] and some of these terms are defined as follows:

- **Roughness** — Roughness consists of surface irregularities that result from the machining process. These irregularities combine to form the surface texture.
- **Roughness height** — This is the height of the irregularities with respect to a reference line. It is also known as the height of unevenness.
- **Roughness width** — This is the distance parallel to the nominal surface between successive peaks or ridges that make up the predominant pattern of the roughness.
- **Roughness width cutoff** — Roughness width cutoff is the greatest spacing of respective surface irregularities to be included in the measurement of average roughness height. This effectively determines the sample length for measurements and should always be greater than the roughness width.
- **Waviness** — Waviness is a form of regular deviation where the wavelength is greater than a specified magnitude (usually about 1 mm). Roughness is a finer irregularity than waviness and is superimposed upon waviness. Waviness may result from workpiece or tool deflection, vibrations or run out.
- **Lay** — The lay of a surface is the direction of the predominant surface pattern and is usually determined by the machining method used to produce the surface. Measurements of a surface are made at right angles to the lay.

Instruments can be used to obtain an enlarged tracing of the surface irregularities; this tracing is known as the *surface profile*. The standard instruments for the determination of surface roughness operate by amplifying the vertical motion of a stylus as it is drawn slowly across the surface; they can produce, in addition, a continuous recording of the profile. In modern instruments the surface profile is digitally sampled and the arithmetical-mean surface-roughness value is calculated continuously over a selected cutoff distance. The cutoff length is selected to eliminate waviness from the measurements.

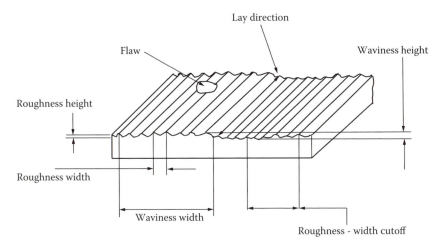

FIGURE 5.12 Basic characteristics of surface roughness measurement. (Adapted from ASME Standard B46.1, 1995.))

The ISO has recommended [19] a series of preferred roughness values and corresponding roughness grade numbers to be used when specifying surface roughness on drawings. The relationship between these numbers and roughness values is shown in Table 5.9.

TABLE 5.9
ISO-Recommended Roughness Values and Grade Numbers for the Specification of Surface Roughness

Roughness Values R_a		Roughness Grade
μm	μin	Numbers
50	2000	N12
25	1000	N11
12.5	500	N10
6.3	250	N9
3.2	125	N8
1.6	63	N7
0.8	32	N6
0.4	16	N5
0.2	8	N4
0.1	4	N3
0.05	2	N2
0.025	1	N1

5.10 TOOL GEOMETRIES FOR IMPROVED SURFACE FINISH

As has been shown the geometric surface roughness developed in machining is dependent on the tool geometry, machine kinematics, and feed rate. This has led to the development of a specific type of edge preparation for cutting tool inserts aimed at improving surface finish or more often to allow increased feed rates for a specified roughness, with associated reductions in cycle time. These inserts are known as *wiper inserts*, and they have become widely used for both turning and milling tools [20–22].

Similar to conventional turning inserts, wiper inserts remove the chip with the leading cutting edge and this leaves the expected geometric surface roughness through the mechanism described in section 5.9.1. However, wiper inserts have an additional radius or flat behind the tool nose that is kept in contact with the workpiece after the initial cut (Figure 5.13). This burnishes the peaks, leaving a smoother surface finish.

These wiper geometries are not effective for all applications. For example they are not suitable for light finishing cuts because they require slightly higher uncut chip thickness values to work well. Also higher feed rates are necessary to take full advantage of the wiper geometry. Figure 5.14 shows the effect of the wiper edge on surface finish for different feed rates. In turning, these inserts are mainly effective on surfaces parallel and perpendicular to axis of rotation of the workpiece.

The overall benefit of wiper inserts can be illustrated by the two case studies shown in Figure 5.15 and Figure 5.16. For the stepped shaft part, doubling the feed rate for the same roughness enables the machining time to be reduced from 68 to 34 seconds. In the second case, keeping the feed rate constant results in an average surface roughness improvement from 8 μm to 4 μm. In both cases an improvement in tool life over conventional inserts also occurs.

5.11 BURR FORMATION IN MACHINING

A surface condition that occurs in most machining operations is a result of the cutting edge exiting the workpiece. A sharp protrusion of material on the workpiece edges can be left, and these protrusions are known as *burrs*. Burrs can cause problems with subsequent assembly and handling operations. In practice deburring operations must be used to remove burrs, and the cost of these additional processes can be significant. It has been stated that the cost of burr removal can be as high as 30% of total manufacturing costs for high precision components such as aircraft engine parts, and for automotive parts of medium complexity 14% of manufacturing costs would be typical [16].

A typical burr in a ductile material results in a significant amount of subsurface damage and deformation associated the formation of the burr. It can be seen that there is a significant amount of subsurface damage and deformation associated with burr formation. Given the range of machining operations and possible

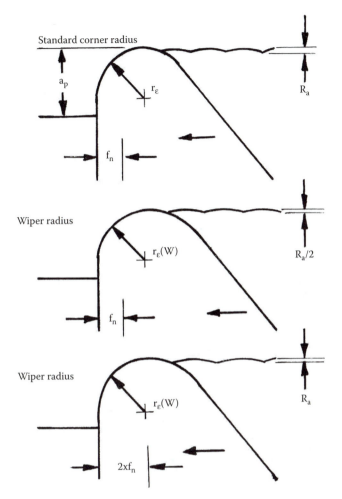

FIGURE 5.13 Typical wiper insert edge geometries and improvement in surface roughness over that for a standard insert. (Adapted from Kennedy, B., *Cutting Tool Eng.*, Vol. 55, 8, August, 2003.)

tool exit conditions that can result, a wide variation of burr types and configurations can be found in practice [16,23]. Substantial differences exist between burr formation in drilling and milling operations for example, but in all cases cutting conditions and tool geometry have a significant influence on burr size and shape. For example in drilling in-feed rate and drill point geometry are significant.

The underlying cause of burr formation is the interaction of the plastic deformation zone necessary for chip removal with the free surfaces as the cutting edge leaves the workpiece. Figure 5.17 illustrates this for the typical roll over burr that is formed in orthogonal cutting as the edge of the workpiece is approached [24]. As the workpiece edge is approached, the final chip material

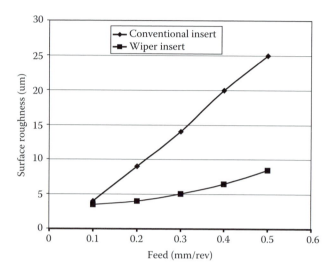

FIGURE 5.14 Effect of feed on the effectiveness of a wiper insert. (Adapted from Kyocera Corp., Wiper Insert-WQ Chip Breaker, Product Brochure, 2002 (www.koyocera.co.jp/ceratip/).)

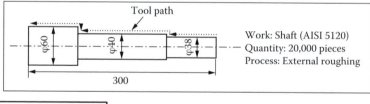

FIGURE 5.15 Productivity improvement from the use of a wiper insert in turning. (Adapted from Kyocera Corp., Wiper Insert-WQ Chip Breaker, Product Brochure, 2002 (www.koyocera.co.jp/ceratip/).)

effectively hinges on the edge to form a burr. For brittle materials the last chip material will break away as the tool leaves the workpiece.

Face milling is used extensively in industry for the machining of mating surfaces on, for example, automotive engine and transmission parts. Burrs on the edges of these surfaces are a particular problem. It has been found [25] that the formation of burrs in face milling is determined by the kinematics of tool exits and also the exit geometry of the cutting edges. Figure 5.18 shows an idealized view of two different exit edge conditions for face milling. If the minor cutting

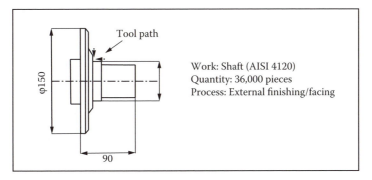

Cutting conditions	Wiper insert benefits
Speed: 180 m/min	1. Same feed rate
Depth of cut: 0.8 mm	2. Improved surface finish (8 μm to 4 μm Ry)
Feed: 0.2 mm/rev	3. Increased tool life (100 pcs. per edge to 150 pcs. per edge)
Wet cutting	

FIGURE 5.16 Surface roughness improvement from the use of a wiper insert in turning. (Adapted from Kyocera Corp., Wiper Insert-WQ Chip Breaker, Product Brochure, 2002 (www.koyocera.co.jp/ceratip/).)

edge, A-B, exits the workpiece before the major cutting edge, B-C, (Figure 5.18a) the chip hinges on the transition surface and a side burr is formed. However, the transition surface is removed by the next tool pass and consequently side burrs are unimportant. If the edge A-B exits after B-C, the chip hinges on the machined surface and an exit burr is formed. This means that, considering the points A, B, and C, the preferred exit order sequence (EOS) is ABC for reduced edge burr formation. Furthermore the six possible EOSs in increasing order of burr size are ABC, BAC, ACB, BCA, CAB, and CBA [25]. The EOS for a particular situation is dependent on tool geometry (rake angle, lead angle, etc.) and cutting conditions (edge approach angle, feed and cutting speed), but careful selection of these can reduce edge burr sizes.

Another strategy for reducing burr formation is to adjust tool paths so that tool exit conditions are avoided. In other words always machine onto the part edge if possible. Figure 5.19 shows three tool exit conditions that may result in edge burrs being formed. Figure 5.20 shows alternative conditions to avoid some of these conditions. This means that tool path planning strategies can be adopted to avoid as much as possible unsuitable edge exits to reduce burr formation and this has been shown to be very effective without significantly impacting machine cycle times [26]. It may not be possible to avoid all tool exits, but the use of tool and cutting conditions to give the best EOS may minimize burr formation in these situations. In addition it has been found that the burr size is determined by the in-plane exit angle and depth of cut (Figure 5.21) [26]. Thus the direction of tool feed can be adjusted to minimize burr height generation for some portions of the machining strategy (Figure 5.22) [27].

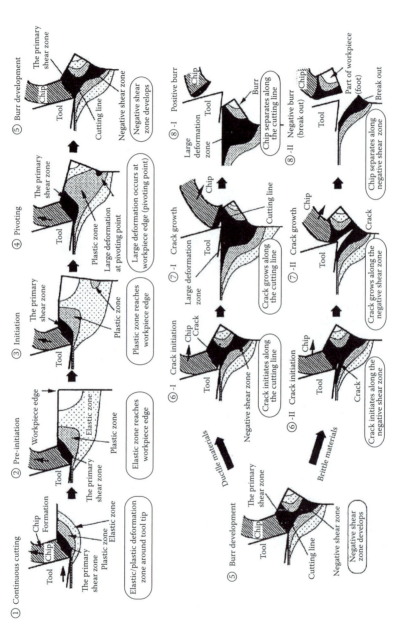

FIGURE 5.17 Edge burr formation during orthogonal cutting. (Adapted from Hashimura, M., Chang, Y.P., and Dornfeld, D.A., ASME Trans., J. Manuf. Sci. Eng., Vol. 121, 1–7, 1999.)

Insert

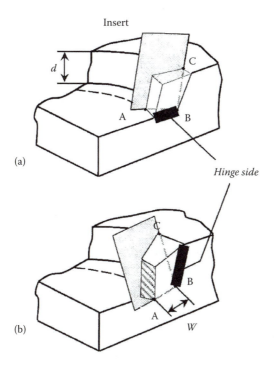

(a)

(b)

FIGURE 5.18 Idealized tool exit conditions in face milling. (a) Minor edge A-B leaves first and chip hinges on B-C to cause a side burr; (b) major edge B-C leaves first and chip hinges on A-B to cause edge burr. (Adapted from Hashimura, M., Hossamontr, J., and Dornfeld, D.A., *Trans. ASME, J. Manuf. Sci. Eng.*, Vol. 121, 13–17, 1999.)

PROBLEMS

1. A turning tool with a sharp corner and a major-cutting-edge angle of 60 degrees is to be used at a feed of 0.05 mm. What minor cutting edge should be provided to obtain an arithmetical mean surface roughness under ideal conditions of 3 μm?

2. What corner radius should be ground on a tool with a rounded corner to give an arithmetical mean surface roughness of 10 μm when turning under ideal conditions with a feed of 0.25 mm?

3. A 150-mm-diameter slab-milling cutter has 10 teeth and is used at a feed speed of 1.5 mm/s. Estimate the "ideal" arithmetical mean surface roughness when the rotational frequency of the cutter is 3s⁻¹.

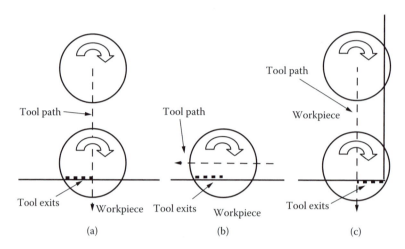

FIGURE 5.19 Tool exit conditions in face milling. (a) Tool enters workpiece along a straight line; (b) tool moves along an edge in up-milling mode; (c) tool encounters an adjacent edge. (Adapted Narayanaswami, R., Dornfeld, D.A., *Trans. ASME, J. Manuf. Sci. Eng.*, Vol. 119, 171–178, 1997.)

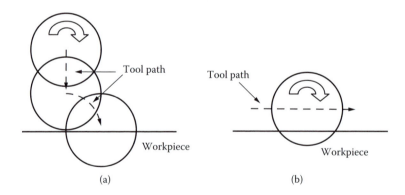

FIGURE 5.20 Alternative tool exit conditions: (a) Tool enters workpiece along a circular arc and avoids condition shown in Figure 5.19a; (b) down milling does not cause tool exits and avoids condition shown in Figure 5.19b. (Adapted Narayanaswami, R., Dornfeld, D.A., *Trans. ASME, J. Manuf. Sci. Eng.*, Vol. 119, 171–178, 1997.)

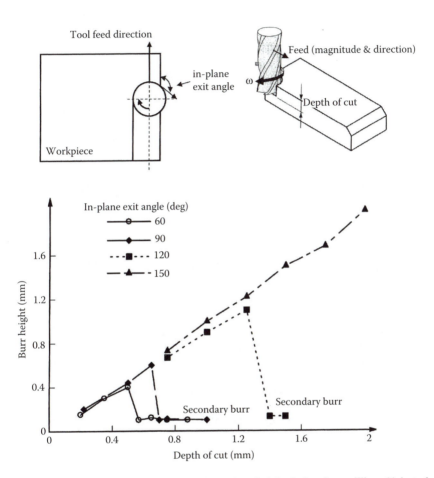

FIGURE 5.21 Effect of in-plane exit angle on burr height during face milling. (Adapted Narayanaswami, R., Dornfeld, D.A., *Trans. ASME, J. Manuf. Sci. Eng.*, Vol. 119, 171–178, 1997.)

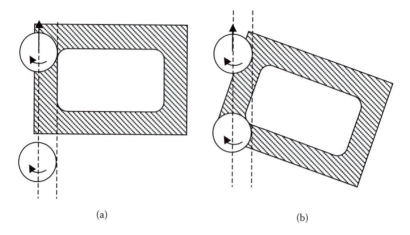

(a)   (b)

FIGURE 5.22 Modifications of part fixturing and/or tool path to avoid in-plane exit angles that cause exit burrs. (a) Original part position; (b) modified part position. (Adapted from Dornfeld, D., Rangerajan, A., Tool Path/Work Fixture Planning for Cycle Time and Part Quality Optimization in Face Milling, CIRP High Performance Machining Workshop, Paris, January 28, 2004.)

REFERENCES

1. Holmes, P.M., Factors affecting the selection of cutting fluids, *Ind. Lubr. Tribology*, Vol. 23, 2, 47–55, 1971.
2. Kuchler, C., Are You Using the Right Cutting Fluid?, *CNC Mach.*, Vol. 6, Issue 20, Winter, 2002.
3. Foltz, G., Cooling Fluids: Forgotten Key to Quality, *Manuf. Eng.*, January, 65–69, 2003.
4. Klocke, F. and Eisenblatter, G., Dry Cutting, *Ann. CIRP*, Vol. 46, No.2, 519–526, 1997.
5. Brockhoff, T. and Walter, A., Fluid Minimization in Cutting and Grinding, *Abrasives*, October, 38–42, 1998.
6. Bowden, F.P., and Tabor, D., *Friction and Lubrication of Solids*, Oxford University Press, London, 1954.
7. Cassin, C. and Boothroyd, G., The Lubricating Action of Cutting Fluids, *J. Mech. Eng. Sci.*, Vol. 7, 1, 67–81, 1965.
8. Waurzyniak, P., Increasing Holemaking Productivity, *Manuf. Eng.*, March, 51–59, 2003.
9. Urdanoff, H., Why Do Coolants Fail?, *Manuf. Eng.*, May, 121–129, 2003.
10. Aronson, B., Why Dry Machining?, *Manuf. Eng.*, January, 33–36, 1995.
11. U.S. Department of Health and Human Services, Occupational Exposure to Metalworking Fluids, NOISH Publication No. 98, January, 1998.
12. Bennett, E.O and Bennett, D.L., Occupational Airway Diseases in the Metalworking Industry, *Tribology Int.*, Vol. 18, 3, 169–176, 1985.
13. Sutherland, J.W., Kulur, V.N., and King, N.C., An Experimental Investigation of Air Quality in Wet and Dry Machining, *Ann. CIRP*, Vol. 49, No. 1, 61–64, 2000.

14. Autret, R. and Liang, S.Y., Minimum Quantity Lubrication in Finish Hard Turning, Proc. of Int. Conf. on Humanoid, Nanotechnology, Information Technology, Communication and Control, Environment and Management 2003 (HNICEM '03), Manila, Philippines, March 27–30, 2003.

15. Makiyama, T., Advanced Near Dry Machining System, 4th NCMS Fall Workshop Series, October, 2002 (also www.horcos.co.jp).

16. Byne, G., Dornfeld, D., and Denekena, B., Advancing Cutting Technology, *Ann. CIRP*, Vol. 52(2), 1–27, 2003.

17. Weinert, K., Inasaki, I., Sutherland, J.W., and Wakabayashi, T., Dry Machining and Minimum Quantity Lubrication, *Ann. CIRP*, Vol. 52/2, 511–537, 2004.

18. ASME Standards, Surface Texture (Surfaces Roughness, Waviness, and Lay), ASME B46.1, 1995.

19. ISO Standards, Technical Drawings-Methods of Indicating Surface Texture, ISO 1302, 1994.

20. Kennedy, B., Take a Bigger Bite, *Cutting Tool Eng.*, Vol. 55, 8, August, 2003.

21. Tisdal, B., Wiper Inserts Boost Turning Productivity, *Machine Shop Guide Web Archive*, March, 2003.

22. Kyocera Corp., Wiper Insert-WQ Chip Breaker, Product Brochure, 2002 (also www.koyocera.co.jp/ceratip/).

23. Ko, S. and Dornfeld, D.A., A Study of Burr Formation Mechanisms, *Trans. ASME, J. Eng. Mater. Technol.*, Vol. 113 (1), pp 75–87, 1991.

24. Hashimura, M., Chang, Y.P., and Dornfeld, D.A., Analysis of Burr Formation in Orthogonal Cutting, *ASME Trans., J. Manuf. Sci. Eng.*, Vol. 121, 1–7, 1999.

25. Hashimura, M., Hossamontr, J., and Dornfeld, D.A., Effect of In-Plane Exit Angle and Rake Angles on Burr Height and Thickness in Face Milling Operations, *Trans. ASME, J. Manuf. Sci. Eng.*, Vol. 121, 13–17, 1999.

26. Narayanaswami, R. and Dornfeld, D.A., Burr Minimization in Face Milling: A Geometric Approach, *Trans. ASME, J. Manuf. Sci. Eng.*, Vol. 119, 171–178, 1997.

27. Dornfeld, D. and Rangerajan, A., Tool Path/Work Fixture Planning for Cycle Time and Part Quality Optimization in Face Milling, CIRP High Performance Machining Workshop, Paris, January 28, 2004.

6 Economics of Metal-Cutting Operations

6.1 INTRODUCTION

Production costs and production rates are of vital interest to the manufacturing engineer. Although in practice a high production rate would probably mean low production costs, it should be pointed out that these two factors must be considered separately and that the manufacturing conditions giving maximum production rate will not be identical to those conditions giving minimum cost of production.

Analysis of production costs and production rates can be a complicated subject, and in many cases the analysis will apply only to the particular operation in question. However, experience gained over the years has led to certain empirical rules or guiding principles for choosing the optimum cutting conditions for a given machining operation, and it is the objective of this chapter to illustrate how these principles can be used.

In the following discussion the *production time* is defined as the average time taken to produce one component, and the *production cost* is defined as the total average cost of performing the machining operation on a component using one machine tool. In general, the production of a component will involve several machining operations using a variety of machine tools (often called *setups*). The total cost will be the sum of production costs for each machine setup. Hence, the total manufacturing costs, apart from the cost of the material, involve many items. For example, the raw material must be brought to the first machine and placed in the machine, and when the first machining process is completed, the component must be removed, stacked, or stored temporarily, transported eventually to the second machine tool, and so on. It is found in practice that, the costs of handling components between machines can be a substantial proportion of the total manufacturing costs. However, it is not proposed here to discuss the problem of material handling but to analyze those factors involved in a particular machining operation. It will be assumed that the components are stacked next to the machine tool ready to be grasped by the operator and that the operator will stack the machined components conveniently near to the machine tool.

Assuming the appropriate tool and cutting fluid were chosen for the machining of a batch of components, the only cutting conditions to be determined are the cutting speed and feed.

For the purposes of this discussion it is necessary to explain that *feed* is the distance moved by the tool relative to the workpiece in the feed direction for each revolution of the tool or workpiece or each stroke of the tool or workpiece.

Confusion may arise in certain multipoint tool operations such as milling where the feed settings on the machine refer to the relative speed between the tool axis and the workpiece in the feed direction (the feed speed).

Thus, if the feed speed in a milling operation is v_f and the rotational frequency of the tool is n_t, the workpiece feed during each revolution of the cutter is given by v_f/n_t and the maximum cutting speed v in a milling operation is given by $d_t n_t$, where d_t is the tool diameter. It now follows that if it is required to double the cutting speed in a milling operation while keeping the feed constant, it will be necessary to double both the rotational frequency of the cutter n_t and the feed speed v_f.

With these considerations in mind it may be stated that in any operation, when either the cutting speed or feed are increased while the other condition is held constant, the actual machining time will be reduced, and the tool-wear rate will increase. Thus, very low speeds and feeds will result in a high production time because of the long machining time. Alternatively, very high speeds and feeds will result in a high production time because of the frequent need to change cutting tools. Clearly, an optimum condition will exist giving minimum production time. Similarly, an optimum condition will arise for minimum production cost. At low speeds and feeds costs will be high because of the cost of using the machine and operator for the longer machining times. At high speeds and feeds costs will be high because of the cost of frequent tool replacement. The manufacturing engineer's problem is how to minimize both the production time and the production cost. Since, in general, these objectives cannot be reached simultaneously, a compromise must be sought.

In this chapter, the appropriate choice of feed will be discussed first. It will be shown that the correct feed to use in roughing operations is the highest the machine tool can withstand in terms of tool forces and power consumption. Next, optimum cutting-speed conditions will be analyzed for those operations involving constant cutting speeds (cylindrical turning, for example), and the optimum spindle-speed conditions will be analyzed for those operations involving variable cutting speeds (facing, for example). Finally, the analyses will be applied to practical machining operations, and the advantages of modern tool materials and tool design will be illustrated.

6.2 CHOICE OF FEED

When a finishing cut is to be taken, the appropriate feed will be that which gives an acceptable surface finish. In this case the choice of feed is in the hands of the designer, who will have specified the surface finish requirements.

To explain how the appropriate feed is chosen in a roughing operation, it is first necessary to consider the relative effects of cutting speed and feed on tool life.

It can be concluded from the work in Chapter 3 that if changes in cutting speed affect neither the geometry of the cutting operation nor the specific energy consumption, the tool temperatures are a function of the thermal number, cva_c/k. Thus, equal changes in speed or feed should affect tool temperatures by the same amount. However, an increase in feed will not affect the relative speed of sliding at the wearing surface of the tool, whereas the speed of sliding will change in proportion to the cutting speed. Since tool wear is a function of both temperature and relative speed of sliding, it can be appreciated that increases in cutting speed will result in a greater reduction in tool life than similar increases in feed.

This conclusion agrees with practical experience and means that if an increased production rate is required in rough machining, it will always be preferable to increase the feed rather than increase the speed. Of course, this procedure will not always be practical since, in general, an increase in feed will increase the tool forces, whereas an increase in cutting speed will not. A limit on feed increase will therefore exist and will depend on the maximum tool force the machine tool is able to withstand. The guiding principle in choosing optimum cutting conditions in a roughing operation is that the feed should always be set at the maximum possible.

6.3 CHOICE OF CUTTING SPEED

Two distinct criteria can be used in choosing the cutting speed for a machining operation: minimum production cost and minimum production time. The optimum cutting speed giving minimum production cost in a constant-cutting-speed operation such as cylindrical turning will first be considered. In this case the time spent by the operator and the machine in producing a batch of components N_b can be separated into three items:

1. The nonproductive time, given by $N_b t_l$, where t_l is the time taken to load and unload each component and to return the tool to the beginning of the cut
2. The total machining time, given by $N_b t_m$, where t_m is the machining time for the component
3. The total time involved in changing worn tools, given by $N_t t_{c_t}$, where t_{c_t} is the tool changing time, and N_t is the number of tools used

Thus, if M is the total machine and operator rate (including overheads), the total machine and operator costs will be

$$M\left(N_b t_l + N_b t_m + N_t t_{c_t}\right) \qquad (6.1)$$

To this cost must be added the cost of the tools used, given by $N_t C_t$, where C_t is the cost of each tool. The average production cost C_{pr}, for each component can now be written

$$C_{pr} = M t_l + M t_m + M \frac{N_t}{N_b} t_{c_t} + \frac{N_t}{N_b} C_t \qquad (6.2)$$

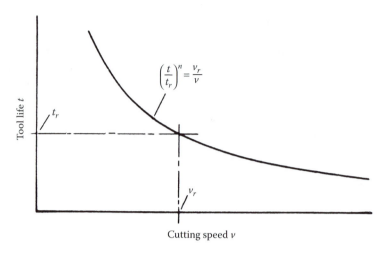

FIGURE 6.1 Taylor's tool-life relation, where n is a constant.

The first item in this expression is the nonproductive cost, which is constant for the particular operation. The second item is the machining cost, which is reduced as the cutting speed is increased at constant feed. The final items are the tool costs, which increase as the cutting speed increases.

To calculate the number of tools used in producing the batch of components, it is necessary to know the relationship between cutting speed and tool life. The work of Taylor [1] showed that an empirical relationship exists between these variables (Figure 6.1), namely,

$$\frac{v}{v_r} = \left(\frac{t_r}{t}\right)^n \qquad (6.3)$$

where

v = cutting speed
t = tool life
n = constant
t_r = measured tool life for a given cutting speed v_r

The value of t_r may be found for a particular workpiece and tool material and a particular feed either by experiment or from published empirical data. The index n depends mainly on the tool material; for high-speed steel $n \sim 0.125$, for carbide $0.25 < n < 0.3$, and for ceramics $0.5 < n < 0.7$. Figure 6.2 gives approximate ranges for the values of v_r for various tool and work materials when the tool life is 60 s.

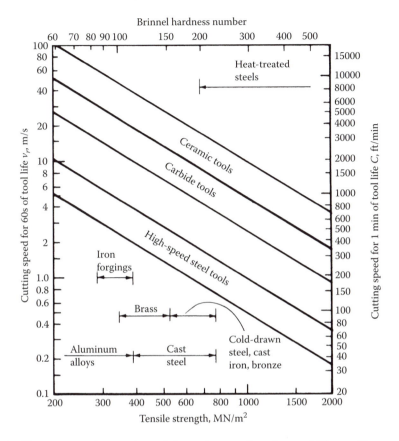

FIGURE 6.2 Approximate values of the cutting speed v_r when tool life $t_r = 60$ s.

The tool life t for a particular situation is therefore given by

$$t = t_r \left(\frac{v_r}{v} \right)^{\frac{1}{n}}$$
(6.4)

Traditionally, the Taylor tool-life equation has been applied in English units in the form

$$vt^n = C$$
(6.5)

where v is measured in feet per minute, t in minutes, and C, the cutting speed for 1 minute of tool life, in feet per minute.

From Equation 6.3, therefore,

$$\frac{v_r}{C} = \left(\frac{1\,\text{min}}{t_r}\right)^n \tag{6.6}$$

Thus, the equation

$$\frac{197v_r}{C} = \left(\frac{60}{t_r}\right)^n \tag{6.7}$$

can be employed to convert previously published data to SI units.

The number of tools N_t used in machining the batch of components is given by $N_b t_m/t$, assuming that the tool is engaged with the workpiece during the entire machining time. Thus,

$$\frac{N_t}{N_b} = \frac{t_m}{t} = \frac{t_m}{t_r}\left(\frac{v}{v_r}\right)^{\frac{1}{n}} \tag{6.8}$$

Finally, the machining time for one component is given by

$$t_m = \frac{K}{v} \tag{6.9}$$

where v is the cutting speed, and K is a constant for the particular operation. In cylindrical turning, for example, the value of K will be given by $d_w l_w/f$ where l_w is the length to be turned, d_w is the diameter of the workpiece, and f is the feed. In general, K can be regarded as the distance moved by the tool corner relative to the workpiece during the machining operation.

The relationship between the production cost and the cutting speed can now be obtained by substitution of Equation 6.8 and Equation 6.9 in Equation 6.2:

$$C_{pr} = Mt_l + Mkv^{-1} + \frac{K}{v_r^{\frac{1}{n}}t_r}\left(Mt_{c_t} + C_t\right)v^{\frac{1-n}{n}} \tag{6.10}$$

To find the cutting speed v_c for minimum cost, Equation 6.10 must now be differentiated with respect to v and equated to zero. Thus,

$$v_c = v_r\left(\frac{n}{1-n}\frac{Mt_r}{Mt_{c_t} + C_t}\right)^n \tag{6.11}$$

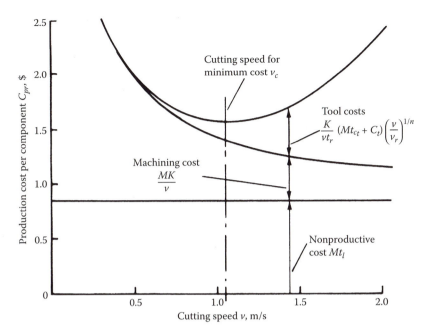

FIGURE 6.3 Variation in production cost per component C_{pr} with cutting speed v for a typical turning operation, where M = \$0.0028/s, C_t =\$2.1, t_{ct} = 240s, n = 0.25, t_l = 300 s, K = 192 m, t_r 60 s, and v_r = 2.75 m/s.

The effect of cutting speed on the cost of production can more clearly be shown in the form of a graph such as that in Figure 6.3. In the figure the three individual cost items represented by the terms in Equation 6.10, the nonproductive cost, the machining cost, and the tool costs, are plotted separately and show how an optimum cutting speed arises for a given set of conditions. It will be noted, from Equation 6.11, that the optimum cutting speed is independent of the batch size and the nonproductive times involved.

To find the cutting speed giving maximum production rate (or minimum production time), it is necessary to follow a similar procedure to that described above. In this case inspection of Equation 6.2 shows that the average production time for one component is given by

$$t_{pr} = t_l + t_m + \frac{N_t}{N_b} t_{c_t} \tag{6.12}$$

Substitution of Equation 6.8 and Equation 6.9 into Equation 6.12 and differentiation of the resulting expression gives the cutting speed v_p, for minimum production time. Thus

$$v_p = v_r \left(\frac{n}{1-n} \frac{t_r}{t_{c_t}} \right)^n \tag{6.13}$$

Comparison of Equation 6.11 and Equation 6.13 shows that application of the criteria for minimum cost and minimum production time yield different conditions, and therefore the final choice of cutting speed requires a compromise. This compromise will be considered later.

6.4 TOOL LIFE FOR MINIMUM COST AND MINIMUM PRODUCTION TIME

In analyzing practical machining operations, it is convenient to employ expressions for the optimum tool life for minimum cost t_c and the optimum tool life for minimum production time t_p. These expressions can be obtained by substitution of Equation 6.11 and Equation 6.13, respectively, into Taylor's tool-life equation, Equation 6.4. Thus

$$t_c = \frac{1-n}{n}\left(t_{c_t} + \frac{C_t}{M}\right) \tag{6.14}$$

and

$$t_p = \left(\frac{1-n}{n}\right)t_{c_t} \tag{6.15}$$

For practical use the factor $(1 - n)/n$ in Equation 6.14 and Equation 6.15 is given the value 7 for high-speed steel, 3 for carbide, and 1 for oxide or ceramic. With these values the approximate expressions for the optimum tool life for various tool materials becomes for

High-speed steel:

$$t_c = 7\left(t_{c_t} + \frac{C_t}{M}\right) \tag{6.16}$$

$$t_p = 7t_{c_t} \tag{6. 17}$$

Carbide:

$$t_c = 3\left(t_{c_t} + \frac{C_t}{M}\right) \tag{6.18}$$

$$t_p = 3t_{c_t} \tag{6.19}$$

Oxide or ceramic:

$$t_c = \left(t_{c_t} + \frac{C_t}{M} \right) \tag{6.20}$$

$$t_p = t_{c_t} \tag{6.21}$$

Finally, the corresponding optimum cutting speeds can be found from

$$v_c = v_r \left(\frac{t_r}{t_c} \right)^n \tag{6.22}$$

and

$$v_p = v_r \left(\frac{t_r}{t_p} \right)^n \tag{6.23}$$

From these equations it can be seen that before optimum conditions can be chosen, it is necessary to know the tool changing time t_c, the cost of providing a sharp tool C_t, the machine and operator costs M, and the cutting speed v_r, for a particular tool life t_r. The following discussion indicates how some of these factors can be estimated.

6.5 ESTIMATION OF FACTORS NEEDED TO DETERMINE OPTIMUM CONDITIONS

The machine and operator costs include the operator's rate of pay plus the overhead associated with his or her employment and the depreciation rate of the machine tool plus the overhead associated with its use. The method of calculating these costs varies from factory to factory, but the following expression would be applicable in most cases:

$$M = W_0 + \left(\frac{O_o}{100} \right) W_0 + M_t + \left(\frac{O_m}{100} \right) M_t \tag{6.24}$$

where W_0, is the operator's wage rate, and M_t is the depreciation rate of the machine tool, with O_0 and O_m being the percent overhead for the operator and machine, respectively. Operator overhead can vary from 100 to 300% and includes the worker's benefits provided by the company, the cost of providing the working facilities, and the cost of the administrators necessary to employ the worker. Machine overhead includes the cost of the power consumed by the machine, the cost of servicing the machine, and possibly the cost of providing the location for the machine.

It is usual when calculating the machine depreciation rate to assume that its initial value can be written off, or amortized, after a certain number of years (amortization period). This period can vary from two to ten years, depending on the type of machine and the tax depreciation rates determined by the Internal Revenue Service. The following expression would generally be used to estimate the machine depreciation rate:

$$M_t = \frac{\text{Initial cost of machine}}{\text{Number of working hours per year} \times \text{amortization period}} \quad (6.25)$$

The method used to estimate tool costs depends on the type of tool used. For tools that can be reground the following expression can be used to estimate the cost of providing a sharp tool:

$$C_t = \text{cost of regrinding} + \frac{\text{Cost of tool}}{\text{Average number of regrinds possible}} \quad (6.26)$$

It should be realized that although a particular tool might theoretically be reground 20 times, the actual number of regrinds will generally be less because the tool becomes damaged or chipped through factors other than gradual wear during machining. In practice the actual number of regrinds might be less than ½ the theoretical number.

For disposable-insert tools, the cost of providing a sharp tool can be estimated from the following equation:

$$C_t = \text{Cost of regrinding} + \frac{\text{Cost of insert}}{\text{Average number of cutting edges used per insert}}$$
$$+ \frac{\text{Cost of tool holder}}{\text{Number of cutting edges used during life of tool holder}} \quad (6.27)$$

Again, it should be realized that the average number of cutting edges used per insert will generally be fewer than the number available, and an effort should be made to estimate the true figure.

The tool-changing time is the time required to remove the tool from the machine, place a new tool in the machine, set the new tool in its correct position, and restart the machining process. If a disposable-insert tool is used, the following expression applies:

$$t_{c_t} = \frac{\text{Time to index insert} \times (\text{average number of cutting edges used per insert} - 1) + \text{time to replace insert}}{\text{Average number of cutting edges used per insert}} \quad (6.28)$$

The cutting speed v_r, for a particular tool life t_r, depends on the tool material, tool shape, work material, and cutting conditions used. Values of v_r can be obtained from values of C (the cutting speed for 1 min of tool life) for various materials and conditions tabulated in machining handbooks. A rough guide to the values is presented in Figure 6.2.

6.6 EXAMPLE OF A CONSTANT-CUTTING-SPEED OPERATION

To illustrate the application of the expressions developed thus far, it will be assumed that a large batch of steel shafts is to be rough-turned to a 76-mm diameter for 300 mm of their length at a feed of 0.25 mm. A brazed-type carbide tool is to be used, and the appropriate constants in Taylor's tool-life equation for the conditions employed are as follows: $n = 0.25$, and $v_r = 4.064$ m/s when $t_r = 60$ s ($C = 800$ ft/min). The initial cost of the machine was $30,000 and is to be amortized over 5 years. The operator's wage will be assumed to be $0.005/s ($18.00/h) and the operator and machine overheads are 100%. Tool-changing and resetting time on the machine is 300 s and the cost of regrinding the tool is $6.00. The initial cost of a tool is $20.00, and, on the average, it can be reground 10 times. Finally, the nonproductive time for each component is 120 s.

The first step in these calculations is to estimate the magnitudes of the relevant factors:

1. *The machine and operator rate M:* If the machine is to be used on an 8-hour shift per day, 5 days per week, and 50 weeks per year, each year will contain 7.2 Ms (7×10^6 s) of working time. The machine depreciation rate (Equation 6.25) is therefore

$$M_t = \frac{30000}{7.2 \times 10^6 \times 5} = \$0.0083/s$$

and thus the machine and operator rate (Equation 6.24) is

$$M = 0.0083 + 0.0083 + 0.005 + 0.005 = \$0.0117/s$$

2. *The cost of providing a sharp tool C_t:* This cost can be found from Equation 6.26 thus:

$$C_t = \$8.00$$

3. *The tool-changing time to t_{c_t}:* This value is given as

$$t_{c_t} = 300 \text{ s}$$

It is now possible to estimate the tool life t_c, and cutting speed v_c, for minimum cost and the tool life t_p and cutting speed v_p for minimum production time. Thus from Equation 6.18 for minimum cost

$$t_c = 3\left(t_{c_t} + C_t/M\right) = 3\left(300 + 8 \times 10^2/1.17\right) = 2.591 \text{ ks (5.12 min)}$$

and the corresponding cutting speed is given by Equation 6.22

$$v_c = v_r\left(t_r/t_c\right)^n = 4.046(60/2591)^{0.25} = 1.57 \text{ m/s (310.7 ft/min)}$$

From Equation 6.19, for minimum production time,

$$t_p = 3t_{c_t} = 900 \text{ s}$$

and the corresponding cutting speed (Equation 6.23) is

$$v_p = v_r\left(\frac{t_r}{t_p}\right)^n = 4.064\left(\frac{60}{900}\right)0.25 = 2.065 \text{ m/s (407 ft/min)}$$

Unless conditions dictate that either the economics of the operation or production time is to be given priority, it is generally considered that any cutting speed between v_c and v_p will result in efficient machining. Since many general purpose machine tools are provided with a limited selection of spindle speeds, a spindle speed giving a cutting speed within this range will be chosen; most probably the spindle speed selected would be the one giving a cutting speed closest to the cutting speed for minimum cost. In pursuing the present example it will be assumed that cutting speeds equal to both v_c and v_p can be obtained.

It is of interest to calculate the cost per component and the production time corresponding to the two criteria of minimum cost and minimum production time.

For minimum cost the time taken to machine one component t_m is given by Equation 6.9, which for turning becomes

$$t_m = \frac{\pi d_w l_w}{vf} = \frac{\pi \times 76 \times 10^{-3} \times 300 \times 10^{-3}}{1.57 \times 0.25 \times 10^{-3}} = 183 \text{ s (3.05 min)}$$

Since the tool life is 2591 s, each tool will produce 14 components, and the ratio N_t/N_b in the cost equation (Equation 6.2) will be equal to 0.0714. Thus, for each component

Nonproductive cost $= Mt_l = 0.0117 \times 120 = \$1.404$

Machining cost $= Mt_m = 0.0117 \times 189 = \$2.141$

Tool cost $= \left(\dfrac{N_t}{N_b}\right)\left(Mt_{c_t} + C_t\right) = 0.0714\left[\left(0.0117 \times 300\right) + 8\right] = \$0.8218$

Finally the total cost C_{pr} is \$4.37.
The production time can be found from Equation 6.12. Thus, for each component

Nonproductive time $= t_l = 120$ s

Machining time $= t_m = 183$ s

Tool changing time $= \left(\dfrac{N_t}{N_b}\right)t_{c_t} = 0.0714 \times 300 = 21.42$ s

Finally, the total production time t_{pr} is 324 s (5.4 min).
The conditions for minimum production time can be obtained in a similar way. In this case the machining time t_m is found to be 139.1sec (2.32 min), the corresponding production cost C_{pr} is \$4.81, and the production time 305 s (5.08 min). In this example it can be seen that use of the cutting speed giving minimum production time rather than minimum cost results in a reduction of the production time of 5.9% and an increase in the production cost of 10.1%. Whether this reduction in production time is worthwhile depends on the way the operation is financed. For example, if it is assumed that the factory is going to receive a fixed price for machining each component and that there is a high demand for the components, it is possible to calculate the profit involved in the operation of the particular machine over a given time period. Suppose the company is to receive \$6.00 for each component; at minimum cost conditions over 1 year (7.2 × 10^6 s) the profit would theoretically be \$36,222. For minimum production time the figure would be \$28,092. Clearly, the minimum cost conditions would be preferable, and although the comparison will depend on many factors other than those considered above, this example illustrates a possible basis for establishing the compromise between minimum cost and minimum production time. The idea of machining at maximum efficiency is pursued in the next section.

6.7 MACHINING AT MAXIMUM EFFICIENCY

It has already been shown that if in a machining operation the minimum cost condition is chosen, production time will be greater than the minimum. Also, if the condition for minimum production time is chosen, the cost of production will be higher than the minimum. To compromise between these two conditions the maximum rate of profit may be considered.

If S denotes the amount of money the machine shop receives for each component machined, the profit per component is given by $S - C_{pr}$, and the rate of profit P_r is given by

$$P_r = \frac{S - C_{pr}}{t_{pr}} \tag{6.29}$$

Substituting for C_{pr} and t_{pr}, from Equation 6.10 and Equation 6.12,

$$P_r = \frac{S - Mt_l - MKv^{-1} - Kv_r^{-1/n}t_r^{-1}\left(Mt_{c_t} + C_t\right)v^{(1-n)/n}}{t_l + Kv^{-1} + Kv_r^{-1/n}t_r^{-1}t_{c_t}v^{(1-n)/n}} \tag{6.30}$$

If Equation 6.30 is now differentiated with respect to v, equated to zero, and combined with Equation 6.3, the tool life t_{ef} for maximum efficiency (maximum rate of profit) is obtained. Thus,

$$t_{ef} = \frac{1-n}{n}\left(t_{c_t} + \frac{t_l C_t}{S}\right) + \frac{C_t K}{n S v_r}\left(\frac{t_{ef}}{t_r}\right)^n \tag{6.31}$$

Unfortunately this equation can only be solved by numerical methods. However, a simple method would be to assume a value of t_{ef}^n equal to 6.5 s^n for carbide tools or 2.5 s^n for high-speed steel tools and substitute this value in the right side of Equation 6.31 to obtain a new value of t_{ef}. This iteration procedure can be continued and will rapidly converge to a sufficiently close approximation.

One of the most interesting features of Equation 6.31 is that the condition for maximum efficiency is independent of the machine rate M. The ability to determine optimum machining conditions without a knowledge of M would be most helpful in practice because it is the uncertainty of overheads and amortization periods that makes M a difficult constant to estimate.

Figure 6.4 shows for a particular case how the rate of profit P varies with cutting speed (or tool life) for various values of the amount S charged for machining each component. Also shown on the graph are the conditions for minimum cost (Equation 6.14), minimum production time (Equation 6.15), and maximum efficiency (Equation 6.31). It can be seen that if a profit is made, the condition for maximum efficiency always lies between the conditions for minimum cost and maximum production time, and that when the profit is zero, the condition for maximum efficiency becomes equal to the condition for minimum

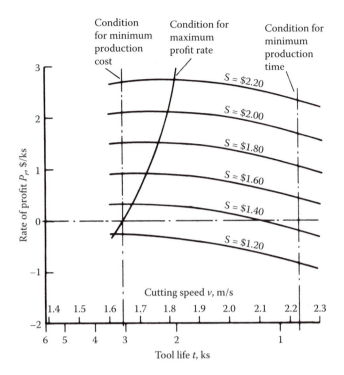

FIGURE 6.4 Effect of cutting speed v and tool life t on rate of profit P_r for a typical turning operation, where $M = \$0.00334/\text{s}$, $C_t = \$2.60$, $t_{c_t} = 300$ s, $n = 0.25$, $t_l = 60$ s, $K = 328$ m, $t_r = 1$ s, and $v_r = 12.2$ m/s.

cost. It can also be seen that unless the profit is very high, the optimum condition for maximum profit lies close to the minimum-cost condition.

Before the equations already developed are used to compare the economics of different tool materials and tool designs, the optimum conditions for facing operations and operations involving interrupted cuts will be analyzed.

6.8 FACING OPERATIONS

Most machine tools are designed to operate at constant spindle speed. In a cylindrical-turning operation constant spindle speed gives a constant cutting speed, and the cutting speeds for minimum cost, minimum production time, and maximum efficiency can be calculated using the equations previously developed. In a facing operation, however, a constant spindle speed results in a variable cutting speed.

A facing operation is shown in Figure 6.5. In this operation the cutting speed varies linearly with the radius of the cut r, the cutting speed is maximum at the periphery of the workpiece and minimum at the end of the operation. The tool-wear rate (the rate of increase of the flank wear-land width) will thus be a

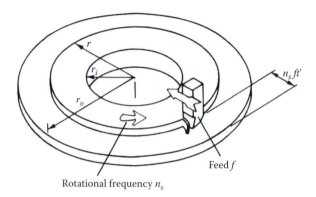

FIGURE 6.5 Facing operation.

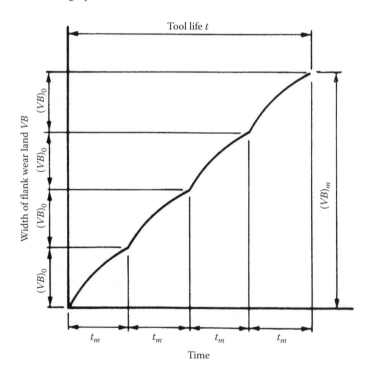

FIGURE 6.6 Tool wear during the facing of several components, where $(VB)_0$ = increase in wear-land per component, $(VB)_m$ = maximum wear-land width, t = tool-life, and t_m = machining time per component.

maximum at the beginning of the cut and decrease as cutting proceeds. Figure 6.6 shows the flank wear-land width plotted against time as a tool is used to face several workpieces. In this graph the increase in tool flank wear-land width (VB) during the machining of each component is $(VB)_0$. Thus,

$$\frac{(VB)_0}{(VB)_m} = \frac{t_m}{t} \qquad (6.32)$$

where $(VB)_m$ is the width of the wear land when the tool must be reground. Assuming the tool wears at a uniform rate at a given cutting speed, Taylor's tool-life equation (Equation 6.4) can be written

$$\frac{(VB)_m}{d(VB)/dt'} \qquad (6\,33)$$

where $d(VB)/dt'$ is the rate of increase of wear-land width. The instantaneous cutting speed v is given by

$$v = 2\pi n_s r \qquad (6.34)$$

where n_s is the rotational frequency of the machine spindle and the instantaneous radius at which cutting is taking place, r, is given by

$$r = r_o - n_s f t' \qquad (6.35)$$

where

f = feed
r_o = outside radius of the workpiece
t' = time

Combination of Equation 6.33 and Equation 6.34 and integration gives

$$\int_0^{(VB)_m} d(VB) = \int_0^{t_m} \frac{(VB)_m}{t_r} \left(\frac{2\pi n_s r}{v_r} \right)^{1/n} dt' \qquad (6.36)$$

Substitution of

$$dr = -n_s f dt' \qquad (6.37)$$

from Equation 6.35, integration and rearrangement yields

$$\frac{(VB)_0}{(VB)_m} = \frac{t_m}{t} = \frac{N_t}{N_b} = \left(\frac{2\pi n_s}{v_r} \right)^{1/n} \frac{n}{f n_s t_r (n+1)} \left(r_0^{(n+1)/n} - r_i^{(n+1)/n} \right) \qquad (6.38)$$

Finally, the machining time t_m is given by

$$t_m = \frac{r_0 - r_i}{n_s f} \tag{6.39}$$

To find the optimum spindle speed n_{s_c} giving minimum cost of production, Equation 6.38 and Equation 6.39 can be substituted in Equation 6.2, which is then differentiated with respect to n_s and equated to zero. Thus,

$$n_{s_c} = \frac{v_r}{2\pi r_0} \left(\frac{1+n}{1-n} \frac{Mt_r}{Mt_{c_t} + C_t} \frac{1 - a_r}{1 - a_r^{(n+1)/n}} \right)^n \tag{6.40}$$

where a_r is equal to r_i/r_0. The tool life t_c for minimum cost is given by combining Equation 6.38, Equation 6.39, and Equation 6.40. Thus,

$$t_c = \frac{1-n}{n} \left(t_{c_t} + \frac{C_t}{M} \right) \tag{6.41}$$

Similarly, the tool life t_p for minimum production time and t_{ef} for maximum efficiency are found to be identical to those for constant-cutting-speed operations. The corresponding spindle speed must, however, be obtained from

$$n_{s;c,p,ef} = \frac{v_{c,p,ef}}{2\pi r_0} \left[\left(1 + \frac{1}{n} \right) \left(\frac{1 - a_r}{1 - a_r^{(1+n)/n}} \right) \right]^n \tag{6.42}$$

where $v_{c,p,ef}$ is the cutting speed for a constant-cutting-speed operation corresponding to a tool life of t_c, t_p, or t_{ef}. In Figure 6.7, Equation 6.42 is plotted for values of n of 0.125, 0.25, and 0.5. It can be seen that when a_r approaches unity, $n_{s;c,p,ef}$ approaches $v_{c,p,ef}/2r_0$.

6.9 OPERATIONS WITH INTERRUPTED CUTS

In all milling, shaping, and planing operations, the cutter tooth is in contact with the workpiece for only a proportion of the machining time. If the tool life t_c for minimum cost, the tool life t_p for minimum production time, and the tool life t_{ef} for maximum efficiency are thought of as the life of the tool while the machine tool is operating, the expressions for these quantities derived earlier will still hold. However, when the cutting speeds that result in these values of tool life are calculated, it is necessary to correct for the proportion Q of the machining time

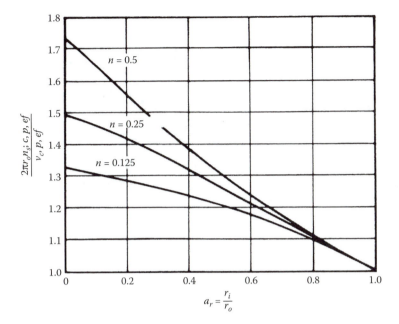

FIGURE 6.7 Effect of radius ratio a_r on the relation between optimum rotational frequency, $n_{s;c,p,ef}$ for a constant spindle-speed operation and the cutting speed $v_{c,p,ef}$ for a constant spindle-speed operation, where r_i = inside radius of surface to be faced and r_o = outside radius of surface to be faced.

t_m during which the cutting edge is engaged with the workpiece. Thus the tool life t_c, t_p, or t_{ef} should be multiplied by Q as shown in Equation 6.43 before the cutting speed is calculated:

$$v_{c,p,ef} = v_r \left(\frac{t_r}{Q t_{c,p,ef}} \right)^n \qquad (6.43)$$

The values of Q for slab milling face milling and side milling can be obtained from the geometries shown in Figure 6.8.

Thus, for slab milling (Figure 6.8a)

$$Q = \frac{\theta}{2\pi} = \frac{1}{4} + \frac{1}{2\pi} \arcsin\left(\frac{2a_e}{d_t} - 1 \right) \qquad (6.44)$$

where a_e is the working engagement, and d_t is the tool diameter. For side milling (Figure 6.8b)

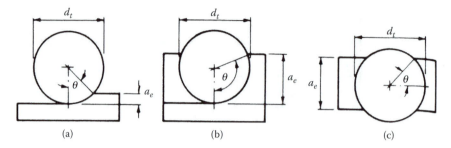

(a)   (b)   (c)

FIGURE 6.8 Determination of proportion Q of machining time when cutting edge is engaged with the workpiece during milling, where a_e = working engagement and d_t = diameter of tool. (a) Slab milling; (b) side milling; (c) face milling.

$$Q = \frac{\theta}{2\pi} = \frac{1}{4} + \frac{1}{2\pi} \arcsin\left(\frac{2a_e}{d_t} - 1\right) \tag{6.45}$$

and for face milling (Figure 6.8c)

$$Q = \frac{\theta}{\pi} = \frac{1}{\pi} \arcsin \frac{a_e}{d_t} \tag{6.46}$$

unless a_e is greater than or equal to d_t when Q is 0.5.

6.10 ECONOMICS OF VARIOUS TOOL MATERIALS AND TOOL DESIGNS

The principal tool materials of interest are high-speed steel, cemented carbide, and ceramics in increasing order of hardness and reducing order of toughness. High-speed steel is still used in the manufacture of cutting tools of complicated shapes (for example, reamers, drills, taps, dies, and various milling cutters). However most turning and boring tools, together with an increasing proportion of milling cutters and drills now use inserts based on cemented carbides and other materials. The individual inserts can be either brazed or more commonly clamped onto a steel body or toolholder. Ceramic tool materials (aluminum oxide, etc.) must also be manufactured in insert form and are mainly used in disposable-insert single-point tools.

For an example economic comparison of different tool materials, only those conditions giving minimum cost will be employed. Three comparisons are made in this section:

1. A rough-turning operation using both a high-speed steel tool and a brazed-carbide tool
2. A finish-turning operation using both a brazed- and a disposable-insert carbide tool and an aluminum oxide disposable-insert tool
3. A milling operation and a shaping operation

The first comparison, presented in Table 6.1, shows that although the optimum tool life for minimum cost was similar for both a high-speed-steel tool and a brazed-carbide tool, the production time using the high-speed-steel tool material was roughly twice that obtained with carbide. It can also be seen that although the cost of providing a sharp carbide tool was seven times the cost of providing a sharp high-speed-steel tool, the lower production time reduced the total production costs by 44%.

TABLE 6.1
Minimum Production Costs and Corresponding Production Times for a Rough-Turning Operation High-Speed Steel and Brazed-Carbide Tools

Production Costs and Times	Type of Tool	
	High-Speed Steel	Brazed Carbide
Proportion of machining time when cutting edge is engaged, Q	1	1
Machine and operator rate M, \$/s	0.01	0.01
Cost of sharp tool C_t, \$	1.10	8.00
Tool changing time t_{c_t}, s	240	240
Tool-life index n	0.125	0.25
Tool life for minimum cost, $t_c[(1/n) - 1]$ $[t_{c_t} + (C_t/M)]$, ks	2.45	3.12
Cutting speed for 1 min tool life v_r, m/s	0.508	2.73
Cutting speed for minimum cost $v_c = v_r (60/Qt_c)^n$, m/s	0.32	1.02
Distance moved by cutting edge relative to workpiece K, m	200	200
Machining time $t_m = K/v_c$, s	626	197
Number of components produced per tool $= t_c/t_m$	4	16
Nonproductive time per component t_l, s	300	300
Tool-changing time per component $= (t_m/t_c) t_{c_t}$, s	61	15
Nonproductive cost per component Mt_l, \$	3.000	3.000
Tool cost per component $= (t_m/t_c) (C_t + Mt_{c_t})$, \$	0.894	0.656
Machining cost per component Mt_m, \$	6.260	1.967
Total production time t_{pr}, s	987	512
Total cost per component C_{pr}, \$	10.15	5.62

TABLE 6.2
Minimum Production Costs and Corresponding Production Times for a Finish-Turning Operation Using Brazed-Carbide, Disposable-Carbide, and Disposable-Ceramic Tool

Production Costs and Times	Type of Tool		
	Brazed Carbide	Disposable Carbide	Disposable Ceramic
Proportion of machining time when cutting edge is engaged, Q	1	1	1
Machine and operator rate M, $/s	0.01	0.01	0.01
Cost of sharp tool C_t, $	8.00	1.00	2.00
Tool changing time t_{c_t}, s	240	60	60
Tool-life index n	0.25	0.25	0.50
Tool life for minimum cost, $t_c[(1/n) - 1]$ $[t_{c_t} + (C_t/M)]$, ks	3.120	0.480	0.260
Cutting speed for 1 min tool life v_r, m/s	6	6	50
Cutting speed for minimum cost $v_c = v_r (60/Qt_c)^n$, m/s	2.23	3.57	24.02
Distance moved by cutting edge relative to workpiece K, m	1000	1000	1000
Machining time $t_m = K/v_c$, s	448	280	42
Number of components produced per tool $= t_c/t_m$	7	2	6
Nonproductive time per component t_l, s	240	240	240
Tool-changing time per component $= (t_m/t_c)t_{c_t}$, s	34	35	10
Nonproductive cost per component Mt_l, $	2.40	2.40	2.40
Tool cost per component $= (t_m/t_c)(C_t + Mt_{c_t})$, $	1.492	0.934	0.416
Machining cost per component Mt_m, $	4.476	2.803	0.416
Total production time t_{pr}, s	722	555	291
Total cost per component C_{pr}, $	8.37	6.14	3.23

Table 6.2 shows an economic comparison between three different tools used in a finish-turning operation. The first tool is a brazed-carbide tool, the second is a disposable-insert carbide tool, and the third is a disposable-insert ceramic tool. The advantages of the disposable-insert tools are immediately evident: lower tool costs and shorter tool-changing time. Comparison of the first two columns, where the same tool material (carbide) is used, shows that disposable-insert tools hold considerable economic advantages: production time is decreased by 19% and production cost by 20%. This reduction of time and cost illustrates why the introduction of disposable-insert-type tools is one of the most important developments in the machining process.

The last column, disposable-ceramic tools, illustrates the advantages of ceramic tools for finishing operations. The much-higher cutting speeds possible with this material considerably reduce machining time and the costs. However,

TABLE 6.3
Minimum Production Costs and Corresponding Production Times for the Machining of a Flat Surface (150 mm long and 80 mm wide) Using a Horizontal Milling Machine and a Shaper

Production Costs and Times	Machining Operation	
	Milling	Shaping
Proportion of machining time when cutting edge is engaged, Q	0.14	0.75
Machine and operator rate M, \$/s	0.012	0.008
Cost of sharp tool C_t, \$	40.00	2.00
Tool changing time t_{c_t}, s	600	120
Tool-life index n	0.125	0.125
Tool life for minimum cost, $t_c[(1/n) - 1]$	27.53	2.59
$[t_{c_t} + (C_t/M)]$, ks		
Cutting speed for 1 min tool life v_r, m/s	1.0	1.0
Cutting speed for minimum cost $v_c = v_r (60/Qt_c)^n$, m/s	0.59	0.65
Machining time $t_m = K/v_c$, s	134.7	190.0
Number of components produced per tool $= t_c/t_m$	204	14
Nonproductive time per component t_l, s	120	120
Tool-changing time per component $= (t_m/t_c)t_{c_t}$, s	3	9
Nonproductive cost per component Mt_l, \$	1.440	0.960
Tool cost per component $= (t_m/t_c) (C_t + Mt_{c_t})$, \$	0.231	0.217
Machining cost per component Mt_m, \$	1.616	1.520
Total production time t_{pr}, s	257.6	318.8
Total cost per component C_{pr}, \$	3.29	2.70

the extremely high cutting speeds required present a difficulty for many general purpose machine tools that are not designed to operate under these conditions.

Table 6.3 compares a slab-milling operation and a shaping operation. Both of these operations involve interrupted cutting, and the tool life required correction before the cutting speed was calculated. This comparison is for the removal of an 18 mm (0.71 in.) deep layer from the surface of a rectangular block 80 mm (3.15 in.) wide and 150 mm (5.9 in.) long. The feed in the shaping operation was limited to 0.13 mm (0.005 in.) because of the maximum cutting force the machine could provide. The feed speed in the milling operation was limited to 1.3 mm/s (3.07 in./min) because of the power available. In the milling operation, a 100 mm (3.94 in.) diameter cutter was employed. The calculation shows that the much-higher tool costs in milling result in a long economic tool life, but that since each cutter produces a large number of components, the tool costs per component are reasonably low. In this comparison the costs of the shaping operation are less than those for milling. This result is probably reasonable for such simple operations as the one used in the example. However, this operation should not be taken as typical: quite complicated shapes can be milled in one pass using "ganged" milling cutters or form milling cutters. Using a shaper in these situations would not be economic, particularly if the batch size were large. In fact, a shaper is not considered to be a production machine tool and is generally only used when very small batches are to be machined because of the relatively long set-up time.

6.11 MACHINABILITY DATA SYSTEMS

The analytical methods described in this chapter can be used to determine recommended cutting conditions for different processes and tool/material combinations. In addition, machining costs can be estimated using a similar approach (this is discussed further in Chapter 13). In all cases data is required on the machinability of a wide range of materials for the effective use of these procedures.

The methods for determining the optimum feeds and speeds described in this chapter are based on the Taylor tool-life relationship (Equation 6.5). In applying this equation, different value of the constants C and n exist for each tool-workpiece material combination utilized and for different cutting fluids used. Establishing the tool-life equation for a particular set of conditions requires lengthy cutting tests and wear measurements, which must be done for the whole range of tool and work materials under consideration. The extent of this task can be reduced by utilizing data found in handbooks (e.g., [2]) or more recently in computer-based machinability data systems [3,4].

Two different types of machinability data systems can be identified: (1) database systems and (2) mathematical model systems.

6.11.1 DATABASE SYSTEMS

Database systems incorporate a database of large quantities of information from laboratory experiments and workshop experience. The database can be accessed to determine the required items for the particular tool-work combination under consideration. Generally these systems include cost information as well as cutting parameter values.

6.11.2 MATHEMATICAL MODEL SYSTEMS

Mathematical model systems go a stage further than database systems and incorporate the types of analysis methods described in this chapter for determining optimum feeds, speeds, and so on. Based on stored machinability data, the constants for the appropriate tool-life equations are determined and the optimum feeds and speeds are then determined using appropriate economic or production rate equations for the particular operation being considered. These recommended cutting conditions are then used as a guide to selection, the general assumption being that the recommended cutting conditions correspond to those for minimum production cost.

6.12 LIMITATIONS OF AVAILABLE MACHINABILITY DATA

The data found in machinability handbooks and data systems is limited and does not cover the full range of materials processed and corresponding cutting conditions. Even for commonly machined materials, a limited range of data exists. For example recommended cutting conditions may be restricted to one or two hardness values for steels and so on. Extension of the available data requires numerous time-consuming machining tests for a wide range of cutting conditions and tool/workpiece material combinations. One approach to dealing with this problem is to determine, from available data, relationships between parameters that enable recommended cutting conditions for other parameter combinations to be determined with a reasonable degree of accuracy.

In order to illustrate this process, relationships for turning free machining carbon steel will be described. A sample of machining parameters for turning is shown in Table 6.4 [5]. Using this information, relationships can be developed to determine the recommended feeds and speeds for a wider range of cutting depths and work material hardness values.

6.12.1 Feed and Depth of Cut in Turning

Examination of the data for most materials shows that a good linear relationship exists between recommended feed and the square root of the depth of cut (back engagement), as is typically shown in Figure 6.9. If the y-intercept is set at zero, then the following relationship can be used to determine the recommended feed:

$$f = a_m \sqrt{a_p} \qquad (6.47)$$

where

f = feed (per rev.)
a_p = depth of cut or back engagement
a_m = constant dependent on materials

6.12.2 Cutting Speed, Depth of Cut, and Hardness in Turning

The data shows that recommended cutting speed is influenced by three variables; depth of cut, work material hardness, and cutting tool material type, as is shown in Figure 6.9. The three lines show how speed is influenced by the work material hardness. In order to determine a suitable relationship between speed, depth of cut, and hardness, it must be assumed that the recommended speed v_1, is known for a depth of cut a_{p_0} and material hardness, R_{c_0}. The following relationship can then be readily derived from data similar to that shown in Figure 6.10 and Figure 6.11:

TABLE 6.4
Selection of Machinability Data — Recommended Cutting Conditions for Different Tool Materials

Material	Hardness	Condition	Depth of Cut (in)	Depth of Cut (mm)	HSS Speed (fpm)	HSS Speed (m/min)	HSS Feed (ipr)	HSS Feed (mm/r)	HSS Tool Material (AISI)	HSS Tool Material (ISO)	Carbide Uncoated Speed Brazed (fpm)	Brazed (m/min)	Indexable (fpm)	Indexable (m/min)	Uncoated Feed (ipr)	Uncoated Feed (mm/r)	Uncoated Grade (C)	Uncoated Grade (ISO)	Coated Speed (fpm)	Coated Speed (m/min)	Coated Feed (ipr)	Coated Feed (mm/r)	Coated Grade (C)	Coated Grade (ISO)
1. Free machining carbon steels, wrought	100 to 150	Hot rolled or annealed	.040	1	200	60	.007	.18	M2, M3	S4, S5	670	205	790	240	.007	.18	C-7	P10	1200	365	.007	.18	CC-7	CP10
			.150	4	150	45	0.15	.40	M2, M3	S4, S5	510	155	600	185	.020	.50	C-6	P20	775	235	.015	.40	CC-6	CP20
			.300	8	120	37	020	.50	M2, M3	S4, S5	400	120	475	145	.030	.75	C-6	P30	625	190	.020	.50	CC-6	CP30
			.625	16	90	27	.030	.75	M2, M3	S4, S5	320	100	370	115	.040	1.0	C-6	P40	—	—	—	—	—	—
Low carbon resulfurized 1116 1117 1118 1119 1211 1212	150 to 200	Cold drawn	.040	1	210	64	.007	.18	M2, M3	S4, S5	680	205	820	250	.007	.18	C-7	P10	1225	375	.007	.18	CC-7	CP10
			.150	4	160	49	0.15	.40	M2, M3	S4, S5	520	160	625	160	.020	.50	C-6	P20	800	245	.015	.40	CC-6	CP20
			.300	8	125	38	020	.50	M2, M3	S4, S5	410	125	495	125	.030	.75	C-6	P30	650	200	.020	.50	CC-6	CP30
			.625	16	100	30	.030	.75	M2, M3	S4, S5	330	100	385	100	.040	1.0	C-6	P40	—	—	—	—	—	—

Source: From Metcut, *Machining Data Handbook*, 3rd. Ed., Metcut Research Associates, Cincinnati, Ohio, 1980. Table 6.5

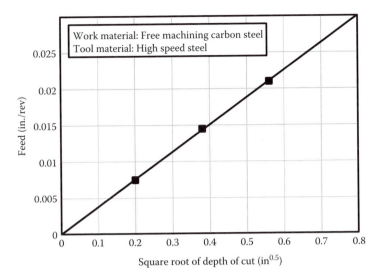

FIGURE 6.9 Typical relationship between recommended feed and depth of cut in turning. (Adapted from Turner, R., Masters Thesis, University of Rhode Island, 1993.)

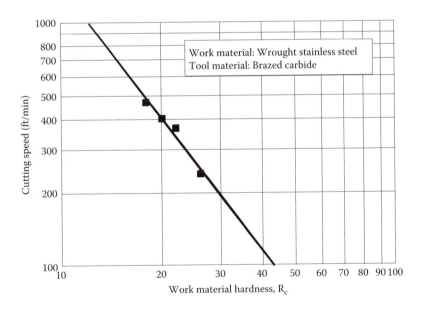

FIGURE 6.10 Typical relationship between recommended cutting speed and workpiece hardness (Rockwell C). (Adapted from Turner, R., Masters Thesis, University of Rhode Island, 1993.)

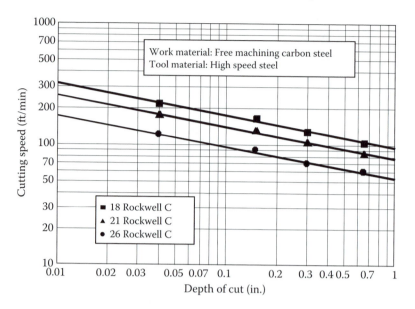

FIGURE 6.11 Typical relationships between recommended cutting speed and depth of cut for different values of workpiece hardness. (Adapted from Turner, R., Masters Thesis, University of Rhode Island, 1993.)

$$v = v_1 \left(a_{p_0} / a_p \right)^p \left(R_{c_0} / R_c \right)^q \qquad (6.48)$$

where v_1 is the recommended speed for a depth of cut of 3.81 mm (0.15 in.) and hardness of 20 Rockwell C, i.e., $a_{p_0} = 3.81$ mm and $R_{c_0} = 20$.

6.12.3 ESTIMATION OF RECOMMENDED FEED AND SPEED

Table 6.5 summarizes the relationships for feed and speed required for turning operations. The appropriate parameters for these relationships have been determined for 21 generic work material types as listed in Table 6.6 and Table 6.7 [6]. Five different cutting tool material types are also included.

Selecting a depth of cut within the specified range and a work material hardness enables the recommended speed for specific turning conditions to be determined.

6.12.3.1 Example

Work material: Free machining carbon steel
Cutting tool: Coated carbide
Depth of cut, a_p: 6.35 mm (0.25 in.)
Material hardness, R_c: 35

TABLE 6.5
Summary of Relationships for Determining Recommended Feeds and Cutting Speeds

Feed Correction

$f = a_m \sqrt{a_p}$ (for $0.20 < a_p \le a_{p_{max}}$) f = feed (mm per rev)

a_p = tool engagement (depth of cut) (mm)

Speed Correction

$v = v_1 (3.81/a_p)^p (20/R_c)^q$ v = cutting speed (m/min)

v_1 = basic cutting speed at a_{p_0} = 3.81 mm and R_{c_0} = 20

$R_c = (Bhn/10.91)$ R_c = Rockwell hardness, C range

Hardness range $20 < R_c < 50$ Bhn = Brinell hardness

Determine v_1, b, and c from Table 6.7.

$v_1 = 228$ m/min (749 ft/min)

$p = 0.32$

$q = 0.43$

Substituting these into Equation 6.48 gives:

$v = 228 (3.81/6.35)^{0.32} (20/35)^{0.43} = 152.2$ m/min (500 ft/min)

The recommended feed is determined by selecting the appropriate depth correction factor, a, from Table 6.6, i.e., $a = 0.19$. Thus from Equation 6.47:

$f = 0.19 (6.35)^{0.5} = 0.48$ mm/rev (0.0188 in./rev)

6.12.4 RELATIONSHIPS FOR OTHER TURNING PROCESSES

For other machining operations based on turning (cutoff, form tool cutting, boring), the recommended cutting conditions are influenced by parameters similar to those for cylindrical turning. The relationships for turning can be used for determining the recommended cutting conditions for boring operations. The depth of cut for cutoff and form tool cutting is equal to the width of the tool, but the recommended cutting speed is independent of the depth of cut. Equation 6.48 can be used to modify the cutting speed to account for the workpiece material hardness by setting the index p equal to zero and then using the values of v_1 and q that correspond to cylindrical turning. The recommended feed is dependent on tool width (depth of cut) and material hardness. The relationship for cutoff operations is:

$$f = f_0 \frac{(w + 0.25)}{0.385} (20/R_c)^q \quad (6.49)$$

TABLE 6.6
Parameters for Determining Recommended Feeds for Different Tool and Workpiece Materials

	a_m for Feed Calculation (Equation 6.47)[b]				Maximum Depth of Cut, $a_{p_{max}}$ (mm)[c]			
	HSS	Carbide[a]	Coated Carbide	Diamond	HSS	Carbide[a]	Coated Carbide	Diamond
Carbon steel	0.19	0.25	0.19	—	7.62	7.62	7.62	—
Free machining carbon steel	0.19	0.25	0.19	—	7.62	7.62	7.62	—
Alloy steel	0.18	0.22	0.16	—	7.62	7.62	7.62	—
Free machining alloy steel	0.18	0.22	0.16	—	7.62	7.62	7.62	—
Stainless steel	0.19	0.22	0.19	—	7.62	7.62	7.62	—
Free machining stainless steel	0.19	0.22	0.19	—	7.62	7.62	7.62	—
Tool steel	0.19	0.19	0.19	—	7.62	7.62	7.62	—
Cast iron	0.19	0.24	0.19	—	7.62	7.62	7.62	—
Titanium alloy	0.13	0.13	—	—	3.81	7.62	3.81	—
Copper alloy	0.19	0.26	—	0.27	7.62	7.62	—	3.18
Zinc alloy	0.25	0.39	—	-	7.62	7.62	—	—
Magnesium alloy	0.25	0.39	—	0.27	7.62	7.62	—	3.18
Aluminum alloy	0.25	0.43	—	0.27	7.62	3.81	—	3.18
Nickel alloy (Monel)	0.22	0.13	—	—	2.54	2.54	3.81	—
High temperature alloy	0.22	0.13	—	—	2.54	3.81	2.54	—
Tungsten alloy	—	0.14	—	—	—	3.81	—	—
Powder metal (Al)	0.09	0.09	—	—	1.52	7.62	—	—
Powder metal (other)	0.09	0.09	—	—	1.52	7.62	—	—
Thermoplastic	0.11	0.11	—	0.27	3.81	7.62	—	3.18
Glass reinforced thermoplastic	0.11	0.11	—	0.27	3.81	7.62	—	3.18
Ceramic and glass	0.06	0.06	—	0.27	3.81	3.81	—	3.18

[a] Brazed and indexable
[b] Feed in mm, for feed in inches divide a_m by 5.04
[c] For inches divide by 25.4

TABLE 6.7
Parameter for Determining Recommended Cutting Speeds for Different Tool and Workpiece Materials

	Basic Cutting Speed (v_1), m/min [b] $R_{c0} = 20$, $a_{p0} = 3.81$ mm					Depth and Hardness Correction Factor p/q			
	HSS	Brazed Carbide	Indexable Carbide	Coated Carbide	Diamond	HSS	Carbide[a]	Coated Carbide	Diamond
Carbon steel	27	94	115	163	—	0.27/0.96	0.27/0.53	0.32/0.43	—/—
Free machining carbon steel	41	140	170	228	—	0.27/0.96	0.27/0.53	0.32/0.43	—/—
Alloy steel	30	98	121	163	—	0.28/0.70	0.26/1.1	0.23/0.91	—/—
Free machining alloy steel	33	119	147	193	—	0.28/0.70	0.26/1.1	0.23/0.91	—/—
Stainless steel	32	107	123	167	—	0.25/1.0	0.25/0.84	0.22/1.1	—/—
Free machining stainless steel	42	131	150	195	—	0.25/1.0	0.25/0.84	0.22/1.1	—/—
Tool steel	24	101	120	158	—	0.25/—	0.28/—	0.23/—	—/—
Cast iron	28	108	122	159	—	0.30/—	0.22/—	0.20/—	—/—
Titanium alloy	50	97	106	—	—	0.11/—	0.39/—	—/—	—/—
Copper alloy	126	271	311	—	135	0.21/—	0.12/—	—/—	0.65/—
Zinc alloy	99	191	236	—	—	0.26/—	0.33/—	—/—	—/—
Magnesium alloy	232	538	610	—	82	0.26/—	0.33/—	—/—	0.59/—
Aluminum alloy	261	506	549	—	384	0.23/—	0.17/—	—/—	0.32/—

TABLE 6.7 (continued)
Parameter for Determining Recommended Cutting Speeds for Different Tool and Workpiece Materials

	Basic Cutting Speed (v_t), m/min [b] $R_{c0} = 20$, $a_{p0} = 3.81$ mm					Depth and Hardness Correction Factor p/q			
	HSS	Brazed Carbide	Indexable Carbide	Coated Carbide	Diamond	HSS	Carbide[a]	Coated Carbide	Diamond
Nickel alloy (Monel)	21	59	67	—	—	0.19/—	0.31/—	—/—	—/—
High temperature alloy	6	20	22	—	—	0.19/—	0.31/—	—/—	—/—
Tungsten alloy	—	40	45	—	—	—/—	0.16/—	—/—	—/—
Powder metal (Al)	53	152	166	—	—	0/—	0/—	—/—	—/—
Powder metal (other)	26	98	106	—	—	0/—	0/—	—/—	—/—
Thermoplastic	126	204	204	—	81	0.12/—	0.1/—	—/—	0.66/—
Glass reinforced thermoplastic	—	134	134	—	81	0.12/—	0.1/—	—/—	0.66/—
Ceramic and glass	12	37	43	—	223	0.11/—	0.03/—	—/—	0.50/—

[a] Brazed and indexable
[b] To convert to ft/min multiply by 3.28

and for form tool cutting is:

$$f = f_0 \frac{\left(2.5 - \sqrt{w}\right)}{1.5}\left(20/R_c\right)^q \qquad (6.50)$$

where w is the width of cut, and f_0 is a known feed similar to the relationships for turning. The index q is for hardness correction and determined in a similar manner to those for turning.

6.12.5 RELATIONSHIPS FOR OTHER MACHINING PROCESSES

A similar approach can be adopted for determining the recommended cutting conditions for milling and drilling operations. For milling the recommended feed, speed is dependent on the cutter diameter, number of teeth, work material hardness, and tool material. The recommended cutting speed depends on the depth of cut, work material hardness, and tool material type. Similarly for drilling and reaming operations, the feed is dependent on tool material type, work material hardness, and drill diameter, but with an addition correction for hole depths greater than two diameters. The recommended cutting speeds for different tool material types depend on the work hardness and hole depth. Further details of the relationships and the associated data can be found in [6].

6.12.6 ACCURACY OF RELATIONSHIPS FOR RECOMMENDED CUTTING CONDITIONS

Relationships between recommended cutting conditions and various machining parameters have been derived for a wide range of materials based on the data presented in [5]. These can then be used to estimate the recommended conditions for combinations of the machining parameters not covered in the available data. The accuracy of these relationships has been checked by comparing the predicted values for combinations of parameters for which data is available with the actual recommended values. It has been found that the overall accuracy of these estimated values is generally better than 10% and in no cases worse than 15%. This is considered acceptable for the purposes of estimating machining costs.

PROBLEMS

1. For a single-shift working in a particular turning operation, the cost of the labor and machine time in resetting a cutting edge is $0.50, the cost of each throwaway insert having four cutting edges is $2.00, and the depreciation of the insert holder per insert is $0.08. The initial cost of the lathe was $30,000, which is amortized for 5 years. Machine running costs including labor and overhead but neglecting depreciation are estimated at $15.00/h.

If the Taylor tool-life index for the carbide inserts is 0.25, calculate, for minimum-cost conditions, the tool life, in seconds (s), and cutting speed, in meters per second (m/s), if the machine is to be used on a:
a. Single-shift, 40 hr/wk basis
b. Double-shift, 80 hr/wk basis
Assuming 50 working weeks per year, a cutting speed of 7.6 m/s, for 60 s tool life and a nonproductive time for loading and unloading a component of 3 mm.

2. A batch of 1000 components is to be rough-turned to an 80-mm diameter for 300 mm of their length using a feed of 0.25 mm per revolution. The Taylor tool-life equation for the particular carbide tool-workpiece combination was found by experiment to be:

$$vt^{0.25} = 28$$

where v is the cutting speed (in m/s), and t is the tool life (in s). If the cost per cutting edge for the particular throwaway-type insert tool is $0.50 (including the tool depreciation and cost of resetting the cutting edge) and the total machine rate, including operator costs, is $20.00/hr, calculate:
a. The tool life (in s) to give minimum production costs
b. The cutting speed (in m/s) that will result in this optimum tool life
c. The total production time (in ks), assuming that the time taken to load and unload a component is 1 mm, that the time taken to reset a cutting edge is 30 s, and that the initial setup time for the batch is 1 hr)
d. The total production cost, in dollars

3. A tool used in a turning operation is of the brazed-carbide-tip type. Its cost is $30.00, and on the average it can be reground 10 times. The cost of regrinding is $2.00. The lathe cost $40,000 and is used for 8 hr per day, 5 days per week, 50 weeks per year. Its cost is to be amortized over 10 years, and the machine overheads are 100%. The operator's wage is $20.00/h, and operator overheads are 100%. The relation between cutting speed and tool life for the conditions employed is:

$$vt^{0.25} = 7$$

where v is the cutting speed (in m/s), and t is the tool life (in s). The tool-changing time is 4 mm, and the handling time per component is 5 mm. The cutting distance per component is 200 m. Calculate the:
a. Minimum production cost, in dollars
b. Minimum production time, in seconds

4. For the conditions of Problem 3, calculate the tool life, in s, and cutting speed (in m/s), for maximum rate of profit and the corresponding production cost and production time. (Assume that the machine shop receives $10.00 for each component.)

5. From tool-life tests with a particular disposable-insert carbide tool and carbon steel workpiece, the tool life was found to be 11.6 ks for a cutting speed of 1 m/s and 363 s for a speed of 2 m/s.
   The cost of the insert is $8, and it has four cutting edges. Indexing time is 10 s, and insert replacement time is 30 s.
   a. For a machine and labor rate of $30/h what is the tool life for minimum cost (ks)? (Note: Neglect tool holder depreciation.)
   b. What cutting speed will give this tool life?

6a. In a certain operation the high-speed-steel drills used cost $2.00 and under the conditions presently used have a life of 1500 s. The time to replace a worn drill is 30 s. If the machine plus operator and overheads cost $30/h, by what percentage should the spindle speed be increased or decreased to maximize production rate?

6b. A turning operation is run under conditions of minimum cost with carbide tools. The tool-changing time is 100 s and the cost of a new tool edge is $2.00. If, as a result of a wage increase, the machine rate, including operator and overheads, rises from $20/h to $30/h, by what percentage should the cutting speed be increased or decreased?

7. The machining time in a particular turning operation is 200 s, and the tool requires changing every five components. The nonproductive time (loading and unloading) is 300 s per component, and the tool changing time is 150 s. The cutting speed is 2.5 m/s, and the tool is a brazed-type carbide ($n = 0.25$) that requires an expenditure of $10.00 for regrinding (including depreciation).
   a. What should the cutting speed be changed to in order to operate at minimum cost? (Assume a machine depreciation rate of $10 per hour, operator wages of $15 per hour, and 100% overheads on operator and equipment.)
   b. How much would be saved per component by operating at minimum cost?

8. At present it takes 30 s to turn a workpiece and the tool must be reground after 80 workpieces. The Taylor tool-life relation for the tool-workpiece combination is:

$$vt^{0.25} = 7$$

where v is the cutting speed (in m/s), and t is the tool life (in seconds).

a. If the loading and unloading of a workpiece takes 20 s (including the time to return the tool to the beginning of the cut) and if it takes 200 s to change a worn tool, what is the present average production time per component?

b. If the cutting speed is tripled, what will the new production time be?

c. If the cutting speed for maximum production rate were used, what would the tool life be (in seconds)?

9. In a slab-milling operation, four milling cutters are "ganged" to produce a shaped surface on a steel workpiece 200 mm in length. The largest cutter has a diameter of 150 mm, and all cutters are of high-speed steel. It takes 5 min to change the cutters, and the tool-life-cutting relationship is:

$$vt^{0.125} = 2.65$$

where v is the cutting speed (in m/s), and t is the tool life (in s). The working engagement (depth of cut) for the largest cutter is 14 mm, and the machine and operator rate is $40.00/h. Two of the milling cutters cost $80.00 each and cost $20.00 each to regrind; the remaining two cutters cost $100.00 each and cost $30.00 each to regrind. All cutters can be reground 20 times on the average. The time to remove and replace a component and return the cutters to the start of the cut is 3 min. Estimate the minimum production cost (in $) and the corresponding production time (in s), when the feed speed is 2 mm/s. Also, estimate the minimum production cost and corresponding production time if the machine is fitted with a special swivel vise that allows a new component to be positioned in the machine while the previous component is being machined. (Assume that this allows the nonproductive time per component to be reduced to 30 s.)

10. At present, on a single-spindle, deep-hole drilling machine one drill can drill six holes in a boiler end plate between regrinds. The drill speed is 25 s^{-1}, and the drill has a carbide tip for which n in Taylor's tool-life relation is 0.25. The machining time per hole is 6 mm. The machine cost $100,000 dollars and is to be amortized over 50,000 hours. The machine overheads are 70%, the operator's wage is $15.00/h, and the operator overheads are $15.00/h. The initial cost of one drill is $50.00; it can be reground 10 times on the average at a cost of $2.00 per regrind. Drill-changing time is 6 min. Neglecting the drill-repositioning time, calculate the:

    a. Cost per hole under present conditions, in dollars

    b. Drill speed to give minimum cost per hole, in reciprocal seconds (s^{-1})

    c. Cost per hole under minimum cost conditions, in dollars

11. A company is to purchase a new three-spindle deep-hole drilling machine at a cost of $250,000. The efficiency with which the three spindles can be employed is 90% because it will not always be possible to drill three holes simultaneously. Tool-changing time per drill will be 2 min, and the remaining costs will be the same as for the old machine described in Problem 6. Calculate, for the new machine:

    a. The total cost per hole, in dollars, if the machine is operated at the present drill speed of 25 s^{-1}

    b. The drill speed, in reciprocal seconds (s^{-1}), to give minimum cost per hole

    c. The total cost, in dollars, per hole under optimum conditions

12. A carbon steel workpiece, with a Rockwell hardness of R_c 30, is to be turned with a carbide tool, and the depth of cut is set at 5 mm.

    a. Estimate the recommended feed and cutting speed for this operation.

    b. The workpiece is hardened to R_c 40 and a coated carbide tool is used. What will be the new recommended feed and cutting speed?

    c. An alternative material for the workpiece is stainless steel, for which a coated carbide tool should be used. Estimate the recommended feed and cutting speed for this case.

REFERENCES

1. Taylor, F.W., On the Art of Cutting Metals, *Trans. ASME*, Vol. 28, 31, 1906.
2. Partons, N.R., ed., *NC Machinability Data Systems*, SME, Dearborn, MI, 1971.
3. Pressman, R.S. and J.E. Williams, *Numerical Control and Computer Aided Manufacturing*, Wiley, New York, 1977.
4. Groover, M.P. and Zimmers, E.W., *CAD/CAM: Computer-Aided Design and Manufacturing*, Pearson Educational, Englewood Cliffs, 1997.
5. Metcut, *Machining Data Handbook*, 3rd. Ed., Metcut Research Associates, Cincinnati, Ohio, 1980.
6. Turner, R., Process Principles and Cost Estimating for Machining and Transfer Lines, Masters Thesis, University of Rhode Island, 1993.

7 Nomenclature of Cutting Tools

7.1 INTRODUCTION

The most important features of a cutting tool are the two cutting edges:

1. The major cutting edge, which usually removes the bulk of metal
2. The minor cutting edge, which mainly controls the final surface finish of the workpiece

and the faces and flanks that intersect to form the cutting edges. Figure 7.1 specifies the important features of a single-point cutting tool and the nomenclature describing these features. The figure also shows the application of a single-point cutting tool in a turning operation.

The geometry and nomenclature of cutting tools, even single-point cutting tools, are surprisingly complicated subjects. It is difficult, for example, to determine the appropriate planes in which the various angles of a single-point cutting tool should be measured; it is especially difficult to determine the slope of the tool face. A number of systems of geometric arrangements and nomenclature exist. To appreciate the advantages and disadvantages of different systems of cutting-tool nomenclature (e.g., the ISO-recommended system, the British maximum rake systems, or the American Standards Association system) it is necessary first to understand how the tool shape affects the direction in which the chip flows.

In much research work consideration is given only to the somewhat artificial geometric arrangement known as *orthogonal cutting*. In the orthogonal arrangement shown in Figure 7.2a the straight cutting edge of the wedge-shaped cutting tool is at right angles to the direction of cutting. Rarely in practice, however, is the cutting edge at right angles to the direction of cutting; most cutting operations involve oblique cutting, where the cutting edge is inclined. In the oblique arrangement shown in Figure 7.2b the cutting edge is inclined at an angle λ_s (known as the cutting-edge inclination) to a line drawn at right angles to the direction of cutting. The cutting edge inclination is measured in the plane of the new workpiece surface and in orthogonal cutting, $\lambda_s = 0$. With oblique cutting, the chip flows up the tool face in a direction forming an angle λ_λ (chip-flow angle) with a line drawn on the face at right angles to the cutting edge. Figure 7.3 shows a development of the chip produced during oblique cutting: the part of the diagram below the line denoting the cutting edge is a view normal to the new workpiece

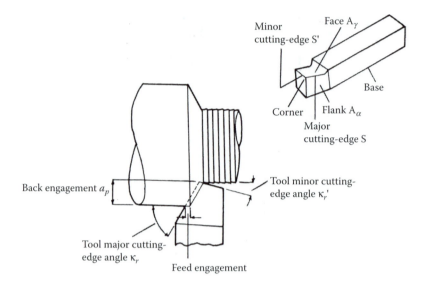

FIGURE 7.1 Terms applied to single-point tools and the single-point cutting operation.

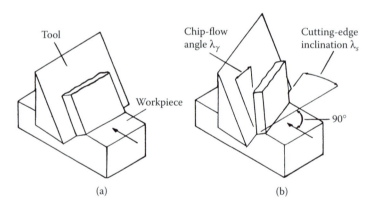

FIGURE 7.2 Basic cutting operations. (a) Orthogonal cutting; (b) oblique cutting.

surface, and that part above the cutting edge is a view normal to the tool face. The diagram shows that if the chip does not change in width during its formation,

$$\lambda_s = \lambda_\gamma \tag{7.1}$$

Experiments have shown that variations from this equality of inclination and chip-flow angles may occur in practice, but for most circumstances Equation 7.1 gives an adequate approximation. It is of interest to note that Equation 7.1 was proposed by Stabler [1] and is commonly known as Stabler's chip-flow law. The

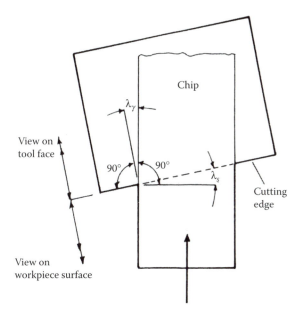

FIGURE 7.3 Demonstration of Stabler's chip-flow law ($\lambda_s = \lambda_\gamma$).

important result of Stabler's law is that the inclination of the cutting edge λ_s is a major parameter determining the direction of chip flow away from the cutting region and is, therefore, one important consideration in the design of practical cutting tools. Thus, any acceptable system of tool nomenclature should include a definition of cutting-edge inclination to give the system physical meaning in relation to the cutting process. A second important consideration in choosing a system of nomenclature related to the grinding of cutting tools is that it should be possible to set the specified tool angles directly onto the grinding vise or jig. With these two important points in mind, the various systems of cutting-tool nomenclature used in the past are now considered.

7.2 SYSTEMS OF CUTTING-TOOL NOMENCLATURE

7.2.1 BRITISH MAXIMUM-RAKE SYSTEM

Figure 7.4 illustrates the old British system (1886) where the slope of the tool face was measured in a plane normal to the tool base and to the tool face, that is, in the direction of maximum slope. This method had the advantage that the specified angles could be set on a grinding vise and the face ground to the specified angle. The difficulty with this system was that the angles specified were quite independent of the position of the cutting edge, and, therefore, complicated expressions or a set of curves had to be used to estimate the direction of chip flow. It has been suggested that the system developed from the idea that the chip flows in the direction of maximum slope of the tool face. Since this idea is not even

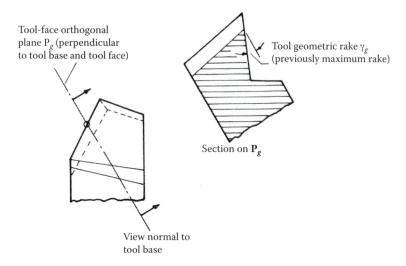

FIGURE 7.4 Early British maximum-rake system.

approximately true, the system had no physical significance in relation to the cutting process.

7.2.2 AMERICAN STANDARDS ASSOCIATION SYSTEM

The American Standards Association (ASA) system (Figure 7.5a) specified the tool face by defining its slope in two orthogonal planes: one parallel to and the other perpendicular to the axis of the cutting tool, both planes being perpendicular to the base of the tool. The two angles thus specified are known as the *tool back rake* and *tool side rake* (Figure 7.5a).

In this system, as in the British maximum-rake system, the angles were specified independently of the position of the cutting edge and, therefore, gave no indication of the behavior of the tool in practice. The advantage of the system was always considered to be the simplicity of its use in the grinding of single-point cutting tools; yet a tool cannot be ground accurately to the back rake and side rake without using equations or curves.

For example, suppose a two-axis vise is used when the tool is being ground. The vise has protractor axes A and B perpendicular to each other and both lie in horizontal planes when set at zero. The tool is mounted in the vise in such a way that the upper of the two axes, A, is parallel to the tool axis. If the tool face is to be surface-ground, the vise must be used to arrange the future tool face in a horizontal plane. If axis B is used first to tilt the tool forward through an angle λ_p to set the tool back rake (Figure 7.5b), the axis A may be used to rotate the tool about its own axis through an angle λ_f equal to the tool side rake. Unfortunately this last operation will also have rotated the plane in which the back rake should be measured; therefore, the back rake setting should be adjusted accordingly. Consideration of Figure 7.5b shows that if the required tool back rake is

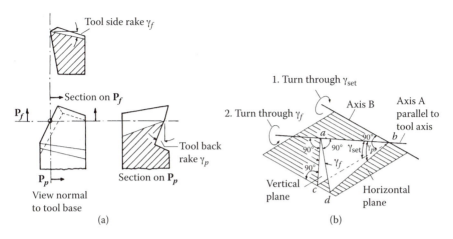

(a)   (b)

FIGURE 7.5 (a) Early American Standards Association (ASA) system; (b) construction to determine tool back-rake compensation owing to the tool side-rake setting, where $ab = ac \, cot\gamma_{set}$, $ab = ad \, cot\gamma_p$, and $ac/ad = cos\gamma_f = tan \, \gamma_{set} \, tan \, \gamma_p$. (Axis A is the upper axis on the vise.)

λ_p, axis B should have been set at λ_{set} which may be obtained from the following expression:

$$\tan \gamma_{set} = \cos \gamma_f \tan \gamma_p \qquad (7.2)$$

where λ_f is the tool side rake.

In practice, this compensation was nearly always ignored, and when the two angles were small, the resulting error was negligible. Nevertheless the system still had the main disadvantage that the angles were independent of the cutting edge.

7.2.3 GERMAN SYSTEM

The German (DIN) system also specified two rake angles, called *back rake* and *side rake* (Figure 7.6), but in the German system the angles were related to the position of the cutting edge. The German back rake was the slope of the cutting edge measured in a plane containing this edge and perpendicular to the tool base; the German side rake was the slope of the tool face measured in a plane that was both perpendicular to the plane in which back rake was measured and perpendicular to the tool base. This system had some physical meaning in relation to the cutting process because both angles were specified in relation to the edge of the tool that performs the cutting operation.

A difficulty arose, however, when the system was used for grinding a cutting tool. The problem is similar to that occurring in the American system. The procedure for grinding a tool to the German system was as follows:

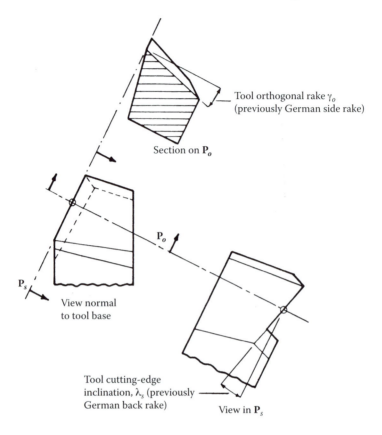

FIGURE 7.6 Early German (DIN) system.

1. The tool was clamped in a universal vise with three orthogonal protractor axes *A, B,* and *C* set at zero.
2. The desired major cutting-edge angle was set on the top vertical axis *A* such that the future cutting edge lay parallel to the third bottom axis *C*.
3. The desired German back rake was set on the second, horizontal axis *B* in order to bring the future cutting edge in a horizontal plane.
4. The next step was to set the German side rake on the bottom axis *C*, but it will be seen that rotation of this axis rotated the tool about the future cutting edge. Thus, the side rake was being measured in a plane normal to the cutting edge and not perpendicular to the tool base. For accurate grinding, the setting on axis *C* should have been adjusted (depending on the setting of axis *B*) using equations or curves in a manner similar to that of the American system.

Clearly, these difficulties can be avoided if the side rake is measured in a plane normal to the tool cutting edge. The side-rake angle could then be set on the third axis *C* without the need for compensations. This idea forms the basis for a system of cutting-tool nomenclature called the *normal-rake system*, which has now been adopted as the new ISO standard and the standard in America, Great Britain, and Germany. It is of interest to note that the normal-rake system of cutting-tool nomenclature was first proposed by Stabler in 1955 [2].

7.3 INTERNATIONAL STANDARD

7.3.1 TOOL-IN-HAND AND TOOL-IN-USE SYSTEMS

The ISO recommendation for the nomenclature of cutting tools [3] first establishes systems of planes that can be used to define the various angles of the faces and flanks of the tool. It should be understood that two systems of angles and planes are required. The first system is the tool-in-hand system of planes and angles and refers to a cutting tool that is held in the hand and is used for the purposes of grinding and sharpening the tool. The second system is the tool-in-use system of planes and angles and refers to the cutting tool being used in a machining operation.

The reason two systems are required is twofold: First, in a simple turning operation as the feed is increased, the effective rake angle increases, and the effective clearance decreases (Figure 7.7); second, it is possible that a cutting tool (particularly a single-point cutting tool) can be held in a machine tool in various orientations, thereby altering the effective angles of the tool. Thus, the tool-in-hand system is defined in relation to the tool base (or, for rotating tools, to the tool axis). The tool-in-use system, on the other hand, is defined in relation to the resultant cutting direction and the direction of feed motion.

Figure 7.8, Figure 7.9, Figure 7.10, and Figure 7.11 show the two systems of planes necessary for defining the tool shape with respect to a selected point on the major cutting edge of a single-point tool. Table 7.1 shows how, using these

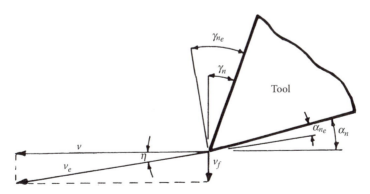

FIGURE 7.7 Effect of resultant cutting-speed angle η on normal rake and normal clearance in orthogonal cutting, where v = cutting speed, v_f = feed speed, v_e = resultant cutting speed, γ_n = tool normal rake, γ_{n_e} = working (effective) normal rake, α_n = tool normal clearance, and α_{n_e} = working (effective) normal clearance.

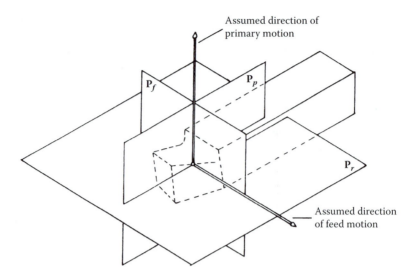

FIGURE 7.8 Tool-in-hand planes P_r, P_f, and P_p for the major cutting edge of a single-point tool, where P_r, the tool reference plane, is parallel to the tool base; P_f, the assumed working plane, is perpendicular to the tool reference plane P_r, perpendicular to the tool axis, and perpendicular to the assumed direction of feed motion; and P_p, the tool back plane, is perpendicular to the tool reference plane P_r and perpendicular to the assumed working plane P_f.

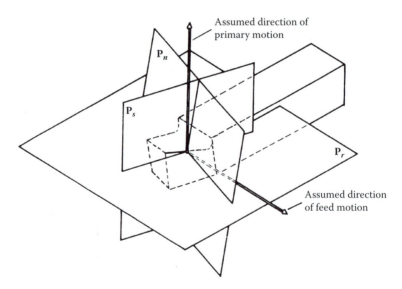

FIGURE 7.9 Tool-in-hand planes P_r, P_s, and P_n for the major cutting edge of a single-point tool, where P_r, the tool reference plane, is parallel to the tool base; P_s, the tool-cutting edge plane, is tangential to the cutting edge and perpendicular to the tool-reference P_r; and P_n, the cutting-edge normal plane, is perpendicular to the cutting edge.

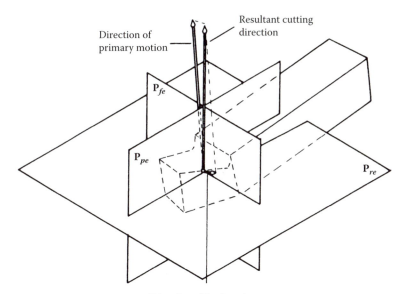

Direction of primary motion

Resultant cutting direction

$\mathbf{P}_{fe}$

$\mathbf{P}_{pe}$

$\mathbf{P}_{re}$

Direction of feed motion

FIGURE 7.10 Tool-in-use planes $\mathbf{P}_{re}$, $\mathbf{P}_{fe}$, and $\mathbf{P}_{pe}$ for the major cutting edge of a single-point tool, where $\mathbf{P}_{re}$, the working reference plane, is perpendicular to the resultant cutting direction; $\mathbf{P}_{fe}$, the working plane, is perpendicular to the working reference plane $\mathbf{P}_{re}$ and contains the directions of the primary and feed motions; and $\mathbf{P}_{pe}$, the working back plane, is perpendicular to the working reference plane $\mathbf{P}_{re}$ and perpendicular to the assumed working plane $\mathbf{P}_{fe}$.

systems of planes, the various angles are defined; it should be noted that the planes and angles in the tool-in-hand system are prefixed by the word *tool* and those in the tool-in-use system are prefixed by the word *working* except for those planes or angles that are the same in either system. For example, it can be seen from the table that the tool normal rake γ_n is the angle between the tool face A_γ and the tool reference plane $\mathbf{P}_r$, measured in the cutting edge normal plane $\mathbf{P}_n$.

Each plane is provided with a symbol consisting of the letter $\mathbf{P}$ with a subscript to denote the plane's identity. In all the planes and angles defined here it is assumed that the selected point of reference lies on the major cutting edge [5]. If, however, the selected point were on the minor cutting edge, the name of the plane or angle would include the words *minor cutting edge* and the symbol would bear a prime. For example $\mathbf{P}'_s$ would be the total minor cutting-edge plane, and γ'_n would be the tool minor cutting-edge normal rake.

The symbol for a plane or angle in the tool-in-use system bears an additional subscript, e, for *effective*. For example, $\mathbf{P}_{s_e}$ would be the working cutting-edge plane, and γ_{n_e} would be the working normal rake.

Figure 7.12 and Figure 7.13 show the various tool and working angles for a single-point cutting tool. The signs of the various angles are all positive in these figures.

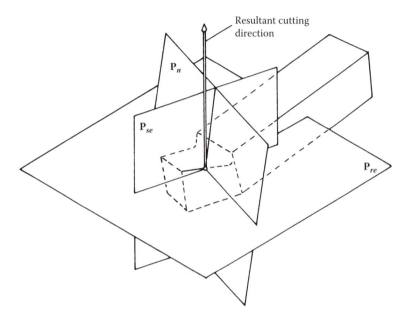

FIGURE 7.11 Tool-in-use planes P_{re}, P_{se}, and P_n for the major cutting edge of a single-point tool, where P_{re}, the working reference plane, is perpendicular to the resultant cutting direction; P_{se}, the working cutting edge plane, is tangential to the cutting edge and perpendicular to the tool-reference P_{re}; and P_n, the cutting-edge normal plane, is perpendicular to the cutting edge.

TABLE 7.1
Tool and Working Angles, Definitions

Angle	Is the Angle Between	and	Measured in Plane
Tool normal rake γ_n	A_γ	P_r	P_n
Working normal rake γ_{ne}	A_γ	P_{re}	$P_{ne}(\equiv P_n)$
Tool normal clearance α_n	A_α	P_s	P_n
Working normal clearance α_{ne}	A_α	P_{se}	$P_{ne}(\equiv P_n)$
Tool cutting-edge angle κ_r	P_s	P_f	P_r
Working cutting edge angle κ_{ne}	P_{se}	P_{fe}	P_{re}
Tool cutting-edge inclination λ_s	S	P_r	P_s
Working cutting-edge inclination λ_{se}	S	P_{re}	P_{se}
Tool-included angle ε_r	P_s	P'_s	P_r
Normal wedge angle $\beta_n(\equiv \beta_{ne})$	A_γ	A_α	$P_{ne}(\equiv P_n)$

A_γ = tool face
A_α = tool flank
S = major cutting edge

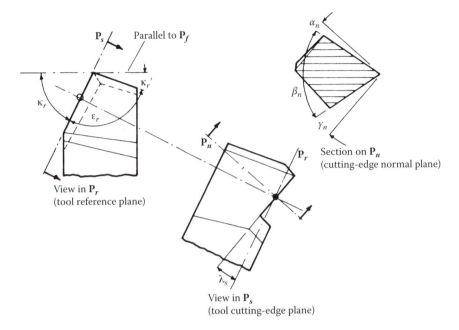

FIGURE 7.12 Tool angles for a single-point tool, where κ_r = tool cutting-edge angle, κ_r' = tool minor cutting-edge angle, λ_s = tool cutting-edge inclination, γ_n = tool normal rake, α_n = tool normal clearance, β_n = normal wedge angle, and ε_r = tool-included angle.

As explained in Chapter 1, in practice the angle between the resultant cutting direction and the direction of primary motion, the resultant cutting speed angle η, is usually very small; therefore if the tool is clamped in the machine in its most natural position (e.g., as in Figure 7.1 for cylindrical turning), the tool-in-use system and tool-in-hand system almost coincide and corresponding angles are almost identical. However, if the tool is oriented differently from its most natural position, the working angles will be different from the tool angles. To relate one system to another it is necessary to define the orientation of the tool in the machine tool, and for this purpose another system of planes and angles, called the setting system of planes and angles, is required.

7.3.2 SETTING SYSTEM

To complete the relationship between the tool and working system, it is necessary to define two sets of angles:

1. Those angles defining the position of the tool in the machine tool (called setting angles)
2. Those angles defining the orientation of the resultant cutting direction and direction of feed motion relative to the machine tool

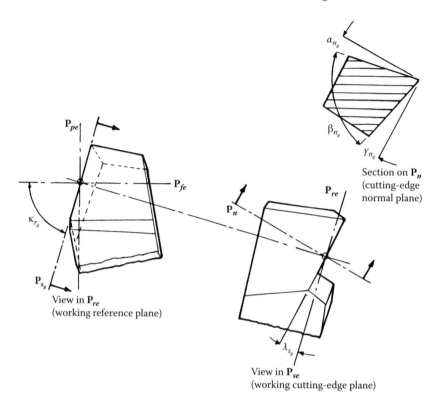

FIGURE 7.13 Working angles for a single-point tool, where κ_{r_e} = working cutting-edge angle, λ_{s_e} = working cutting-edge inclination, γ_{n_e} = working normal rake, α_{n_e} = working normal clearance, β_{n_e} = normal wedge angle (= β_n), and ε_{r_e} = working included angle.

In the present discussion only the first set of angles will be defined; the definitions will provide sufficient information to obtain the working angles from the tool angles when the resultant cutting speed angle η is assumed to be zero. Neglecting the effect of the resultant cutting speed angle η simplifies the mathematical analysis, but it should be remembered that for situations where the angle η is significant (say, greater than 1 degree), more complicated equations must be employed.

The setting angles are defined in Figure 7.14 for a single-point tool. The figure also indicates the signs of the angles determined from the right-hand-screw rule. The axes X, Y, and Z correspond to the machine-tool axes defined and illustrated for various machine tools in Chapter 1.

7.3.3 Mathematical Relationships Between Tool and Working Systems

Equation 7.3, Equation 7.4, Equation 7.5, Equation 7.6, Equation 7.7, Equation 7.8, Equation 7.9, Equation 7.10, and Equation 7.11 allow the working angles to be

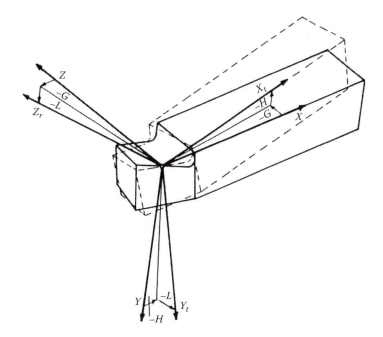

FIGURE 7.14 Setting planes and angles, where X, Y, and Z are the machine axes as defined in Chapter 1; X_t, Y_t, and Z_t are the tool-in-hand axes and correspond to the intersections of the tool-in-hand planes as follows, X_t ($\mathbf{P}_r$, $\mathbf{P}_p$), Y_t ($\mathbf{P}_p$, $\mathbf{P}_f$), Z_t ($\mathbf{P}_f$, $\mathbf{P}_r$), as well as to the machine axes when the tool is in its most "natural" position on the machine tool; G is the plane-setting angle and is positive if it has the effect of decreasing κ; H is the elevation-setting angle and is positive if it has the effect of increasing the angle λ; L is the roll-setting angle and is positive if it has the effect of decreasing the angle γ. (Note: For the right-hand tool shown, the signs of the setting angles obey the right-hand-screw rule.)

calculated when a single-point tool is held in a machine tool in any orientation and when it can be assumed that the resultant cutting-speed angle η is zero (Figure 7.7). Each working angle is calculated in three steps, each step corresponding to moving the tool from its most natural position through each of the three setting angles in turn. Thus, in each group of equations the suffix G, H, or L refers to the effect of the appropriate setting angle. It should be noted that the resulting angles are the true working angles only when the working plane $\mathbf{P}_{fe}$ coincides with the machine plane P_{yz} and when the effect of feed is negligibly small, that is, $\eta = 0$. Also, under certain conditions (e.g., when drilling or turning with a high feed), the effect of the resultant cutting speed angle η must be taken into account.

The equations for calculating the cutting-edge angle are

$$\kappa_{r_L} = \arctan \frac{\sin \kappa_r}{\left(\cos \kappa_r \cos L\right) - \left(\tan \lambda_s \sin L\right)} \tag{7.3}$$

$$\kappa_{r_H} = \arctan \frac{\left(\sin \kappa_{r_L} \cos H\right) - \left(\tan \lambda_{s_L} \sin L\right)}{\cos \kappa_{r_L}} \tag{7.4}$$

$$\kappa_{r_G} = \kappa_{r_H} - G = \kappa_{r_e} \tag{7.5}$$

for the cutting-edge inclination

$$\lambda_{s_L} = \arcsin\left[\left(\sin \lambda_s \cos L\right) + \left(\sin \kappa_r \cos \lambda \sin L\right)\right] \tag{7.6}$$

$$\lambda_{s_H} = \arcsin\left[\left(\sin \lambda_{s_L} \cos H\right) + \left(\sin \kappa_{r_L} \cos \lambda_{s_L} \sin H\right)\right] \tag{7.7}$$

$$\lambda_{s_G} = \lambda_{s_H} = \lambda_{s_e} \tag{7.8}$$

and for the normal rake

$$\gamma_{n_L} = \gamma_n - \arctan \frac{\sin \kappa_r \sin L}{\left(\cos \lambda_s \cos L\right) - \left(\cos \kappa_r \sin \lambda \sin L\right)} \tag{7.9}$$

$$\gamma_{n_H} = \gamma_{n_L} + \arctan \frac{\cos \kappa_{r_L} \sin H}{\left(\cos \lambda_{s_L} \cos H\right) - \left(\sin \kappa_{r_L} \sin \lambda_{s_L} \sin H\right)} \tag{7.10}$$

$$\gamma_{n_G} = \gamma_{n_H} = \gamma_{n_e} \tag{7.11}$$

The effect of setting angles on normal clearance is numerically the same as for normal rake except the change is in the opposite sense.

7.3.4 EXAMPLE

A lathe tool is to be placed in a machine such that the plane setting angle G is −20 degrees, the elevation setting angle H is −10 degrees, and the roll setting angle L is zero. Also, the tool-in-hand angles (Figure 7.12) are as follows:

1. The tool major-cutting-edge angle κ_r 30 degrees
2. The tool major-cutting-edge inclination λ_s is 5 degrees
3. The tool major-cutting-edge normal rake γ_n is 10 degrees
4. The tool major-cutting-edge normal clearance α_n is 5 degrees

From Equation 7.3, Equation 7.6, and Equation 7.9 when $L = 0$,

$$\kappa_{r_L} = \kappa_r = 30°$$

$$\lambda_{s_L} = \lambda_s = 5°$$

$$\gamma_{n_L} = \gamma_n = 10°$$

$$\alpha_{n_L} = \alpha_n = 5°$$

From Equation 7.4 and Equation 7.5

$$\kappa_{r_e} = \kappa_{r_G} - G$$

$$= \arctan\left[\frac{\left(\sin 30° \cos 10°\right) + \left(\tan 5° \sin 10°\right)}{\cos 30°}\right] + 20°$$

$$= 30.3° + 20° = 50.3°$$

From Equation 7.7 and Equation 7.8

$$\lambda_{s_e} = \lambda_{s_H}$$

$$= \arcsin\left[\left(\sin 5° \cos 10°\right) - \left(\sin 30° \cos 5° \sin 10°\right)\right] = 0$$

From Equation 7.10 and Equation 7.11

$$\gamma_{n_e} = \gamma_{n_H}$$

$$= 10° + \arctan\left[\frac{-\cos 30° \sin 10°}{\left(\cos 5° \cos 10°\right) + \left(\sin 30° \sin 5° \sin 10°\right)}\right]$$

$$= 10° - 8.9° = 1.1°$$

and finally since the normal rake has been reduced by the setting angles, the normal clearance would be increased by the same amount. Thus,

$$\alpha_{n_e} = \alpha_n + 8.9° = 5° + 8.9° = 13.9°$$

7.3.5 CALCULATION OF TOOL ANGLES FROM WORKING ANGLES

It is often desired to calculate the tool angles when the required working angles and the setting angles are specified. To calculate the tool angles when the effect of feed is neglected and when the working plane $\mathbf{P}_{fe}$ coincides with the machine plane $\mathbf{P}_{yz}$) the appropriate equations are as follows:

For the cutting-edge angle

$$\kappa_{r_G} = \kappa_r - G \tag{7.12}$$

$$\kappa_{r_H} = \arctan \frac{\left(\sin \kappa_{r_G} \cos H\right) - \left(\tan \lambda_{s_G} \sin H\right)}{\cos \kappa_{r_G}} \tag{7.13}$$

$$\kappa_{r_L} = \arctan \frac{\sin \kappa_{r_H}}{\left(\cos \kappa_{r_H} \cos L\right) - \left(\tan \lambda_{s_H} \sin L\right)} = \kappa_r \tag{7.14}$$

for the cutting-edge inclination

$$\lambda_{s_G} = \lambda_{s_e} \tag{7.15}$$

$$\lambda_{s_H} = \arcsin \left[\left(\sin \lambda_{s_G} \cos H\right) + \left(\sin \kappa_{r_G} \sin \lambda_{s_G} \sin H\right) \right] \tag{7.16}$$

$$\lambda_{s_L} = \arcsin \left[\left(\sin \lambda_{s_H} \cos L\right) + \left(\cos \kappa_{r_H} \cos \lambda_{s_H} \sin L\right) \right] = \lambda_s \tag{7.17}$$

and for the normal rake

$$\gamma_{n_G} = \gamma_{n_e} \tag{7.18}$$

$$\gamma_{n_H} = \gamma_{n_G} + \arctan \frac{\cos \kappa_{r_G} \sin H}{\left(\cos \lambda_{s_G} \cos H\right) - \left(\sin \kappa_{r_G} \sin \lambda_{s_G} \sin H\right)} \tag{7.19}$$

$$\gamma_{n_L} = \gamma_{n_H} - \arctan \frac{\sin \kappa_{r_H} \sin L}{\left(\cos \lambda_{s_H} \cos L\right) - \left(\cos \kappa_{r_H} \sin \lambda_{r_H} \sin L\right)} \tag{7.20}$$

The effect of setting angles on normal clearance is numerically the same as for normal rake except the change is in the opposite sense.

All of the equations relating the tool and working systems were derived by Verma [4].

PROBLEMS

1. A tool bit is to be mounted in the standard straight tool holder shown in Figure 7.15. The shank of the tool holder is to be mounted horizontally and perpendicular to the spindle axis of a lathe with the tool corner at the same level as the spindle axis. Calculate the tool angles for the bit to give the following working angles:
   a. A working major cutting-edge angle κ_{r_e} of 45 degrees
   b. A working major cutting-edge inclination λ_{s_e} of 0
   c. A working major cutting-edge normal rake γ_{n_e} of 5 degrees
   d. A working major cutting-edge normal clearance α_{n_e} of 3 degrees

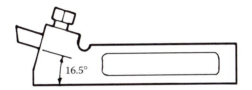

FIGURE 7.15 Standard straight tool holder.

2. If the tool holder of Problem 1 had been a right-hand tool holder with the front part bent to the left through an angle of 10 degrees in the plan view, what tool angles for the bit would be required?

REFERENCES

1. Stabler, G.V. The fundamental geometry of cutting tools, *Proc. Inst. Mech. Eng.,* vol. 165, 14, 1951.
2. Stabler, G.V. The basic nomenclature of cutting tools, *J. Inst. Prod. Eng.,* Vol. 34, 264, 1955.
3. International Standards Organization, Geometry of the Active Parts of Cutting Tools — General Terms, Reference Systems, Tool and Working Angles, Draft International Standard ISO/DIS 3002, 1973.
4. Verma, D.K. Nomenclature of Cutting Tools, M.S. Thesis, Department of Mechanical Engineering University of Massachusetts, Amherst, 1969.

8 Chip Control

8.1 INTRODUCTION

Before the advent of carbide tools and the use of higher cutting speeds, continuous chips produced in metal cutting did not present a serious problem. At low cutting speeds, these chips usually have a natural curl and tend to be brittle. However, cutting speeds have now increased to such an extent that chip control has become necessary. In turning operations, where the tool is continuously removing metal for a long period, a continuous chip can become entangled with the tool, the workpiece, or the machine-tool elements. This type of chip can be hazardous to the operator and, unless controlled properly, can result in mechanical chipping of the cutting edge. If coolant is being applied, the chip may interfere with the flow of coolant, causing alternate heating and cooling of the cutting edge. The resulting thermal stresses can reduce the life of cemented-carbide tools.

The handling of continuous chips in bulk can present a major economic problem. A handling characteristic of chips can be expressed by the bulk ratio, the total volume occupied by the chips divided by the volume of solid chip material. Unbroken, continuous chips have a bulk ratio of approximately 50, tightly wound chips a bulk ratio of approximately 15, and well-broken chips a bulk ratio of approximately 3. The volume occupied by well-broken chips is therefore about 1/17 the volume of unbroken chips, a considerable advantage when it comes to handling and disposal.

Figure 8.1 shows some of the various forms of chips that may be produced in a cutting operation, together with the first two digits of a proposed international standard-coding system [1]. The dangers and difficulties that result from the formation of long chips has necessitated the study and development of various means to control chips and break them into small coils or fragments as they are produced. At the present time all tool manufacturers offer cutting tool inserts with a variety of chip control or chip breaking features built into rake face of the tool.

8.2 CHIP BREAKERS

The controlling and breaking of chips is accomplished by a chip breaker, which is defined as a modification of the face to control or break the chip, consisting of either an integral groove or an integral or attached obstruction.

There are basically two kinds of chip breakers: the *groove type* and the *obstruction type*. These chip breakers together with some of the important dimensions are illustrated in Figure 8.2 and Figure 8.3. Modern chip control devices are generally more complex than those shown, but the essential features are

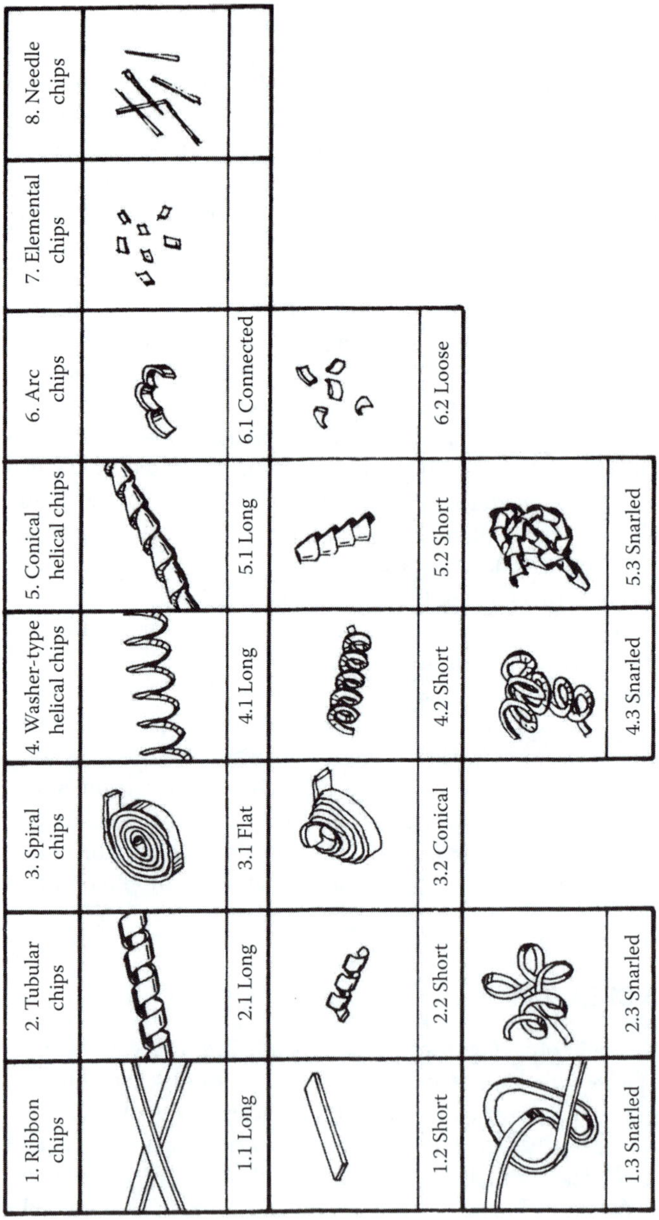

1. Ribbon chips	2. Tubular chips	3. Spiral chips	4. Washer-type helical chips	5. Conical helical chips	6. Arc chips	7. Elemental chips	8. Needle chips
1.1 Long	2.1 Long	3.1 Flat	4.1 Long	5.1 Long	6.1 Connected		
1.2 Short	2.2 Short	3.2 Conical	4.2 Short	5.2 Short	6.2 Loose		
1.3 Snarled	2.3 Snarled		4.3 Snarled	5.3 Snarled			

FIGURE 8.1 Chip forms produced in machining operations.

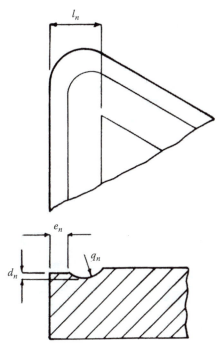

FIGURE 8.2 Groove-type chip breaker, where d_n = chip-breaker-groove depth, e_n = chip-breaker land width, l_n = chip-breaker distance, and q_n = chip-breaker groove radius.

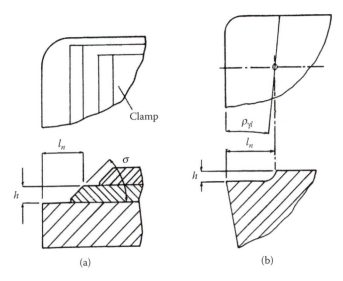

FIGURE 8.3 Obstruction-type chip breakers, where h = chip-breaker height, l_n = chip-breaker distance, σ = chip-breaker wedge angle, and $\rho_{\gamma l}$ = chip-breaker angle. (a) Attached; (b) integral.

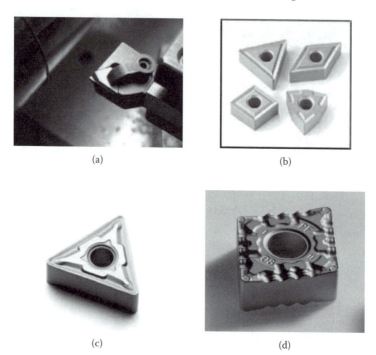

(a)   (b)

(c)   (d)

FIGURE 8.4 Chip-breaking devices for turning tools. (a) Clamp-on obstruction chip breaker; (b) two dimensional chip-breaking grooved inserts; (c) simple three-dimensional chip-breaking grooved inserts; (d) complex three-dimensional chip-breaking grooved inserts.

similar. Figure 8.4 shows examples of different types of chip breaking or chip control features available for turning tools.

A chip breaker acts by controlling the radius of the chip and directing the chip in such a way that it breaks into short lengths. In addition to an appropriate chip-breaker design, it is necessary to have the correct tool geometry so that the chip will follow the proper path across the tool face. It was seen in Chapter 7 that the direction of chip flow is controlled mainly by the working major cutting-edge inclination. When this angle is zero, the condition is known as orthogonal cutting, and the chip will usually form a spiral, as shown in Figure 8.5a. Under these conditions up curl predominates. For orthogonal conditions and with the chip initially tightly curled, the radius of chip curvature will be forced to increase gradually as cutting proceeds. This gradual increase in chip-curvature radius will impose gradually increasing stresses in the chip, eventually causing breakage and resulting in spiral chips. If the chip does not have a natural curl (the case at high cutting speeds) and no chip breaker is present, straight or ribbon chips are produced that can become snarled if the cutting process is continuous. With a chip breaker present, the chip will be curled, but because of the restrictions imposed on its path, it will strike the transient surface of the workpiece and

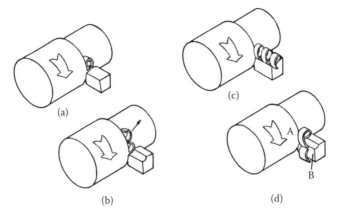

FIGURE 8.5 Production of various chip forms in turning. (a) Orthogonal cutting, flat spiral chips; (b) orthogonal cutting, loose-arc chips; (c) oblique cutting, long tubular chips; (d) oblique cutting, connected-arc chips.

continuously break into small fragments (Figure 8.5b). These fragments, which take the form of loose arc chips, often fly violently from the cutting region and present a hazard to the machine operator.

In the more common situation of oblique cutting (when the working major cutting-edge inclination is not zero) long chips are usually tubular, washer-type helical, or conical helical (Figure 8.1) and can become snarled. Under these conditions side curl predominates. In the production of tubular chips (Figure 8.5c) the helix angle of the chip will be equal to the chip-flow angle and therefore approximately equal to the working major cutting-edge inclination. However, in many situations the cutting speed can vary considerably along the major cutting edge of the tool (in drilling, for example), and washer-type or conical-helical-type chips are produced. To provide chip breaking under any of these conditions the radius of the chip must be controlled by a chip breaker; in addition, the working major cutting-edge angle and the working major cutting-edge inclination must be such that the free end of the chip strikes one of the surfaces on the workpiece or the tool flank. The most common situation is shown in Figure 8.5d, where the free end of the chip clears the transient surface and strikes the work surface at some point A. The free end of the chip is then carried down by the rotation of the workpiece and eventually strikes the tool flank at B. Further chip production, with the free end effectively fixed at B, results in an increasing radius of chip curvature, increasing stresses, and eventually, fracture of the chip. This process results in the production of arc-type chips (Figure 8.1). Under these circumstances the radius of chip curvature is critical. If this radius is too large, the free end of the chip may not strike the tool flank but rather coil around the tool. If this radius is too small, the chip may clear the tool face and form a tubular or helical chip. In the latter case, the chip may eventually break because of its weight and motion.

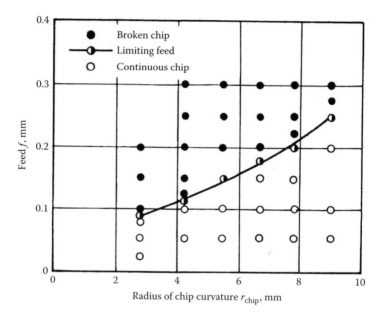

FIGURE 8.6 Effects of radius of chip curvature and feed on chip breaking for S45C at a cutting speed of 1.83 m/s. (After Nakayama, K, *Bull. Japanese Soc. Mech. Eng.*, Vol. 5, 17, 142, 1962.)

Hendriksen [2] classified chips into three categories: over broken, efficiently broken, and under broken. Generally, connected arc chips are considered to be efficiently broken. In turning operations with a proper chip-former geometry and tool shape, the type of broken chip produced can usually be controlled by the feed. A change in feed affects both the chip thickness and radius of chip curvature, the basic rule being that to change from an under broken to an efficiently broken chip, the feed should be increased. Several early analyses [3–5] were made to determine the conditions under which various types of chip breaking will occur. The principal parameters were found to be the radius of chip curvature, the thickness of the chip, and the properties of the workpiece material. However, limited success was achieved in correlating these analyses with experimental data such as that obtained by Nakayama [4] (shown in Figure 8.6). This figure illustrates how the feed and radius of chip curvature affect the efficiency of the chip-breaking operation.

Since these early studies, chip breaking and chip control have been investigated in some detail [6–8]. For a given tool/work combination, the conditions under which efficient chip breaking occurs can be shown in the form of chip breaking charts. These show, for different feeds and depths of cut (tool engagements), the regions in which different types of chip and in particular broken chips occur. Figure 8.7 shows a typical experimental chip breaking chart. This was

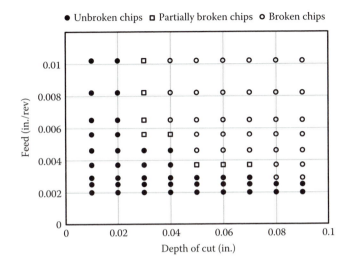

FIGURE 8.7 Typical experimental chip breaking chart. (Adapted from Zhou, L., Rong, Y., Li, Z., and Yang, J.A., *Int. J. Adv. Manuf. Technol.*, Vol. 22, 336–343, 2003.)

obtained by cutting at a range of feeds and depths of cut and observing the types of chips produced. As typically can be observed, chip breaking does not occur at low feeds and low depths of cut.

Different chip breaking charts will exist for various cutting speeds and tool geometries, for a given work material. The development of chip breaking charts such as that shown in Figure 8.7 requires a large number of cutting tests and inspection of the chips collected. However, more automated ways of generating these charts have been developed [9] based upon programming a CNC machine tool to follow a range of feed/depth of cut combinations. Cutting force measurement signals are then processed to compare with stored signatures obtained for different chip breaking conditions to automatically determine the chip breaking chart.

In general chip breaking charts take the typical form shown in Figure 8.8 [8]. There exist critical feeds and depths of cut below which chip breaking does not occur and these define the chip breaking limit curve. The chip breaking limit curve can be divided into three parts:

1. Region AB. This region is two-dimensional up curl dominated. Above the limit curve, up curl broken chips are formed. On the unbroken side mainly snarl type chips occur.
2. Region CD. This region is side curl dominated. On the unbroken side of the limit curve mainly side curl spiral continuous chips are formed.
3. Region BC. This is a transition region and more complex three-dimensional chip curling occurs.

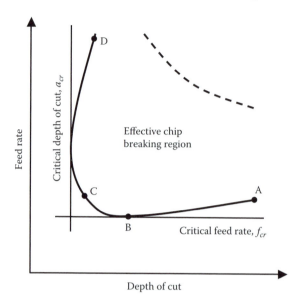

FIGURE 8.8 General form of chip breaking charts; f_{cr} = critical feed below which no chip breaking occurs; a_{cr} = critical back engagement or depth of cut below which no chip breaking occurs.

8.3 PREDICTION OF RADIUS OF CHIP CURVATURE

Prediction of the actual chip-breaking performance of various tool designs is obviously desirable. In general chip breaking is determined by the radius of curvature of the chip. If the chip is tightly curled, breaking will usually occur. Some success has been experienced [10] in predicting the radius of chip curvature when a chip breaker is employed.

For any of the obstruction-type chip breakers, it is only necessary to realize that the chip starts to curl away from the tool face at the end of the chip-tool contact region and that the chip then maintains a constant radius of curvature until it clears the chip breaker. From the geometry of Figure 8.9 it can be deduced that for an integral obstruction-type chip breaker, the radius of chip curvature, r_{chip} at the defined point is given by

$$r_{chip} = \frac{\left(l_n - l_f\right)^2}{2h} + \frac{h}{2} \tag{8.1}$$

where

l_n = chip-breaker distance
l_f = length of chip-tool contact
h = chip-breaker height

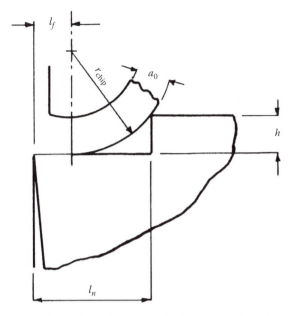

FIGURE 8.9 Construction to determine radius of chip curvature for an integral obstruction type chip breaker, where l_f = contact length between the chip and tool, l_n = chip-breaker distance, h = chip-breaker height, a_0 = chip thickness, and r_{chip} = chip-curvature radius.

Creveling, Jorden, and Thomsen [11] investigated the length of chip-tool contact and concluded that

$$\frac{l_f}{a_0} = K \tag{8.2}$$

where K is a constant for a given material, and a_0 is the chip thickness.
Hence, Equation 8.1 becomes

$$r_{chip} = \frac{\left(l_n - Ka_0\right)^2}{2h} + \frac{h}{2} \tag{8.3}$$

These authors found that for steel K equals unity, and although K is probably not equal to unity for all conditions and materials, it seems unlikely that the value of K will vary considerably. For practical purposes, therefore, it is reasonable to assume that l_f/a_0 equals unity and Equation 8.3 becomes

$$r_{chip} = \frac{\left(l_n - a_0\right)^2}{2h} + \frac{h}{2} \tag{8.4}$$

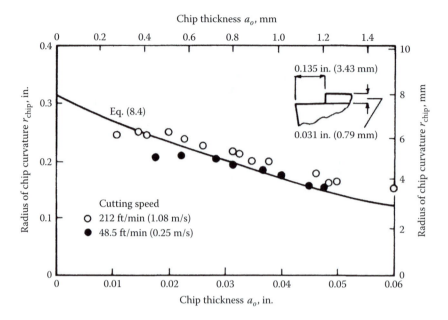

FIGURE 8.10 Effect of chip thickness on radius of chip curvature when 85/15 brass is cut. (After Trim, A.R. and Boothroyd, G. *Int. J. Prod. Res.,* Vol. 6, 3, 227, 1968.)

Figure 8.10 shows that Equation 8.4 agrees closely with experimental results. For an attached obstruction-type chip breaker (Figure 8.11), the equation for the radius of chip curvature can be developed in a similar way:

$$r_{chip} = \left[\left(l_n - l_f \right) - \left(h \cot \sigma \right) \right] \cot \frac{\sigma}{2} \tag{8.5}$$

where σ is the chip-breaker wedge angle, and when l_f equals a_0 this equation becomes

$$r_{chip} = \left[\left(l_n - a_0 \right) - \left(h \cot \sigma \right) \right] \cot \frac{\sigma}{2} \tag{8.6}$$

For a groove-type chip breaker, no chip control is obtained unless the chip-breaker land width e_n is less than the length of chip-tool contact l_f. When e_n is less than l_f, the chip will try to flow down into the groove and conform to the chip breaker groove radius q_n (Figure 8.12). If, however, q_n is too small, the chip

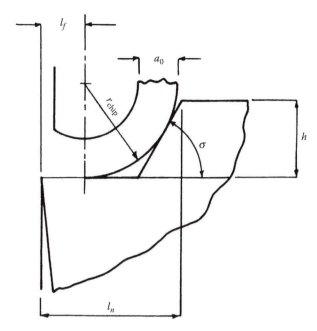

FIGURE 8.11 Construction to determine radius of chip curvature for an attached obstruction-type chip breaker, where l_f = contact length between the chip and tool, l_n = chip-breaker distance, h = chip-breaker height, a_0 = chip thickness, r_{chip} = chip-curvature radius, and σ = chip-breaker wedge angle.

will not follow the groove contour until tool wear is significant. Thus, if l_f is assumed equal to a_0 and the choice of chip breaker groove radius q_n is appropriate,

$$r_{chip} = q_n \qquad (8.7)$$

when e_n is less than a_0.

Equation 8.4, Equation 8.6, and Equation 8.7 allow the radius of chip curvature to be estimated for a chip breaker of given dimensions, and the chip thickness a_0 always appears as a significant parameter. It can therefore be appreciated why both r_{chip} and a_0 have a controlling influence on the efficiency of chip breaking.

8.4 PREDICTION OF CHIP BREAKING PERFORMANCE

Investigations into the analytical prediction of chip breaking performance have been carried out [8,12,13], including finite element based analyses to predict the chip-tool and chip-work contacts that lead to chip breaking. Some success with 2-D chip curl has been achieved [13].

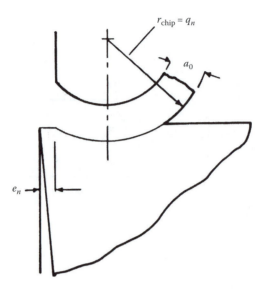

FIGURE 8.12 Chip radius of curvature when $e_n < a_0$ for a groove-type chip breaker, where e_n = chip-breaker land width, a_0 = chip thickness, and r_{chip} = chip-curvature radius.

For tools with two-dimensional chip breaking features, analytical expressions for the critical feed and critical depth of cut for the chip breaking chart have been derived [8]. From the analysis of up-curl, the critical feed, f_{cr}, can be represented by

$$f_{cr} = \frac{\varepsilon_B}{\alpha_c} \frac{r_c w_n K_R}{\gamma_n \sin \kappa_r} \left(1 - 2 \frac{l_f}{w_n} \cos \gamma_{n_e} \right) \tag{8.8}$$

in which

$$K_R = \frac{r_{chip}}{r_l - r_{chip}} \tag{8.9}$$

and where
    ε_B = fracture strain of the chip
    α_c = proportion of chip thickness to neutral surface in bending
    r_c = cutting ratio
    w_n = width of chip breaking groove
    κ_r = cutting edge angle
    γ_{n_e} = normal rake angle
    l_f = chip/tool contact length

r_{chip} = chip flow radius
r_l = chip breaking radius

Similarly from the analysis of side curl chips, the critical value of the tool back engagement (depth of cut), a_{cr}, can be derived as

$$a_{cr} = \frac{\varepsilon_B r_{sc} \cos \delta}{\alpha_c} - \left(\frac{\pi}{2} - 1\right) r_\varepsilon, \text{ for } a_p \geq r_\varepsilon \qquad (8.10)$$

or

$$a_{cr} = \left(\cos \frac{57.3 \varepsilon_B r_{sc} \cos \delta}{\alpha_c r_\varepsilon}\right) - r_\varepsilon, \text{ for } a_p < r \qquad (8.11)$$

where

r_{sc} = radius of side curl chips
δ = cross section related parameter
r = tool nose radius.

If it is assumed that ε_B and r_c are mainly determined by the cutting speed and workpiece material and all other parameters are mainly determined by the tool insert/chip breaking groove geometry, then the critical feed can be represented by

$$f_{cr} = f_r K_{fT} K_{fv} K_{fm} \qquad (8.12)$$

where

f_r = critical feed under representative cutting conditions
K_{fT} = cutting tool (insert) effect coefficient
K_{fv} = cutting speed effect coefficient
K_{fm} = workpiece material effect coefficient

Similarly, the critical tool back engagement (depth of cut) can be expressed as

$$a_{cr} = a_r K_{dT} K_{dv} K_{dm} \qquad (8.13)$$

where

a_r = critical back engagement (depth of cut) under representative cutting conditions
K_{dT} = cutting tool (insert) effect coefficient
K_{dv} = cutting speed effect coefficient
K_{dm} = workpiece material effect coefficient

The various coefficients in Equation 8.12 and Equation 8.13 can be determined by a limited range of cutting tests and then the equations used to predict the limits of chip breaking charts for other tool geometries and cutting conditions. This approach has been used for a web-based chip breaking prediction system that appears to estimate chip breaking conditions with some success [8].

8.5 TOOL WEAR DURING CHIP BREAKING

As a tool wears, a crater forms on its face. The shape of this crater conforms to the shape of the curled chip, and thus the presence of a chip breaker will affect the wear process. Conversely, as a crater forms on a plain tool, the crater can eventually act as a chip breaker. Figure 8.13 shows the results of some experiments [14] where the progress of tool wear was studied for three tools under identical cutting conditions. In each test, the shape of the worn face of the tool was observed at intervals of 4 min (240 s).

In the first test (Figure 8.13a) a plain tool was used. Initially the chip was a long ribbon chip that became snarled. Gradually a crater formed that caused the chip to curl, and after only 8 min (480 s) this crater started to act as a chip breaker.

In the second test (Figure 8.13b) an attached obstruction-type chip breaker was employed. It can be seen that in addition to controlling the radius of chip curvature and the breaking of the chips, this obstruction-type chip breaker caused the rate of wear on the tool face to decrease substantially. However, as wear progressed, the radius of chip curvature decreased, until after 11 min (660 s) the chips became over broken.

Finally, in the third test (Figure 8.13c) a groove-type chip breaker was employed. Again after about 11 min (660 s) chip breaking became unsatisfactory, and the chips were over broken because of the reduced radius of chip curvature. In this test the rate of wear was comparable to that of the plain tool but the effect of the groove-type chip breaker was to eliminate the initial period where no chip breaking occurred.

It appears from these tests that the obstruction-type chip breaker causes a reduced rate of wear on the tool face. However, since satisfactory chip breaking occurred for the first 11 min (660 s) only, this amount of time became the effective tool life, even though a wear land of only 0.01 in. (0.25 mm) in width had developed on the tool flank.

It can be concluded that the effective life of a cutting tool can be shortened considerably using a chip breaker, simply because the chip-breaking action may become ineffective before the tool itself is significantly worn.

PROBLEMS

1. A single-point cutting tool is fitted with an integral obstruction-type chip breaker and is provided with positive, working major cutting-edge inclination so that a chip of helical form is produced. The chip-breaker

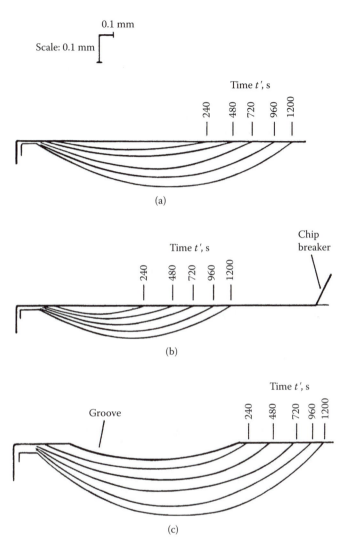

FIGURE 8.13 History of crater wear during turning for a work material of AISI 1045 hot-rolled steel and a tool material of carbide (Kennametal-KH4), where the cutting speed is 550 ft/min (2.8 m/s), the feed is 0.014 in. (0.356 mm), and the major-cutting-edge angle is 75 degrees. (a) Plain tool; (b) attached obstruction-type chip breaker; (c) groove-type chip breaker. (After Reinhart, L.E. and Boothroyd, G. Proc. 1st North American Metal-working Research Conf., Hamilton, Ontario, Vol. 2, 13, May 1973.)

height is 0.5 mm and the chip-breaker distance is 1.5 mm. The unde-formed chip thickness is 0.05 mm and the cutting ratio is 0.2. If the chip width is 2 mm, estimate the minimum working major cutting-edge inclination that will ensure that the successive chip coils will not overlap. (Assume that the chip-tool contact length is equal to the chip thickness.)

2. An integral obstruction-type chip breaker has a chip-breaker height of 1 mm and a chip-breaker distance of 4 mm. It is found that the chip is broken satisfactorily when the chip thickness is 0.8 mm. What chip-breaker distance should be provided for an attached obstruction-type chip breaker to give the same performance as the integral obstruction-type chip breaker if the chip breaker height is 2 mm and the chip-breaker wedge angle is 45 degrees? (Assume that the chip-tool contact length is equal to the chip thickness.)

REFERENCES

1. International Standards Organization, Tool Life Testing with Single-Point Turning Tools, ISO 3685, 1993.
2. Hendriksen, E.K., Chip Breaking — A Study of Three Dimensional Chip Flow, ASME Paper 53-S-9.
3. Okushima, K., Hoshi, T., and Fujinawa, T., On the Behavior of Chip in Steel Cutting, Part 2, In the Case of the Parallel Type Chip Breaker, *Bull. Japanese Soc. Mech. Eng.*, Vol. 3, 10, 199, 1960.
4. Nakayama, K., A Study of Chip Breaker, *Bull. Japanese Soc. Mech. Eng.*, Vol. 5, 17, 142, 1962.
5. Subramanian, T.L. and Bhattacharyya, A., Mechanics of Chip Breakers, *Int. J. Prod. Res.,* Vol. 4, 1, 1965.
6. Jawahir, I.S. and van Lutterfeld, C.A., Recent Developments in Chip Control Research and Applications, *Ann. CIRP*, Vol. 42, 2, 659–693, 1993.
7. Jawahir, I.S., On the Controllability of Chip Breaking Cycles and Modes of Chip Breaking in Metal Machining, *Ann. CIRP*, Vol. 39, 1, 47–51, 1990.
8. Zhou, L., Rong, Y., Li, Z., and Yang, J.A., Development of Web-based Machining Chip Breaking Prediction Systems, *Int. J. Adv. Manuf. Technol.*, Vol. 22, 336–343, 2003.
9. Andreasen, J.L. and De Chiffre, L., An Automatic System for Elaboration of Chip Breaking Diagrams, *Ann. CIRP*, Vol. 47, 1, 35–40, 1998.
10. Trim, A.R. and Boothroyd, G., Action of Obstruction Type Chip Former, *Int. J. Prod. Res.,* Vol. 6, 3, 227, 1968.
11. Creveling, J.M., Jorden, T.F., and Thomsen, E.G., Some Studies of Angle Relationships in Metal Cutting, *Trans. ASME,* Vol. 79, 1, 127–137, 1957.
12. Lutterfelt, C.A. van, Childs, T.H.C., Jawahir, I.S, Klocke, F., and Venuvinod, P.K., Present Situation and Future Trends in Modeling of Machining Operations, *Ann. CIRP*, Vol. 47 (2), 587–626, 1998.
13. Morusich, T.D., Brand, C.J., and Theile, J.D., A Methodology for Simulation of Chip Breakage in Turning Processes using an Orthogonal Finite Element Model, Proc. 5th CIRP Int. Workshop on Modeling of Machining Operations, 139–148, West Lafayette, Indiana, May 20–21, 2002.
14. Reinhart, L.E. and Boothroyd, G., Effect of Chip Forming Devices on Tool Wear in Metal Cutting, Proc. 1st North American Metalworking Research Conf., Hamilton, Ontario, Vol. 2, 13, May 1973.

9 Machine Tool Vibrations

9.1 INTRODUCTION

The machine, cutting tool, and workpiece form a structural system having complicated dynamic characteristics. Under certain conditions, vibrations of the structural system may occur, and as with all types of machinery, these vibrations may be divided into three basic types:

1. Free or transient vibration: resulting from impulses transferred to the structure through its foundation, from rapid reversals of reciprocating masses such as machine tables, or from the initial engagement of cutting tools. The structure is deflected and oscillates in its natural modes of vibration until the damping present in the structure causes the motion to die away.

2. Forced vibrations: resulting from periodic forces within the system such as unbalanced rotating masses or the intermittent engagement of multi-tooth cutters (milling), or transmitted through the foundations from nearby machinery. The machine tool will oscillate at the forcing frequency, and if this frequency corresponds to one of the natural frequencies of the structure, the machine will resonate in the corresponding natural mode of vibration.

3. Self-excited vibrations: usually resulting from a dynamic instability of the cutting process. This phenomenon is commonly referred to as *machine tool chatter* and, typically, if large tool-work engagements are attempted, oscillations suddenly build up in the structure, effectively limiting metal removal rates. The structure again oscillates in one of its natural modes of vibration.

It is important to limit vibrations of the machine tool structure as their presence results in poor surface finish, cutting-edge damage, and irritating noise. The causes and control of free and forced vibrations are generally well understood, and the sources of vibration can be removed or avoided during operation of the machine. Chatter vibrations are less easily controlled and metal removal rates are frequently limited because the operator must stop the machine to change the cutting conditions, which often means reducing the depth of cut or feed rate. The causes of machine tool chatter are generally well understood, and the vibrations occur because of the basic interaction between the cutting process and the machine tool structure [1–3].

9.2 FORCED VIBRATIONS

A machine structure that is subjected to a periodic force will vibrate at the forcing frequency. Several basic results can be illustrated by first considering a single-degree-of-freedom system.

9.2.1 SINGLE-DEGREE-OF-FREEDOM SYSTEM

For the one-degree-of-freedom system shown in Figure 9.1, the equation of motion is

$$m_e \frac{d^2 x}{dt'^2} + c_d \frac{dx}{dt'} + S_e x = F_{max} \cos \omega_f t' \qquad (9.1)$$

where

x = displacement
t' = time
m_e = equivalent mass
c_d = damping force per unit velocity (viscous damping constant)
S_e = restoring force per unit displacement (spring stiffness)
F_{max} = peak value of the external harmonic force
ω_f = angular frequency of the external harmonic force

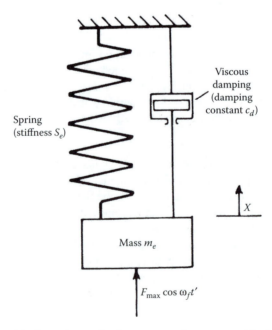

FIGURE 9.1 Model of one-degree-freedom mass-spring system with viscous damping.

The steady-state vibration of this system is given by

$$x = \frac{F_0 \cos\left(\omega_f t' - \phi_f\right)}{\sqrt{4\xi^2\omega_f^2 - \left(\omega_n^2 - \omega_f^2\right)^2}}$$ (9.2)

where

$F_0 = F_{max}/m_e$
ω_n = the natural angular frequency
ξ = the damping ratio

Equation 9.2 represents a motion of angular frequency ω_f at an amplitude given by:

$$a_0 = \frac{F_0}{\sqrt{4\xi^2\omega_f^2 - \left(\omega_n^2 - \omega_f^2\right)^2}}$$ (9.3)

and lagging the disturbing force by a phase angle ϕ_f where

$$\phi_f = \tan^{-1}\frac{2\xi\omega_n}{\left(\omega_n^2 - \omega_f^2\right)}$$ (9.4)

Resonance occurs when ω_f equals ω_n and the amplitude at resonance is $F_0/2\xi\omega_n$. These results are shown in Figure 9.2, where Equation 9.3 and Equation 9.4 are plotted for various values of the damping ratio.

It can be seen from these results that to minimize the amplitude of vibration of a damped mass-spring system, the damping should be as large as possible and the natural frequency of the system should be significantly less than the frequency of the disturbing force.

Forced vibrations in machine tools are most often caused by cyclic variations in the cutting forces. Such variations will occur in side or face milling, for example, where the frequency of the forced vibration equals the product of the tool rotational frequency and the number of teeth on the tool.

Figure 9.3 shows the variation in spindle torque during a slab-milling operation for cutters having various numbers of teeth. It can be seen that as the number of teeth increases, the frequency of the torque variations increases and the peak torque decreases.

The machine tool designer will be able to estimate from data such as that presented in Figure 9.3 the magnitudes and frequencies of the disturbing forces likely to occur during the operation of the machine tool. With this information, the designer will be better able to design the machine so that the natural frequencies of the various parts of the structure do not approach the forcing frequencies.

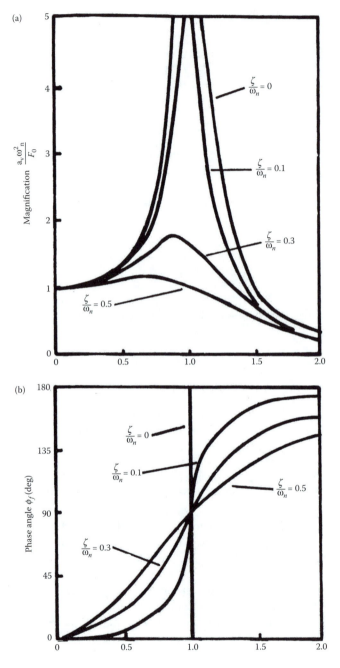

FIGURE 9.2 Response curves for the system in Figure 9.1, where a_v = amplitude of vibration, F_0 = peak value of disturbing force divided by mass, ω_f = angular forcing frequency, ω_n = angular natural frequency of the system, ξ = damping ratio, and ϕ_f is the phase angle between the displacement and the disturbing force.

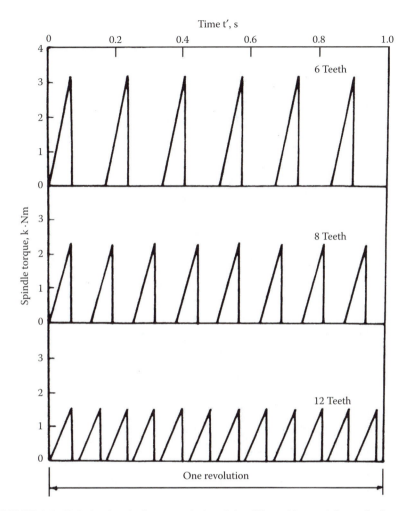

FIGURE 9.3 Variation in spindle torque during slab milling with a straight-toothed cutter, where the cutter diameter d_t = 100 mm, the work engagement (depth of cut) a_e = 5 mm, the back engagement (width of cut) a_p = 200 mm, the rotational frequency of the cutter n_t = 1 s^{-1} and feed speed v_f = 17 mm/s.

The periodic forces in Figure 9.3 are not sinusoidal and have significant harmonics at multiples of the tooth engagement frequencies. These harmonics will also have to be considered and avoided if resonances are not to occur.

The response curves for the single-degree-of-freedom system shown in Figure 9.2 can be plotted in polar form on a single diagram (Figure 9.4). As can be seen, the resulting curve is approximately circular, with the resonant frequency occurring when the phase lag of the displacement is around 90 degrees. Such a frequency response diagram is referred to as the *harmonic response locus* or

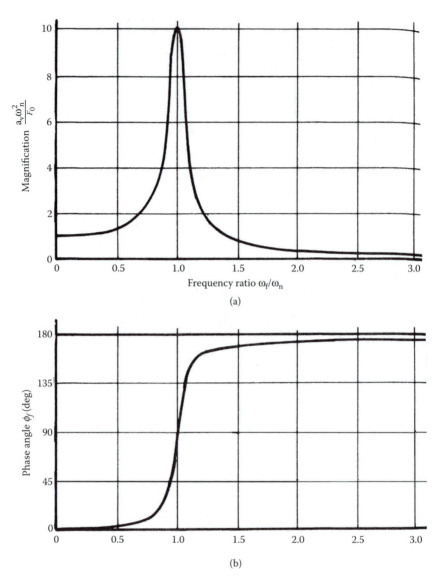

FIGURE 9.4 Relation between resonance and phase curves and harmonic response locus. (a) Resonance curve; (b) phase curve; (c) harmonic response locus of a system with one-degree-of-freedom with viscous damping corresponding to $\xi = 0.05$.

harmonic receptance of the system. This representation of the frequency response is useful when considering complex structures with many degrees of freedom and in particular for the analysis of machine tool chatter [1–3].

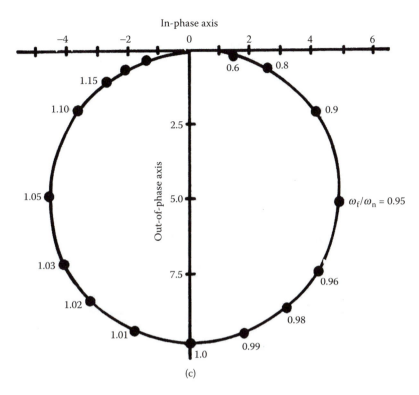

FIGURE 9.4 (continued).

9.2.2 COMPLEX STRUCTURES

It is possible to simulate the behavior of complex structures by considering numerous masses, springs, and damping elements connected together, although in practice choosing the appropriate values for each element is difficult. Structural analysis programs based on the finite element method can also be used to predict resonant frequencies and associated mode shapes for complex structures [4]. However, the allocation of appropriate damping for each structural element throughout the structure is difficult.

If the structure is excited with a periodic force, it will behave in a similar manner to the single-degree-of-freedom system considered in section 9.2.1. As the frequency of the exciting force is varied, the structure will resonate with large-amplitude vibrations as before but with a number of important differences:

1. A complex structure will exhibit several resonances or natural frequencies, and the frequency response curve will have several peaks corresponding to each resonance. For example, Figure 9.5 shows the resonance curves for a machine tool structure, and Figure 9.6 shows

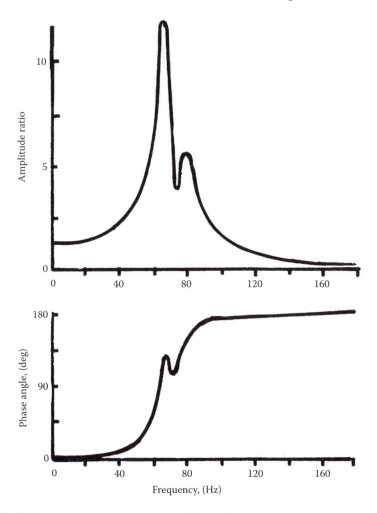

FIGURE 9.5 Frequency response of a machine tool structure.

the corresponding harmonic response locus. Each resonance produces a loop of approximately circular form in the curve.

2. At each resonance the contribution of the various elements of the structure to the overall response will vary. Some parts will be moving with large amplitudes at some frequencies, but hardly moving at all for others. Each resonance will have a corresponding mode of vibration or mode shape that the structure adopts. The modes of vibration can be represented graphically by considering the motion of various points on the structure and indicating the amplitude and relative phase of the motion of these points on a picture of the structure. Figure 9.7 shows some of the mode shapes for a horizontal-milling machine structure [5]. The important modes of vibration are those that result in relative

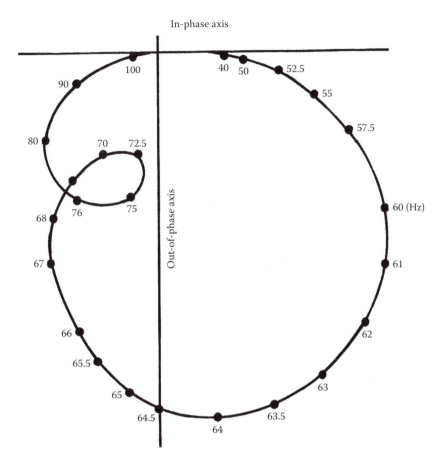

In-phase axis

Out-of-phase axis

90 100 40 50 52.5 55 57.5 80 70 72.5 68 76 75 67 66 65.5 65 64.5 64 63.5 63 62 61 60 (Hz)

FIGURE 9.6 Polar frequency response locus corresponding to Figure 9.5.

displacements of the tool and workpiece, as these will have a detrimental effect on the surface finish of the workpiece. If two modes of vibration occur at frequencies close to each other, these are closely coupled modes of vibration.

3. A machine tool is a three-dimensional structure. Consequently the points of application and the direction of the exciting force and the measured vibration are important. For the single-degree-of-freedom system discussed in section 9.2.1, the direction and point of application of the exciting force, F_{max}, corresponded to the measured vibration, x. The resulting frequency response is referred to as a *direct response*, since these two directions correspond. However, in a complex three-dimensional structure, it is obviously possible to apply the exciting force at a particular point in one direction and then to measure the displacement at another point in a different direction. The resulting frequency response curve will again exhibit resonances as before and

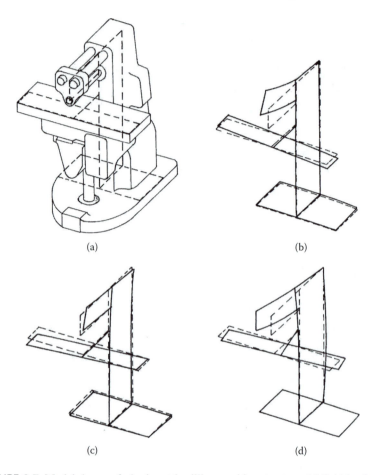

(a)   (b)

(c)   (d)

FIGURE 9.7 Modal shapes of a horizontal-milling machine structure. (a) Grid for describing modal shapes; (b) distortion for 97-Hz mode; (c) distortion for 107-Hz mode; (d) distortion for 112-Hz mode.

is referred to as a *cross-frequency response*. This is important because forces applied at certain points and directions may result in only some of the modes of vibrations being excited.

9.3 SELF-EXCITED VIBRATIONS (CHATTER)

9.3.1 Interaction of the Cutting Process and Machine Structure

The basic cause of chatter is the dynamic interaction of the cutting process and the machine tool structure. A schematic representation of an orthogonal machining

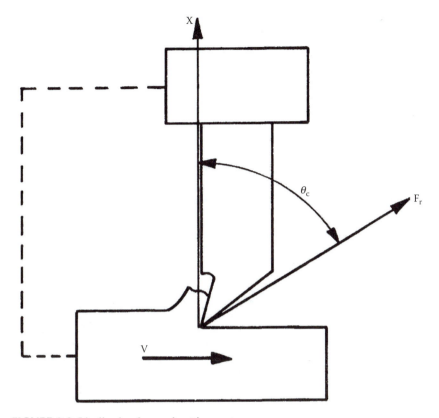

FIGURE 9.8 Idealized orthogonal cutting system.

operation is presented in Figure 9.8, which depicts a cutting tool and workpiece connected by an elastic structure. During cutting, a force F_r is generated between the tool and workpiece, which acts at an angle to the cut surface. The magnitude of this cutting force depends largely on the tool-work engagement and depth of cut. The cutting force strains the structure elastically and can cause a relative displacement of the tool and workpiece, which alters the tool-work engagement (undeformed chip thickness). Thus, during cutting the machine tool behaves as a closed-loop feedback system, as shown in Figure 9.9. A disturbance in the cutting process (e.g., because of a hard spot in the work material) will cause a deflection of the structure, which may alter the undeformed chip thickness, in turn altering the cutting force. There is a possibility for the initial vibration to be self-sustaining (unstable) and build up, with the machine oscillating in one of its natural modes of vibration.

The closed-loop cutting-process structural system will become unstable if a mechanism exists for transferring energy into the structure to sustain the vibration. Two major effects can cause instability: the regenerative effect, which is the dominant cause, and the mode-coupling effect.

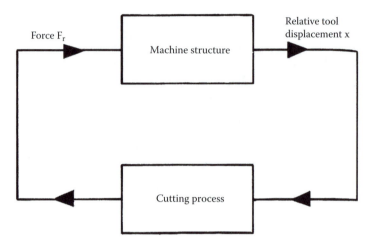

FIGURE 9.9 Closed-loop representation of the machine tool structure and cutting process system.

9.3.1.1 Regenerative Instability

Regenerative instability can occur when successive passes of the cutting tool overlap, that is, when the tool at any instant is removing a surface that was cut on the previous pass of the tool or revolution of the workpiece (Figure 9.10). A disturbance in the cutting process will cause a wavy surface to be generated on the workpiece. During the next pass of the tool the undeformed chip thickness will depend on the current motion of the tool and the wave cut on the surface during the previous pass. Depending on the phase between these waves, the force variation may increase after successive passes of the tool and the vibration will build up, until limited by some nonlinearity such as the tool leaving the workpiece during part of the vibration cycle.

9.3.1.2 Mode-Coupling Instability

Mode-coupling instability can occur when successive passes of the tool do not overlap, such as in screw cutting, and results from a particular motion of the tool relative to the workpiece in the presence of closely coupled modes of vibration of the structure. The characteristics of the structure can be such that a free vibration of the structure causes the tool to follow an elliptical path relative to the workpiece when disturbed, as illustrated in Figure 9.11. The cutting force F_r, does work during this periodic motion when F_r opposes the motion. It will put work into the system during half of the cycle and take work out during the other half. Over the lower part of the elliptical path, the undeformed chip thickness is greater and consequently the value of F_r is higher. Thus, more work may be put into the system over the lower part of the cycle elliptical path than is taken out in the upper part, resulting in a net input of energy and causing the amplitude of vibration to increase.

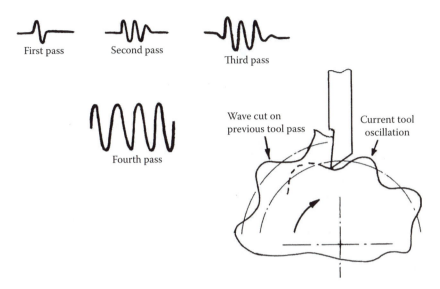

First pass

Second pass

Third pass

Fourth pass

Wave cut on previous tool pass

Current tool oscillation

FIGURE 9.10 Regenerative instability.

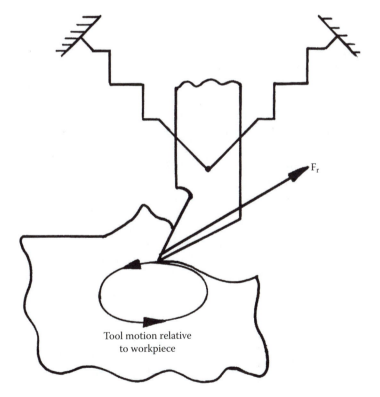

F_r

Tool motion relative to workpiece

FIGURE 9.11 Mode coupling instability.

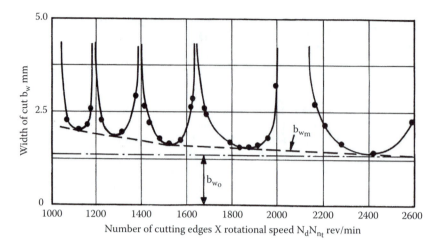

FIGURE 9.12 Experimental stability chart of a horizontal milling machine, feed speed = 0.79 mm/rev and depth of cut = 5.24 mm. (Adapted from Sadek, M.M. and Tobias, S.A., *Proc. Inst. Mech. Eng. (London)*, Vol. 185, 319, 1971.)

9.3.2 STABILITY CHARTS

The stability of a machine tool can be represented graphically in the form of a chart, which depicts the effect on stability of the width of cut, drill diameter, and so on, and the rotational speed of the tool or workpiece. A typical experimental stability chart for a horizontal-milling machine is shown in Figure 9.12, with the unstable ranges shown as lobed areas. This chart was obtained by cutting at different widths of cut and rotational speeds of the cutter and noting the conditions at which instability occurs and vibrations build up. This type of stability chart, with the unstable regions forming lobes, is typical of that obtained when chatter is caused by the regenerative effect. A number of important features should be noted:

1. The lobes lie on an envelope forming a boundary below which the machine is stable for all rotational speeds.
2. The stability of the machine increases at low speeds.
3. The envelope b_{w_0} of the unstable regions has a horizontal asymptote, b_{w_0} which represents a width of cut below which the machine is always stable.
4. A judicious selection of cutting speed can increase the chatter-free width of cut, that is, points lying between the lobes of the stability chart. Although not important for cutting at conventional speeds, the presence of regions of increased stability at higher speeds has significant relevance with the current trend towards a greater use of high-speed machining.

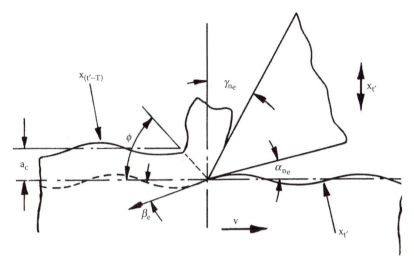

FIGURE 9.13 Effect of oscillations of the tool in the normal direction during orthogonal cutting.

9.3.3 ANALYSIS OF MACHINE TOOL CHATTER

An analysis of machine tool instability is based on the condition at the threshold of stability, assumed to be the point at which a disturbance in the cutting process neither grows nor decays in magnitude, that is, a small amplitude harmonic vibration persists. Considering the closed-loop system of Figure 9.9, this requires a representation of the response of the machine structure and the response of the cutting process itself to such vibrations, in other words:

1. A transfer function between a harmonic force input and a relative displacement output of the structure
2. A transfer function for the cutting process between a vibrational displacement input and a force output

For simplicity, the analysis will consider initially an orthogonal turning operation, as indicated in Figure 9.13.

9.3.3.1 Cutting Process

Under steady-state conditions the cutting force F_r remains reasonably constant and acts at an angle $_c$ to the normal to the cut surface (Figure 9.8). When the tool oscillates relative to the workpiece, some of the cutting-process parameters (undeformed chip thickness, relative tool geometry, etc.) undergo cyclic variations and a cyclic force variation dF_r is superimposed on the mean force F_r. The magnitude of dF_r is dependent on the nature of the tool motion and the way it affects the parameters governing the cutting force. The cutting-force variation is determined

TABLE 9.1
Dynamic Variation of Cutting Parameters Resulting from the Normal Motion x_t'

Parameter	Dynamic Value
In-phase element:	
Uncut chip thickness a_c	$da_c = x_{t'} - x_{(t'-T)}$
Out-of-phase elements:	
Cutting speed v	Negligible change due to $\dot{x}_{t'}$ in the normal direction
Tool rake angle γ_{n_e}	$d\gamma_{n_e} = -\dot{x}_{t'}/v$
Tool clearance angle α_{n_e}	$d\alpha_{n_e} = \dot{x}_{t'}/v$
Instantaneous cutting direction, β_e	$d\beta_{n_e} = \dot{x}_{t'}/v$
Free surface slope δ_e	$d\delta_e = \dot{x}_{t'}/v - \dot{x}(t' - T)/v$
Shear angle ϕ	$d\phi = f(\dot{x}_{t'})/v$

Source: Adapted from Knight, W.A., Das, M.K., and Sadek, M.M., A Critical Assessment of Cutting Force Models in the Analysis of Machine-Tool Instability, *Proc. 11th Int.,* MTDR Conf., Pergamon Press, New York, 1970..

mainly by relative oscillations normal to the cut surface, and this case is illustrated in Figure 9.13. Oscillations parallel to the cut surface and torsional oscillations of the drive system cause changes in the cutting speed, and these have been shown to have little effect on the onset of instability [6], except for processes in which the build up of vibrations is relatively slow such as in grinding [7]. If the situation shown in Figure 9.13 is considered, it can be seen that the depth of cut is dependent on both the current motion of the tool x, and the wave left on the surface during the previous tool pass $x_{(t'-T)}$. Note that x is the displacement of the tool perpendicular to the mean cutting direction, t is time, and T is the time for one pass of the tool or revolution of the workpiece. The presence of a vibration normal to the mean cutting direction causes variations in a number of the parameters known to influence the cutting force, as indicated in Table 9.1 [8]. The actual influence of each of these parameter variations on the cutting force is difficult to determine. Several analyses of the dynamic cutting process have been attempted [9–11] and these are mostly modifications of the basic shear-plane model discussed in Chapter 2. However, examination of the entries in Table 9.1 indicates that the cutting-force variation should be divided into a portion that is in phase with the displacement x_t and another portion that is out of phase with x_t, dependent on the derivative of x_t.

The in-phase force variation can be expressed as

$$dF_1 = k_1 \left(x_{t'} - x_{(t'-T)} \right) \tag{9.5}$$

where k_1 is the chip-thickness coefficient or cutting stiffness, which is proportional to the width of cut, drill diameter, and so on. This expression applies when successive passes of the cutting tool completely overlap. For cases when successive passes do not overlap completely, an overlap factor μ_c can be introduced, which can have a value of from zero to one, and Equation 9.5 becomes

$$dF_1 = k_1\left(x_{t'} - \mu_c x_{(t'-T)}\right) \tag{9.6}$$

Several of the parameters in Table 9.1 depend on the derivative of the tool motion and the cutting speed, but the individual contribution of each parameter is difficult to determine. However the effect of all these parameter variations can be combined into one out-of-phase force variation of the following form:

$$dF_2 = k_2 \frac{2\pi}{\Omega} \dot{x}_{t'} \tag{9.7}$$

where k_2 is a constant, and Ω is the rotational speed of the workpiece (rad/s). The total force variation due to the tool oscillation relative to the workpiece is

$$dF_r = dF_1 + dF_2 = k_1\left(x_{t'} - \mu_c x_{(t'-T)}\right) + k_2 \frac{2\pi}{\Omega} \dot{x}_{t'} \tag{9.8}$$

which can be used to analyze the conditions at the limit of stability.

9.3.3.2 Machine Structure

At the threshold of stability, the transfer function for the structure is its frequency response. Since for most cutting operations, the cutting force acts in a direction different from the normal to the cut surface, the relevant frequency response is the cross response between these two directions. This may be represented as a polar diagram (Figure 9.14) and is called the *operative receptance*. This polar diagram represents the in-phase and out-of-phase components of the response at each frequency and can be expressed as a complex number. Thus, at each frequency

$$x = \left(a_h + ib_h\right)F \tag{9.9}$$

where $i = \sqrt{-1}$, x and F are varying harmonically, and a_h and b_h depend on the magnitude of the vibrational frequency ω_f.

The required operative receptance depends on the cutting process and the particular cutting conditions (depth of cut, tool rake angle, etc.). Figure 9.15 shows the cutting force directions and normal to the cut surface for several machining processes. For multipoint operations such as milling, a mean direction for the force F, must be used because the number of teeth engaged with the workpiece varies as the tool rotates. Also, the effective normal direction to the

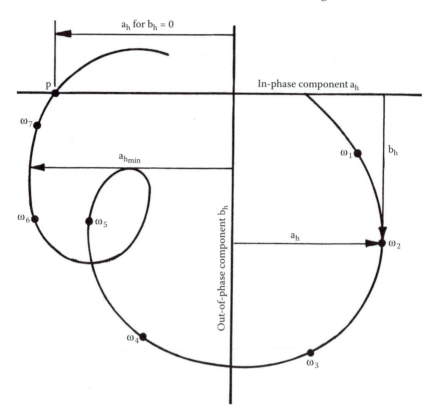

FIGURE 9.14 Cross-frequency response locus between cutting force direction and the normal to the cut surface (operative receptance locus).

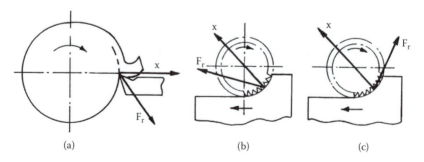

FIGURE 9.15 Forcing and response directions in some machining processes. (a) Orthogonal turning; (b) up-milling; (c) climb-milling.

cut surface must be known and is assumed to pass through the center of the arc joining the points of contact between the tool and workpiece [12]. As a result the direction of the normal to the cut surface is different for different values of the depth of cut in slab milling.

For the analysis of machine tool instability, the cross receptance between the applicable cutting force direction and the normal to the cut surface must be known, for conditions similar to those existing at the threshold of stability. Determination of this frequency response will be discussed later.

9.3.3.3 Stability Analysis

At the threshold of stability $x = dF_r (a_h + ib_h)$ or

$$x = -\left[k_1 \left(x_{t'} - \mu x_{(t'-T)} \right) + k_2 \frac{2\pi}{\Omega} \dot{x}_{t'} \right] \left(a_h + ib_h \right) \tag{9.10}$$

with the negative sign introduced because an increase in the displacement, x, causes a decrease in the cutting force, F_r.

At the limit of stability x is harmonic, that is, $x = Xe^{i\omega t'}$, and therefore in Equation 9.10

$$Xe^{iwt'} = -\left[k_1 \left(Xe^{i\omega t'} - \mu_c Xe^{i\omega(t'-T)} \right) + k_2 \frac{2\pi}{\Omega} i\omega Xe^{i\omega t'} \right] \left(a_h + ib_h \right)$$

or

$$1 = -(a_h + ib_h) \left[k_1 \left(1 - \mu_c Xe^{-i\omega T} \right) - k_2 \frac{2\pi}{\Omega} i\omega \right] \tag{9.11}$$

However, $e^{-i\omega T} = \cos(\omega T) - i \sin(\omega T)$ and, therefore, expanding and equating real and imaginary parts

$$1 = b_h \left(k_1 \mu_c \sin \omega T + k_2 \frac{2\pi}{\Omega} \right) - a_h k_1 \left(1 - \mu_c \cos \omega T \right)$$

and

$$0 = a_h \left(k_1 \mu_c \sin \omega T + k_2 \frac{2\pi}{\Omega} \right) + b_h k_1 \left(1 - \mu_c \cos \omega T \right)$$

which after replacing T by β_c and k_2/k_1 by C_c, assumed a constant these become

$$k_1 = \frac{1}{b_h \mu_c \sin \beta_c + \beta_c C_c b_h - a_h \left(1 - \mu_c \cos \beta_c \right)} \tag{9.12}$$

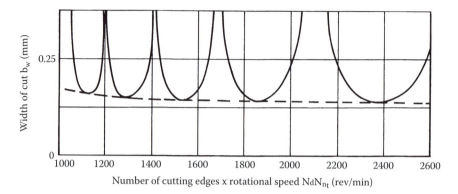

FIGURE 9.16 Predicted stability chart corresponding to Figure 9.12, $C_c = 0.001$ (From Sadek, M.M. and Tobias, S.A., *Proc. Inst. Mech. Eng.* (*London*), Vol. 185, 319, 1971.)

and

$$0 = a_h \left(\mu_c \sin \beta_c + C_c \beta_c \right) + b_h \left(1 - \mu_c \cos \beta_c \right) \qquad (9.13)$$

These two equations define the limit of stability for a given frequency in terms of critical values of the cutting stiffness k_1 to which the width of cut is directly proportional.

Equation 9.12 and Equation 9.13 must be solved numerically to give a stability chart. The procedure for this is as follows:

1. For a particular frequency, obtain the values of a_h and b_h from the frequency response locus (Figure 9.14).
2. From Equation 9.13, obtain the solutions for the phase angle β_c. There are a large number of these solutions and for each one $\beta_c = \omega T = 60/n_w$ where n_w is the rotational frequency of the workpiece.
3. Substituting each solution for β_c into Equation 9.12 gives a corresponding critical value of k_1.
4. Repeat these calculations for other values of the frequency ω.

The stability chart (k_1 versus n_w) may then be plotted and will be of the form shown in Figure 9.16.

9.3.3.4 Modification for Multiedge Cutting Operations

The preceding equations were derived for a single-point turning operation but can be readily modified for multipoint cutting tools by replacing k_1 by $N_d k_1$ and setting $\beta_c = \omega T/N$, where N is the number of cutting edges, and N_d is the average

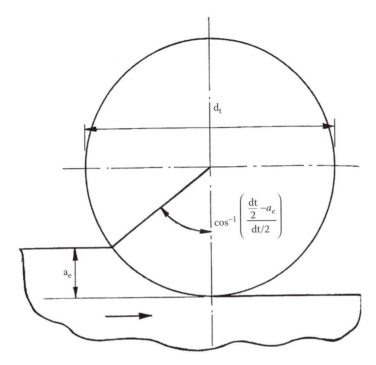

FIGURE 9.17 Arc of contact during slab milling, N_d = average number of teeth in contact, N = number of teeth on the cutter, d_t = tool diameter, and a_e = the work engagement (depth of cut).

number contacting the workpiece at any time. Thus, for a drilling operation, $N = N_d = 2$. For a slab-milling operation, the value of N_d depends on the tool radius and the depth of cut a_e (Figure 9.17):

$$N_d = \frac{N}{2\pi}\cos^{-1}\frac{\left(d_t - a_e\right)}{d_t/2} \tag{9.14}$$

9.3.3.5 Special Cases

Two special cases of this analysis exist, from which some general conclusions can be drawn. First, a simplification can be made by assuming $C_c = 0$, $\mu_c = 1$, which means that the increased damping from the cutting process is ignored at low speeds and successive cuts by the tool completely overlap. In this case Equation 9.10 becomes

$$x_{t'} = -\left[k_1\left(x_{t'} - x_{(t'-T)}\right)\right]\left(a_h + ib_h\right) \tag{9.15}$$

or

$$\frac{x_{t'}}{x_{(t'-T)}} = \frac{k_1\left(a_h + ib_h\right)}{1 + k_1\left(a_h + ib_h\right)}$$

which has a magnitude of 1 at the threshold of stability and gives:

$$k_1 = \frac{1}{2a_h} \text{ or } \infty \tag{9.16}$$

Equation 9.13 becomes

$$a_h \sin\beta_c + b_h\left(1 - \cos\beta_c\right) = 0 \tag{9.17}$$

This result allows some important conclusions to be drawn, which can be shown to apply also to the more general case when k_2 and therefore C_c has a finite value.

1. Since the chip-thickness coefficient must be positive, instability can occur only for frequencies for which a_h is negative, that is, frequencies corresponding to the left-hand portion of the polar frequency response diagram (Figure 9.14).
2. The stability chart for this case is similar to Figure 9.16, except the increase in stability at low speeds is not present (Figure 9.18). The horizontal asymptote in both stability charts is the same.
3. The lowest value of k_1 corresponds to the negative value of a_h of greatest magnitude. This gives the lowest point on each lobe of the stability chart and defines the horizontal asymptote in the stability charts. This is important because if cutting conditions are chosen such that k_1 is below this asymptote, the machine will be stable for all cutting speeds. This result can form the basis for a comparative dynamic acceptance test for machine tools.

The second special case is where $C_c = 0$ and $\mu_c = 0$. For this condition Equation 9.12 and Equation 9.13 give $k_1 = -1/a_h$ and $b_h = 0$, which means that instability can occur only when a_h is negative and b_h is zero. This instability occurs at the frequency at which the locus crosses the negative in-phase axis (point F, Figure 9.14). This stability condition is independent of cutting speed; thus, the stability chart is a horizontal line at $k_1 = -1/a_h$.

In general, for mode-coupling instability, the operative response locus must cross the negative in-phase axis, and this can occur only with a cross-frequency response locus for a structure with closely coupled modes of vibration. Also, mode-coupling instability cannot occur for cutting processes for which the cutting

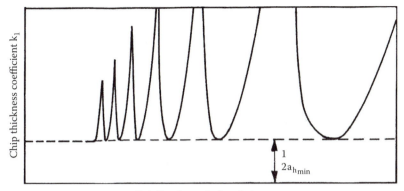

Rotational speed of workpiece, n_ω

FIGURE 9.18 Form of the stability chart when $C_c = 0$ and $\mu_c = 1$.

force direction corresponds to the normal to the cut surface, since the operative response locus is then a direct-frequency response, which will not cross the negative in-phase axis. This means, for example, that this type of instability is not possible for a drilling operation.

The value of k_1 and therefore the corresponding width of cut for mode-coupling instability is much greater than for regenerative instability. Thus, this type of instability will not be significant for conditions where successive passes of the cutting tool overlap each other.

9.4 DETERMINATION OF FREQUENCY RESPONSE LOCI

The analysis of machine tool stability requires knowledge of the cross-frequency response between the cutting-force direction and the normal to the cut surface. This response should be experimentally determined for the conditions prevailing at the threshold of stability, that is, when a small amplitude harmonic force exists between the tool and workpiece and a large mean cutting force is applied. In addition, the spindle of the machine should be rotating.

To determine the required cross-frequency response directly involves applying a harmonic force in the direction of the cutting force and measuring the response in the direction of the normal to the cut surface. A different setup would therefore be required for each possible direction of the cut force and direction of the normal to the cut surface. Fortunately, it is possible to deduce the cross response for any given direction of the applied force and direction of the measured response from experimentally measured responses in only a few directions. For example, consider the slab-milling process shown in Figure 9.19 [13]. To determine the cross-frequency response between the direction of force F_r, and the normal to the cut surface x, it is necessary to measure the direct-frequency response in two directions (u and v), preferably orthogonal to each other and the cross response between these two directions.

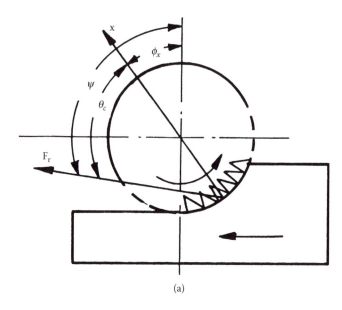

(a)

FIGURE 9.19 (a) Orientation of the cutting force F_r and effective normal to the cut surface x in up-milling; (b) orientation of two electromagnets u and v of the vibration exciter with the directions of F_r and x (After Sadek, M.M. and Tobias, S.A., *Proc. Inst. Mech. Eng.* (*London*), Vol. 185, 319, 1971.); (c) force components in the u and v directions; (d) displacements in the $u, v, R,$ and x directions.

The direct- and cross-receptance loci for the u and v directions are $R_{uu}, R_{vv}, R_{uv},$ and $R_{vu},$ respectively. By Maxwell's theorem of reciprocity, $R_{uv} = R_{vu}.$ Figure 9.19c shows the force diagram in slab milling. The forces in the v and u directions are

$$F_v = \frac{F \sin(\alpha + \psi)}{\sin 2\alpha} \qquad (9.18)$$

$$F_u = \frac{F \sin(\psi - \alpha)}{\sin 2\alpha} \qquad (9.19)$$

The responses (deflections) in the v and u directions due to the force F become:

$$\delta_v = R_{vv} F_v + R_{uv} F_{u_{vu}} \qquad (9.20)$$

and

$$\delta_u = R_{uu} F_u + R_{vu} F_v \qquad (9.21)$$

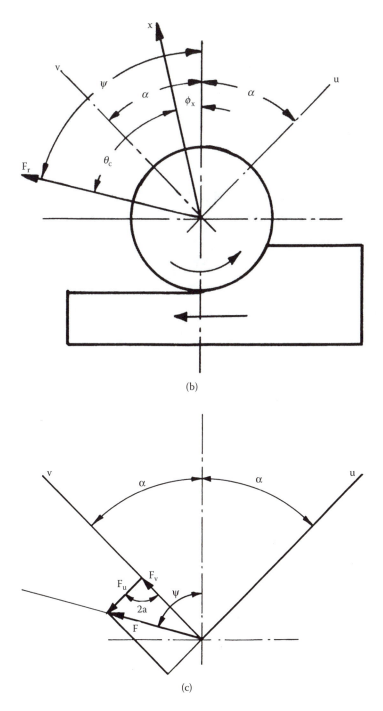

(b)

(c)

FIGURE 9.19 (continued).

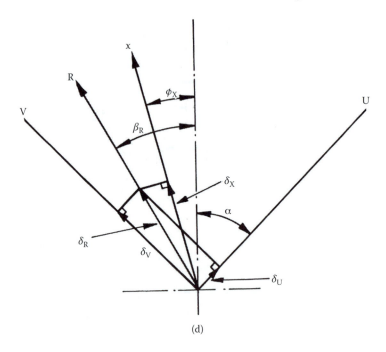

(d)

FIGURE 9.19 (continued).

In the experimental set-up, these deflections will be sensed by transducers mounted in the u and v directions and will arise because of a resultant displacement in the direction R (Figure 9.19d):

$$\delta_R = \frac{\delta_v}{\cos\left(\alpha - \beta_r\right)} = \frac{\delta_u}{\cos\left(\alpha + \beta_r\right)} \qquad (9.22)$$

The response sensed by a transducer in the direction x is given by

$$\delta_x = \delta_v \cos\alpha \cos\beta_r \qquad (9.23)$$

From Equation 9.22

$$\frac{\delta_v}{\cos\alpha\cos\beta_r + \sin\alpha\sin\beta_r} = \frac{\delta_u}{\cos\alpha\cos\beta_r - \sin\alpha\sin\beta_r} \qquad (9.24)$$

Substitution of Equation 9.18, Equation 9.19, Equation 9.20, Equation 9.21, and Equation 9.24 into Equation 9.23 gives:

$$\delta_x = \frac{F}{\sin^2 2\alpha} \left[\begin{array}{l} R_{uu}\sin\left(\psi + \alpha\right)\sin\left(\alpha - \phi_x\right) \\ +R_{uv}\left\{\sin\left(\psi + \alpha\right)\sin\left(\alpha - \phi_x\right) + \sin\left(\psi - \alpha\right)\sin\left(\alpha - \phi_x\right)\right\} \\ +R_{vv}\sin\left(\psi + \alpha\right)\sin\left(\alpha + \phi_x\right) \end{array} \right] \quad (9.25)$$

The cross receptance between the F and x directions is δ_x/F, and it can be determined from Equation 9.25.

In this way the operative receptance can be computed for any directions of F and x. For a three-dimensional cutting process such as face milling, similar equations can be developed, but direct and cross receptances for three directions must then be measured.

Measurement of the required frequency responses under appropriate conditions can be readily carried out using electromagnetic vibrators of the type shown in Figure 9.20. These exciters consist essentially of two U-shaped laminated electromagnets mounted on a base of nonmagnetic light alloy, which can be rigidly clamped either to the table of the milling machine or to the saddle of the lathe, with their forcing axes 30 or 45 degrees to an axis of symmetry. The magnet acts on a rotor that replaces the workpiece on a lathe or the tool on a horizontal-milling machine.

The relative vibration induced by the magnetic force between the magnet poles and the rotor (armature) can be measured by a capacitance transducer rigidly

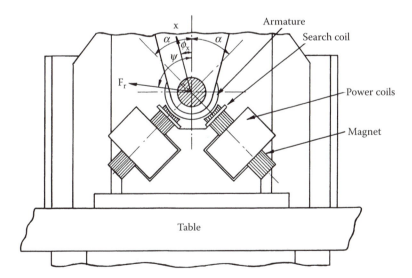

FIGURE 9.20 Electromagnetic exciter for milling machines. (From Sadek, M.M. and Tobias, S.A., *Proc. Inst. Mech. Eng. (London)*, Vol. 185, 319, 1971.)

mounted midway between the poles of each magnet, with its axis parallel to the forcing axis of the magnet. These transducers are also used for setting up the clearance between the poles and armature. Search coils placed around each arm of the U-shaped magnets, near their poles, are used for measuring the exciting-force component.

The main advantages of this form of vibration exciter are:

1. Low weight and high rigidity to minimize the effect of the mass and flexibility of the exciter on the response of the machine tool structure
2. Ability to provide a harmonic exciting force superimposed on a constant load simulating the mean cutting thrust
3. Ability to excite the machine with the spindle rotating without physical contact between the vibrator and the rotating member

The conditions existing at the threshold of stability are well simulated by these exciters. The necessary direct- and cross-frequency response data can be obtained directly in one setting by vibrating each set of magnets in turn. The required frequency responses can be determined by other forms of instrumentation, including impulse response measurements processed through fast Fourier analyzers or similar.

9.5   DYNAMIC ACCEPTANCE TESTS FOR MACHINE TOOLS

The analysis of regenerative instability described earlier showed that the horizontal asymptote to the stability charts is given by Equation 9.16 and depends only on the maximum negative in-phase magnitude of the operative receptance. This receptance can form the basis for a coefficient of merit for the dynamic assessment of machine tools, which is particularly useful for the comparison of the performance of similar machines [12]. Thus, from Figure 9.14, the coefficient of merit can be defined as

$$COM = \frac{1}{2} a_{h_{\min}} \tag{9.26}$$

To compare the performance of alternative machines of a similar type, it is necessary to determine the operative frequency response for each machine as described earlier and then to compute the coefficient of merit for each machine. An advantage of this approach is that machines can be compared for a wide range of cutting conditions from a single set of frequency response loci. For example, in horizontal milling the relative stability can be determined for different depths of cut and directions of cutting. This is achieved by computing the operative receptances for the appropriate cutting force and normal to the cut surface directions and determining the coefficient of merit. Figure 9.21 shows values of the coefficient of merit determined for a typical horizontal-milling machine for different

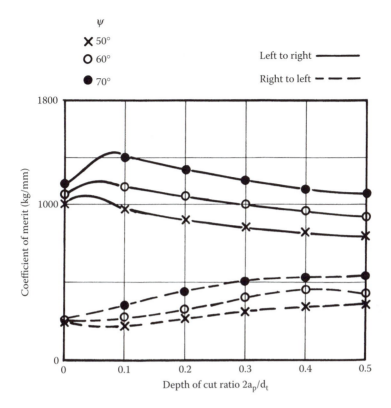

FIGURE 9.21 Variation of coefficients of merit with different cutting conditions for horizontal milling. (From Sadek, M.M. and Tobias, S.A., *Proc. Inst. Mech. Eng.* (*London*), Vol. 185, 319, 1971.)

depths of cut and when the machine table is fed in opposite directions. This result shows that this particular machine is much more stable when the table is fed from left to right. In fact, the width of cut for unconditional stability is increased by more than a factor of 2.

9.6 IMPROVING MACHINE TOOL STABILITY

During the operation of a machine tool, particularly under numerical control, it is not always possible to avoid the conditions at which chatter vibrations may build up. A means of improving the stability of a machine may therefore be necessary, and several possibilities are available.

9.6.1 STRUCTURAL ALTERATIONS

More damping and/or stiffness may be added to the structure to effectively reduce the magnitude of $a_{h_{\min}}$ on the operative response locus (Figure 9.14). Only those modes of vibration causing significant relative separation of the tool and work-piece need be considered. However, determination of where to apply additional damping most effectively may require careful analysis of the structure.

It may be possible, by redesign of portions of the structure, to alter the principal directions of the major modes of vibration so they do not coincide with the cutting force and/or the normal to the cut surface directions. In machines such as vertical-milling machines, the direction of the normal to the cut surface and cutting force changes continually during operation of the machine, so that this approach may not be effective. However, in lathes, these directions remain rela-tively constant. Redesign of the tailstock of an engine lathe (Figure 9.22) to alter the relationship between the principal directions of the modes of vibration and the normal to the cut surface has been found to result in improvements in stability [14].

9.6.2 Vibration Absorbers

A vibration absorber is a mass-spring system added to the structure and tuned so that its natural frequency coincides with that of the major vibration mode, thus reducing the effect of vibration in the major mode. Vibration absorbers can be considered for machine tools, but may be difficult to use if several major modes of vibration exist. For machines with a single dominant mode of vibration, these devices may be successful. Figure 9.23 shows the effect of adding an absorber to the overarm of a horizontal-milling machine [15]. This resulted in considerable improvement in stability for the particular cutting conditions used.

9.6.3 Modification of the Regenerative Effect

Regenerative instability occurs when the phasing between successive waves cut on the surface of the workpiece by a vibrating tool reaches critical values. Consequently, a means of disturbing this phasing can improve stability. This can be done in several ways, as discussed in the following sections.

9.6.3.1 Variable-Pitch Milling Cutters

Several investigations of the use of milling cutters with variable pitch between successive teeth have been made [16–18], and this technique can lead to improve-ments in stability. Figure 9.24 shows the effect on stability charts of irregular tooth pitch for slab milling [16]. Such cutters result in different phasing between the waves cut by successive teeth on the cutter and can destroy the basic regen-erative effect.

Another way to achieve irregular pitch for helical slab milling cutters is to use different helix angles on successive teeth. Figure 9.25 shows improvements in stability obtained using such cutters [18].

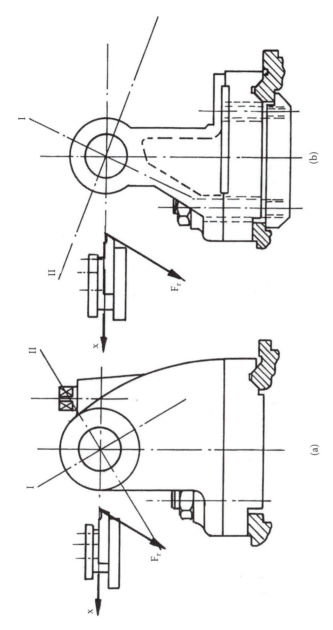

FIGURE 9.22 (a) Redesign of a lathe tailstock to reorient principal directions of vibration; (b) improvements in the width of cut for instability. (Adapted from Tlusty, J., *Proc. 7th Int. MTDR Conf.*, Pergamon Press, New York, 1966, 13.)

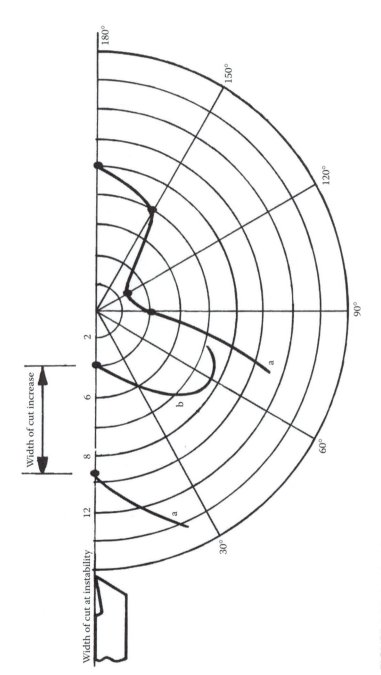

FIGURE 9.22 (continued).

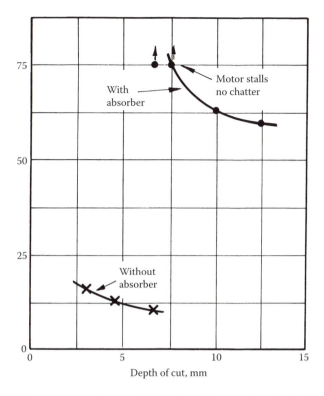

FIGURE 9.23 Improvements in the stability of a milling machine fitted with a vibration absorber. (From Stone, B.J. and Andrew, C., *Proc. 9th Int. MTDR Conf.*, Pergamon Press, New York, 1968.)

9.6.3.2 Superimposed Speed Variation

A second method to alter the phasing between successive waves cut on the work surface is to superimpose a variation in speed on the steady spindle rotation. For machines with continuously variable spindle speed under program control, the speed can be controlled to fluctuate appropriately and thereby reduce the tendency for regenerative instability to occur. This approach can be effective for processes with a slow buildup of vibrations, such as some grinding operations and has also been applied to other machining operations [19–20].

9.6.4 ACTIVE FORCE CONTROL

Several attempts have been made to control chatter vibrations by means of active feedback control, generally in connection with lathe turning. This is achieved by sensing vibrations of the tool or workpiece and then feeding this signal to a vibrator, usually electro hydraulic, connected to the tool post. The signal is modified so that the force applied opposes the motion of the tool and damps

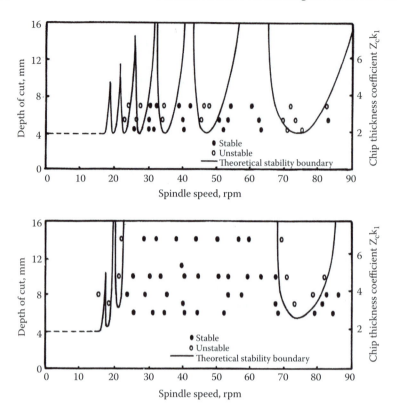

FIGURE 9.24 Stability charts for regular and irregular tooth pitch milling cutters. (From Opitz, H., Dregger, E.U., and Roese, H., *Proc. 7th Int. MTDR Conf.*, Pergamon Press, New York, 1966.)

down any vibrations. This fairly complex instrumentation has been used effectively under laboratory conditions.

PROBLEMS

1. A simple lathe structure can be represented as a single-degree-of-freedom system aligned along the direction of the normal to the cut surface, with an equivalent mass m_e, damping coefficient c_d, and stiffness S_e. If the direction of the cutting force is inclined at an angle θ_c, to the normal to the cut surface, show that the limit of stability for fully regenerative chatter is defined by the following equations:

$$\frac{k_1}{S_e} = \frac{-\left(1 - \left(\omega/\omega_n\right)^2\right)}{\left(1 - \cos\beta_c\right)\cos\theta_c}$$

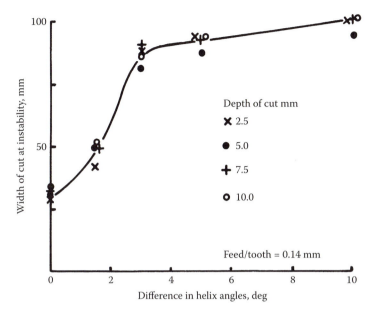

FIGURE 9.25 Width of cut at stability boundary for various differences in helix angles on helical milling cutters (rotational speed = 122 rpm). (From Opitz, H., Dregger, E.U., and Roese, H., *Proc. 7th Int. MTDR Conf.*, Pergamon Press, New York, 1966.)

and

$$\frac{k_1}{S_e} = \frac{-2\omega\xi}{\omega_n \sin\beta_c \cos\theta_c}$$

where $\beta_c = \omega T$, T = the period between successive passes of the tool, ω_n is the natural frequency of the system, and ξ is the damping ratio. It should be assumed that the cutting force varies only due to variations in the undeformed chip thickness and k_1 is the chip thickness coefficient (or cutting stiffness). If $\omega_n = 50$ Hz, $\xi = 0.5$, and $\theta_c = 60$ degrees, plot the highest speed unstable region of the stability chart in terms of k_1/S_e and β_c.

2. Figure 9.26 shows the cross-frequency response (receptance) locus for a lathe between the cutting-force direction and the normal to the cut surface. It is assumed that variations in cutting force occur only because of variations in the undeformed chip thickness.

   a. Prepare a sketch of the stability chart for fully regenerative cutting conditions in terms of the critical values of the chip-thickness coefficient, indicating some of the cutting speeds for the minima on the chart and also the frequency range in which instability will occur.

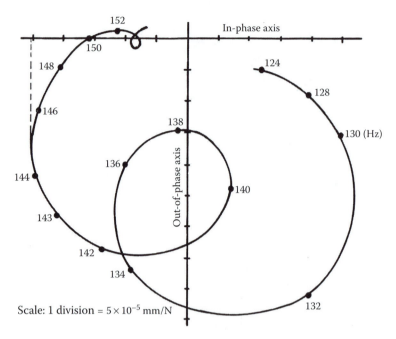

FIGURE 9.26 Operative frequency response locus for a milling machine tool structure.

b. What will be the form of the stability chart for machining under conditions in which no overlapping of successive cuts occurs?

c. If, instead of a turning operation, the process were a slab-milling operation with a cutter having 20 teeth with an average of three engaged with the workpiece at any time, what difference would this make to the stability chart?

3. A cylindrical workpiece of diameter d_w is to reduced in diameter to d_a using an end-milling tool of diameter d_t. The workpiece is mounted on the table of the milling machine as shown in Figure 9.27. The machine structure behaves as a single-degree-of-freedom dynamic system in the x direction as shown. The direct frequency response receptance of the structure has a maximum negative in phase component of R_a. Assuming that the normal to the cut surface and resultant cutting force act midway along the tool arc of contact, show that the limit of stability at any position of the tool during the cut is given by:

$$k_1 = \frac{1}{2R_a \cos(\theta_c - \alpha + \psi)\cos(\alpha - \psi)}$$

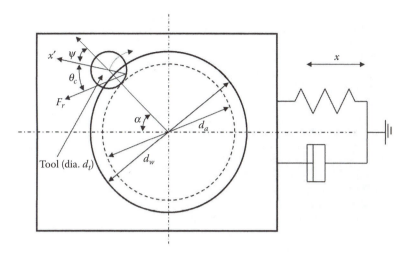

FIGURE 9.27

For a particular workpiece $d_w = 100$ mm, $d_a = 85$ mm and $d_t = 25$ mm. Determine how the limit of stability changes as the tool passes around the workpiece in terms of $k_1 R_a$. Assume that the angle between the cutting force and the normal is the cut surface, θ_c, is equal to $60°$.

REFERENCES

1. Tobias, S.A., *Machine-Tool Vibrations,* Blackie and Son, London, 1965.
2. Tlusty, J. and Polacek, K.M., Experiences with Analysing Stability of Machine-Tool against Chatter, *Proc. 9th Int. MTDR Conf.,* Pergamon Press, New York, 1968.
3. Welbourne, D.B. and Smith, J.K., *Machine-Tool Dynamics: An Introduction,* Cambridge University Press, New York, 1970.
4. Baker, J.R. and Rouch, K.E., Use of Finite Element Structural Models in Analyzing Machine Tool Chatter, *Proc. CIRP Int. Workshop on Modeling of Machining Operations,* Atlanta, 151–160, May 19, 1998.
5. Andrew, C., Chatter in Horizontal Milling, *Proc. Inst. Mech. Eng. (London),* Vol. 179, 877, 1968.
6. Knight, W.A., Torsional Vibrations and Machine-Tool Stability, Ph.D. thesis, University of Birmingham, Birmingham, England, 1967.
7. Drew, S.J., Mannan, M.A., and Stone, B.J., Torsional Vibration Effects in Grinding, *Ann. CIRP,* Vol. 49 (1), 249–252, 2000.
8. Das, M.K., Knight, W.A., and Sadek, M.M., A Critical Assessment of Cutting Force Models in the Analysis of Machine-Tool Instability, *Proc. 11th Int. MTDR Conf.,* Pergamon Press, New York, 1970.
9. Das, M.K., The Relationship Between the Static and Dynamic Cutting of Metals, *Int. J. Machine Tool Design and Res.,* Vol. 7, 1965.
10. Nigm, M.M., Sadek, M.M., and Tobias, S.A., Prediction of Dynamic Cutting Coefficients from Steady State Cutting Data, *Ann. CIRP,* Vol. 21 (1), 98, 1972.

11. Sarnicola, J.F. and Boothroyd, G., Machine-Tool Chatter: Factors Which Influence Cutting Forces During Wave Generating and Regenerative Chatter, *Proc. 2nd NAMR Conf.*, Madison, Wisconsin, May 1974.

12. Sadek, M.M. and Tobias, S.A., Comparative Dynamic Acceptance Tests for Machine Tools Applied to Horizontal Milling Machines, *Proc. Inst. Mech. Eng. (London)*, Vol. 185, 319, 1971.

13. Choudhury, I., Sadek, M.M., and Tobias, S.A., Determination of the Dynamic Characteristics of Machine Tool Structures, *Proc. Inst. Mech. Eng. (London)*, Vol. 184, 943, 1970.

14. Tlusty, J., New Tailstock Design Improves Stability of Centre Lathes, *Proc. 7th Int. MTDR Conf.*, Pergamon Press, New York, 1966, 13.

15. Stone, B.J. and Andrew, C., Vibration Absorbers for Machine-Tools, *Proc. 9th Int. MTDR Conf.*, Pergamon Press, New York, 1968.

16. Opitz, H., Dregger, E.U., and Roese, H., Improvement of the Dynamic Stability of the Milling Process by Irregular Tooth Pitch, *Proc. 7th Int. MTDR Conf.*, Pergamon Press, New York, 1966.

17. Slavicek, J., The Effect of Irregular Tooth Pitch on Stability of Milling, *Proc. 6th Int. MTDR Conf.*, Pergamon Press, New York, 1965.

18. Stone, B.J., The Effect on Regenerative Chatter Behaviour of Machine Tools of Cutters with Different Approach Angles on Adjacent Teeth, MTIRA Report No. 34, MTIRA, England, July, 1970.

19. Jayaram, S., Kapoor, S.G., DeVor, R.E., Analytical Stability Analysis of Variable Spindle Speed Machining, *Trans. ASME, J. Manuf. Sci. Eng.*, Vol. 122 (3), 391–397, 2000.

10 Grinding

10.1 INTRODUCTION

Significant developments have recently been made in the grinding of metals. Whereas a few years ago, grinding was considered only a finishing process, the development of new grinding wheels has led to such improvements in performance that in many cases grinding can now be economically used for the bulk removal of metal. In addition, processes such as creep-feed grinding have been developed for removing relatively large amounts of stock material in a single pass. In the future an increasing proportion of the total amount of metal removal will be accomplished by grinding. It is believed, however, that grinding will never replace conventional metal cutting as the principal method of machining components. In fact the development of more advanced cutting tool materials, for turning operations particularly, has resulted in the ability to effectively machine hardened steels and other difficult to cut materials by conventional machining (see Chapter 4). Hard machining, as this process is sometimes referred to, therefore now competes with grinding for many applications [1].

10.2 THE GRINDING WHEEL

A grinding wheel essentially consists of a large number of abrasive particles, called *grains*, held together by a suitable agent called the *bond* (Figure 10.1). It may be regarded as a multipoint cutting tool with a cutting action similar to that of a milling cutter except that the cutting points are irregularly shaped and randomly distributed over the active face of the wheel. Cutting speeds in grinding are considerably higher than for other conventional machining operations.

Those grains at the surface of the wheel that perform the cutting operation are called *active grains*. In peripheral grinding, each active grain removes a short chip of gradually increasing thickness in a way that is similar to the action of a tooth on a slab-milling cutter (Figure 10.2). Because of the irregular shape of the grains, however, there is considerable interference, or plowing action, between each active grain and the new work surface. Each active grain is a small cutting tool with a large negative rake angle (-60 degrees or larger) (Figure 10.3), and this results in very low values for the equivalent shear angle during the removal of each chip. The plowing action results in progressive wear, causing the formation of worn areas on the active grains. As grinding proceeds, the number and size of these worn areas increase, thus increasing the interference or friction, resulting in an increase in the force acting on the grain. Eventually this force becomes large enough either to tear the worn grain from the bond of the wheel and thus expose a new unworn grain or to fracture the worn grain to produce

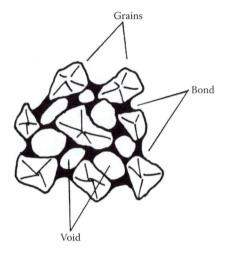

FIGURE 10.1 Grinding wheel structure.

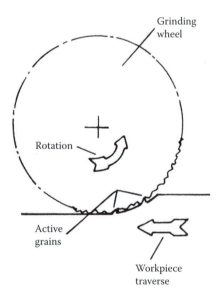

FIGURE 10.2 Action of a grinding wheel.

new cutting edges. Thus, a grinding wheel has a self-sharpening characteristic, and the force a grain can withstand before being torn from the wheel or fractured is a most important factor when grinding-wheel performance is considered.

A wheel consisting of relatively tough grains strongly bonded together will only exhibit the self-sharpening characteristic to a small degree and will quickly develop a glazed appearance during grinding. This glazed appearance is caused

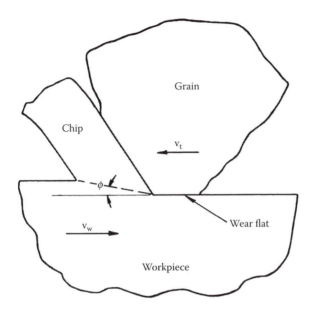

FIGURE 10.3 Chip formation by an abrasive grain, characterized by negative rake angle, small shear angle, and wear flat on the grain.

by the relatively large worn areas that develop on the active grains. These worn areas result in excessive friction and the consequent overheating of the workpiece. Grinding under these conditions is inefficient and necessitates "dressing" the wheel at frequent intervals to remove worn grains from the wheel surface. Dressing is usually carried out by passing a diamond-tipped dressing tool across the wheel surface while the wheel rotates. Dressing with a diamond-tipped tool removes or fractures the worn grains at the wheel periphery, thus generating a new and sharpened cutting surface. The need for frequent dressing to remove worn grains characterizes a hard wheel.

The opposite characteristic, where the grains in the wheel are torn out of the structure or fracture very rapidly during grinding, describes a soft wheel. Under the circumstances characterizing a soft wheel, the volume of wheel removed becomes significant compared to the volume of metal removed; the wheel will rapidly lose its shape and dressing will soon be necessary. Rapid fracturing represents inefficient grinding conditions, and thus the object in selecting a wheel for a particular operation is to achieve a compromise between the two extremes just described for hard and soft wheels. Several factors combine to determine whether a wheel behaves hard or soft and some of these factors are discussed below.

10.2.1 GRAIN TYPE

Table 10.1 shows the hardness of various materials used for grains in grinding wheels, together with the hardness of some other materials for comparison. The

TABLE 10.1
Knoop Hardness for Various Materials and Abrasives

Common glass	300–500	Titanium carbide	1800–3200
Hardened steels	700–1300	Silicon carbide[a]	2100–3000
Tungsten carbide	1800–2400	Boron carbide	2800
Aluminum oxide[a]	2000–2400	Cubic boron nitride[a]	4000–5000
Titanium nitride	2000	Diamond[a]	7000–8000

[a] Used commonly as grinding abrasives

hardest type of grain used in grinding wheels is diamond, and naturally this type of grain is used for grinding very hard materials such as cemented carbides. This is followed by cubic boron nitride (CBN), which has now taken over some of the applications for which diamond was previously used. The use of CBN wheels is increasing particularly in the automotive industry and for difficult-to-grind materials [2]. However, grains of aluminum oxide or silicon carbide are more commonly used in the manufacture of general purpose grinding wheels. Silicon carbide grains are the tougher of the two types, a property that makes them more efficient abrasives when used in grinding such materials as the hard cast irons. On the other hand, silicon carbide grains dull more rapidly than aluminum oxide grains when used in grinding steels. The general rule is to select aluminum oxide for surface grinding of steels of all kinds, including the hard stainless steels, annealed malleable iron, and tough bronzes and to select silicon carbide for surface grinding of gray iron, chilled iron, brass, soft bronze, aluminum, hard facing alloys, and the cemented carbides. A detailed discussion of the various types of abrasive in general use can be found in [3].

10.2.2 GRAIN SIZE

Grain size is determined mainly by the surface-finish requirement: the smaller the grain, the smoother the surface obtained. The grain size may also affect the relative hardness of the wheel. The size of grain is designated by the mesh of the screen through which it just passes. Thus, a grain of size 20 grit passes through a mesh of 20 openings per inch. Grains are sorted into various sizes by a mechanical sifting machine.

10.2.3 BOND

Six principal types of bond are used for grinding wheels: vitrified, silicate, shellac, resinoid, rubber, and metallic.

A vitrified bond is a clay melted to a glasslike consistency. The abrasive grains are mixed with clay, formed into the shape of the required wheel and then baked to form the bonded wheel structure. A vitrified bond is strong and rigid and is the most common type of bond used for grinding wheels. A silicate bond is essentially water glass, (sodium silicate) hardened by baking and the process for forming the wheel is similar to that used for vitrified bonds. A silicate bond holds the grains more loosely than a vitrified bond, and therefore the wheel behaves softer. A resinoid bond is a thermosetting resin and a wide range of properties can be obtained by different mixes of the compound. It is strong but fairly flexible and is sometimes used for manufacturing large grinding wheels. A rubber bond consists of fairly hard vulcanized rubber; it can be used for the manufacture of thin, flexible wheels, used particularly for cutoff machines and similar applications. Resin and rubber bonded wheels are often reinforced with fiber glass and other materials. A shellac bond is used for grinding wheels employed in producing very smooth finishes on hard surfaces. A metallic bond is used for grinding wheels with diamond or CBN grains and cannot be dressed in the usual way. Metallic-bonded wheels are produced by sintering compacted mixtures of the grains and metallic powders, usually of bronze, but other metals including cast iron can be used.

For most grinding wheels the whole of the wheel is made as one piece from the grit and bond mixture. This is usually the case for aluminum oxide and silicon carbide wheels. However, for metallic bonded CBN and diamond wheels, the abrasive is usually a thin layer (typically less than 5 mm) of abrasive bonded onto a steel disk to form the wheel structure.

10.2.4 STRUCTURE

The structure of a wheel is determined by the proportion and arrangement of the abrasive grains and bond. Some wheels have an open (porous) structure, while others have more tightly packed grains with the spaces between grains filled with the bonding material; this is known as a closed structure (Figure 10.4) [3]. A wheel with an open structure provides for easier swarf, or chip, removal but will tend to behave soft; on the other hand, a wheel with a closed structure will tend to clog with chips (swarf) and will behave hard thus requiring more frequent dressing to ensure efficient grinding.

10.2.5 DESIGNATION OF GRINDING WHEELS

Grinding wheels can be manufactured with a wide range of characteristics. For this reason a standard marking system is available to aid in the selection of the appropriate wheel for a given application.

Table 10.2 shows the ISO standard marking system for grinding wheels that use aluminum oxide or silicon carbide abrasives. It can be seen that the grinding-wheel code consists of seven alphanumeric symbols referring to type of abrasive, nature of abrasive, grain size, wheel grade (relative hardness), structure, nature of bond, and finally a manufacturer's record. In general, the following guidelines can be used for the selection of a grinding wheel with aluminum oxide or silicon carbide grains:

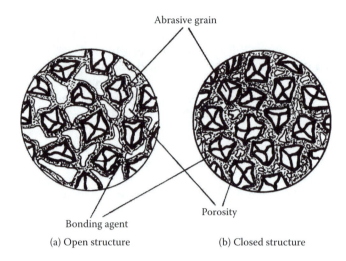

Abrasive grain

Porosity

Bonding agent

(a) Open structure   (b) Closed structure

FIGURE 10.4 Grinding wheel structure. (a) Open structure; (b) closed structure.

1. Choose aluminum oxide for steels and silicon carbide for carbides and nonferrous metals.
2. Choose a hard-grade wheel for soft materials and a soft-grade wheel for hard materials.
3. Choose a large grit for soft and ductile materials and a small grit for hard and brittle materials.
4. Choose a small grit for a good finish and a large grit for a maximum metal removal rate.
5. Choose a resinoid, rubber, or shellac bond for a good finish and a vitrified bond for a maximum metal removal rate.
6. For surface speeds greater than 32 m/s, do not choose a vitrified bond.

Table 10.3 shows the standard marking system for diamond and CBN wheels. There is also a similar standard marking system for coated abrasives (abrasive belts and papers), but this is less widely used in practice than the systems for grinding wheels (Table 10.2 and Table 10.3).

Grinding wheels are also marked with a maximum rotational speed to be used. This is necessary because excessive speeds can result in centrifugally induced hoops stresses that are sufficient to cause a wheel to burst. When a wheel bursts, fragments are ejected at the peripheral velocity of the wheel and therefore at velocities as high as 100 meters per second or more. As a quality control test, grinding wheel manufacturers must conduct standard bursting tests in which sample wheels are rotated at increasing speeds until bursting takes place. The designated maximum rotational speed is significantly less than the expected bursting speeds.

TABLE 10.2
ISO Standard Marking System for Aluminum Oxide and Silicon Carbide Grinding Wheels[a]

Sequence	1	2	3	4	5	6
	Nature of Abrasive	Grain Size	Grade	Structure	Nature of Bond	Manufacturer's Record
Type of Abrasive	A	36	L	5	V	23
51						

Manufacturer's symbol indicating exact type of abrasive (optional)

Aluminum Oxide — A
Silicon Carbide — C

Coarse	Medium	Fine	Very Fine
10	30	70	220
12	36	80	240
14	46	90	280
16	54	100	320
20	60	120	400
24		150	500
		180	600

Grade Scale
ABCDEFGHIJKLMOPQRSTUVWXYZ
Soft   Medium   Hard

Dense to open
0 — 8
1 — 9
2 — 10
3 — 11
4 — 12
5 — 13
6 — 14
7 — Etc.

V — vitrified
S — silicate
R — rubber
RF — rubber reinforced
B — resinoid
BF — resinoid reinforced
E — shellac
Mg — magnesia

Manufacturer's private marking to identify wheel (optional)

[a] The sample grinding wheel designation shown is 51A36L5V23.

TABLE 10.3
ISO Standard Marking System for CBN and Diamond Grinding Wheels[a]

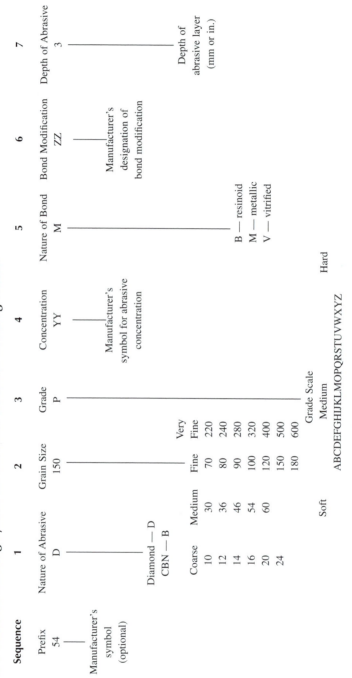

Sequence	1	2	3	4	5	6	7
	Nature of Abrasive	Grain Size	Grade	Concentration	Nature of Bond	Bond Modification	Depth of Abrasive
Prefix	D	150	P	YY	M	ZZ	3
54							

Manufacturer's symbol (optional)

Diamond — D
CBN — B

Manufacturer's symbol for abrasive concentration

Manufacturer's designation of bond modification

Depth of abrasive layer (mm or in.)

B — resinoid
M — metallic
V — vitrified

Coarse	Medium	Fine	Very Fine
10	30	70	220
12	36	80	240
14	46	90	280
16	54	100	320
20	60	120	400
24		150	500
		180	600

Grade Scale
Soft   Medium   Hard
ABCDEFGHIJKLMOPQRSTUVWXYZ

[a] The sample grinding wheel designation shown is 54D150PYYMZZ3.

10.3 EFFECT OF GRINDING CONDITIONS ON WHEEL BEHAVIOR

It has been seen that wheels of different types manufactured from different types of grain behave in different ways, and the choice of wheel type will depend on the material to be ground, the surface finish required, and the metal removal rate. However, grinding conditions also play an important part in determining whether a wheel will behave hard or soft.

Figure 10.5 shows the approximate shape of the layer of material removed by a single grain during plunge grinding. The average length of a chip l_c during grinding is given approximately by

$$l_c = \frac{d_t}{2} \sin \theta \tag{10.1}$$

where d_t is the diameter of the wheel and

$$\cos \theta = \frac{(d_t/2) - f}{d_t/2} = 1 - \frac{2f}{d_t} \tag{10.2}$$

where f is the infeed. Now

$$\sin^2 \theta = 1 - \cos^2 \theta = \frac{4f}{d_t} - \frac{4f^2}{d_t^2} \tag{10.3}$$

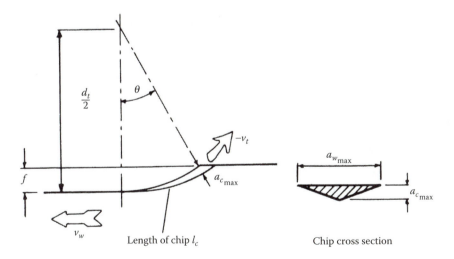

FIGURE 10.5 Geometry of chip removal in plunge grinding, where d_t = diameter of wheel, f = infeed, v_t = surface speed of wheel, v_w = surface speed of workpiece, $a_{c_{max}}$ = maximum undeformed chip thickness, and $a_{w_{max}}$ = maximum chip width.

by substitution of Equation 10.2. Substituting Equation 10.3 in Equation 10.1 and neglecting the second-order term,

$$l_c = \left(f d_t \right)^{0.5} \tag{10.4}$$

The metal removal rate Z_w, during grinding is given by

$$Z_w = f a_p v_w \tag{10.5}$$

where a_p is the back engagement (width of the grinding path), and v_w is the surface speed of the workpiece.

If the chips have the triangular cross section shown in Figure 10.5, the average volume V_0 of each chip is given by

$$V_0 = \frac{1}{6} a_{w_{max}} a_{c_{max}} l_c \tag{10.6}$$

where $a_{c_{max}}$ is the maximum undeformed chip thickness, and $a_{w_{max}}$ is the maximum width of the chip and can be expressed by

$$a_{w_{max}} = r_g a_{c_{max}} \tag{10.7}$$

where r_g is the grain aspect ratio.

The number of chips produced per unit time N_c is given by

$$N_c = v_t a_p C_g \tag{10.8}$$

where v_t is the surface speed of the wheel, and C_g is the number of active grains per unit area on the wheel surface.

Finally, since $V_0 N_c$ is equal to Z_w, from Equation 10.4, Equation 10.5, Equation 10.6, and Equation 10.7,

$$a_{c_{max}}^2 = \frac{K v_w}{v_t} f^{0.5} \tag{10.9}$$

where K is given by $6/(C_g r_g d_t^{0.5})$ and is a constant for the particular grinding wheel.

Since an increase in $a_{c_{max}}$ will result in an increase in the maximum force acting on each active grain during grinding, it follows that any change in grinding conditions tending to increase $a_{c_{max}}$ increases the self-sharpening process, and the grinding wheel tends to behave softer. Thus from Equation 10.9 the following changes in grinding conditions would be expected to make a wheel behave soft [4]:

1. An increase in the work surface speed v_w
2. An increase in the infeed f
3. A decrease in the wheel surface speed v_t

10.4 DETERMINATION OF THE DENSITY OF ACTIVE GRAINS

If it is desired to evaluate the constant K in Equation 10.9, it is necessary to measure the number of active grains per unit area of wheel surface C_g. In a method used by Backer, Marshall, and Shaw [5] the grinding wheel was rolled over a glass plate covered with a layer of carbon black. The resulting picture on the glass plate was enlarged photographically, and C_g was determined by counting the number of marks per unit area.

A device [6] that may be used to measure the number of active grains passing one point on the workpiece surface during one revolution of the grinding wheel employs a thermocouple located at the surface of the workpiece. The thermo-couple wire is insulated and passes through the workpiece. As each active grain passes, a thermocouple junction is formed between the wire and the workpiece, and a pulse is obtained from the high temperature developed. The pulses can be displayed on an oscilloscope screen and counted.

A more direct method was employed by Grisbrook [7]. The surface of the grinding wheel was viewed on a projection microscope, and the number of cutting points passing a line on the projection screen was counted as the wheel was rotated a given amount. The number of active grains determined in this manner was found to agree with the number of chips produced during one revolution of the wheel in grinding.

10.5 TESTING OF GRINDING WHEELS

An important factor in determining the characteristics of a grinding wheel is the force required to break the bond between grains. Much effort has been made to devise a method that will determine bond forces so that the operating performance of a grinding wheel may be predicted. A typical method [6] is to pass a sintered-metal carbide or diamond chisel over the wheel surface in such a way that it tears a layer of grains from the bond. The forces required to separate a layer of grains from the bond are taken as a measure of the strength of the bond, thus giving a guide to the behavior of the wheel in practice.

Grinding wheels must be tested to establish hardness as designated in the standard coding system outlined above. Hardness can be measured in several different ways. One method is to drill the wheel with a hard spade-type drill with a constant force. The depth of penetration in a given time is a measure of wheel hardness. A second method is to use an air/abrasive jet to break the bond. The depth of penetration of the jet erosion in a standard period of time is used to determine the equivalent wheel hardness. A third method used is to measure the resonant frequency of the isolated wheel after a sharp blow with a rubber hammer. The resulting frequency can be related to the wheel hardness [3].

10.6 DRESSING AND TRUING OF GRINDING WHEELS

After a grinding wheel has been mounted on the spindle and properly balanced, two important tasks must be carried out before the wheel can be used: truing and dressing of the wheel surface. *Truing* is the process of ensuring that the radial and axial run out of the wheel surface is eliminated or minimized. *Dressing* is the process of sharpening the wheel to expose active sharp grits and for form grinding shaping the wheel surface to the required shape for the particular operation to be carried out. Redressing of the wheel will usually be required during production as the wheel becomes dull or clogs. For most grinding wheel applications, truing and dressing are carried out simultaneously, and then the whole processes is usually just referred to as dressing. However, for metallic bonding wheels, truing is done separately and is achieved by accurately mounting the wheel so that any run out is minimized.

A number of different methods are available for the dressing of grinding wheels. The most common way to dress grinding wheels is single point dressing, in which a diamond pointed dresser, consisting of a diamond tip mounted onto a steel holder, is moved across the surface of the wheel while the wheel is rotating (Figure 10.6) This action removes grits from the surface of the wheel to expose sharp active grains. The dressing tool is usually mounted at an angle to the wheel surface as shown in Figure 10.6. The feed rate of the dressing point across the

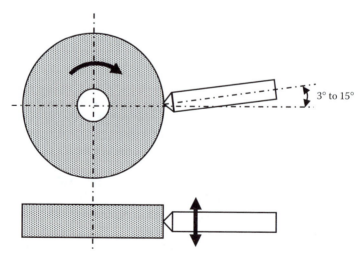

FIGURE 10.6 Single point dressing of a grinding wheel.

wheel surface influences the subsequent cutting action of the wheel; high feeds result in a coarsely dressed surface more suitable for rough grinding, and low feeds produce a finely dressed surface more appropriate for finish grinding (see section 10.7.5). Configurations of dressing tool other than that shown in Figure 10.6 are also used [3].

For form grinding the surface of the wheel must be accurately shaped to the surface to be machined. In this case the single point dressing tool must be moved under automatic control to form the surface shape. Various mechanisms, including the use of numerical control (NC), have been developed to achieve the required motions [3].

Another commonly used method for wheel dressing is crush dressing. In this process a roller of hardened steel, tungsten carbide, or other hard material is pressed into the surface of the wheel while the wheel rotates slowly. The dressing roller may be separately driven or be rotated by the contact with the grinding wheel surface. The dressing roller crushes grits from the surface and dresses the wheel. Crush dressing is generally much faster than single point dressing, but for form grinding the shaped rollers required may be expensive to produce, and the process is then more suitable for large quantity product.

In a process similar to crush dressing, diamond dressing rolls are used in which the shaped roll surface is coated with a layer of diamond grits to aid the dressing process. In this case the wheel material is removed mainly through the abrasive action of the diamond grits on the dressing roll rather than by the crushing action. Shaped diamond dressing rolls are more expensive to produce than conventional crush dressing rolls, but the dressed surface is generally more accurately produced.

Metallic bonded wheels require specialized dressing techniques. Some wheels are manufactured with matrices that can be crush dressed. In other cases the most effective process for dressing is electro-discharge machining (EDM) (see Chapter 14). An electrode corresponding to the required wheel surface shape is prepared. The electric discharge erodes away the matrix material, and the current density can be varied to produce both course and finely dressed surfaces on the wheel.

10.7 ANALYSIS OF THE GRINDING PROCESS

In the previous discussion it was shown that wheel wear is an essential phenomenon in grinding and provides the necessary self-sharpening characteristic. This variation in wheel surface means that wheel-workpiece engagements (width of grinding path and feed) will change as cutting proceeds. In addition, because the engagements are so small compared to the grinding machine itself, structural deflections arising from the grinding forces can significantly affect the values of these engagements. In other words, machining conditions will be affected by the machining process, a phenomenon that cannot be neglected in the grinding process.

Figure 10.7 shows an idealized model of the cylindrical plunge-grinding process. The stiffness of the wheel-support system in the infeed direction is represented by a spring of stiffness S_t, the stiffness of the work-support system by a spring of stiffness S_w, and finally the stiffness of the wheel and work are

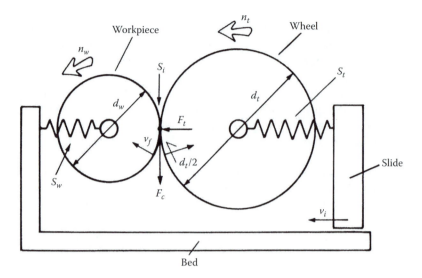

FIGURE 10.7 Idealized model of cylindrical plunge grinding, where S_w = stiffness of workpiece support, S_t = stiffness of wheel support, S_i = combine stiffness of workpiece and wheel, n_w = rotational frequency of workpiece, n_t = rotational frequency of wheel, d_w = diameter of workpiece, d_t = diameter of wheel, v_i = machine infeed speed, v_f = feed speed, F_c = cutting force, and F_t = thrust force.

designated by S_i. The wheel support is approaching the workpiece at the machine infeed rate of v_i, the radius of the wheel is being reduced at a rate $\dot{d}_t/2$, and the radius of the workpiece is being reduced at a rate $\dot{d}_t/2$ or v_f. Under steady-state conditions, the force components generated at the wheel-workpiece contact zone are the cutting force F_c and the thrust force F_t.

It can be appreciated from the figure that at steady state, the feed speed v_f resulting from a given machine infeed rate v_i depends on the stiffness of the machine tool, workpiece, and wheel and on the rate $\dot{d}_t/2$ at which the wheel wears. Thus,

$$v_f = v_i - \dot{d}_t/2 \qquad (10.10)$$

To obtain quantitative relationships for the forces generated, it is necessary to know the effect of the grinding conditions on the grinding forces. Figure 10.8 shows a relationship between the thrust force per unit width of cut and the feed. (It should be emphasized that the feed f is the advance of the wheel surface per revolution of the workpiece.) It can be seen that three distinct regions occur [8]:

1. When the thrust force per unit back engagement (width of cut) is very low, only rubbing occurs, where the grains rub on the work surface with essentially no material removal.

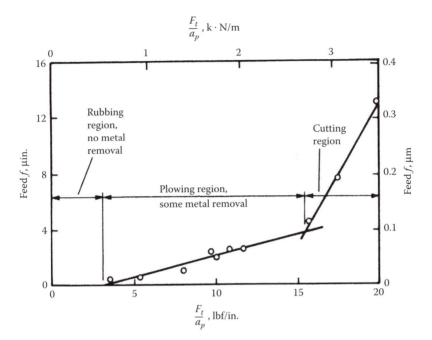

FIGURE 10.8 Effect of thrust force per unit width of cut on feed during internal, plunge, climb grinding of AISI 4150 at Rockwell C 53-55 using an A80L8V wheel with an equivalent wheel diameter of 2.09 in. (53 mm), where the wheel surface speed is 7700 ft/min (30 m/s), the workpiece surface speed is 540 ft/min (2.74 m/s), F_t = thrust force, a_p = back engagement (width of cut), and f = feed. (After Hahn, R.S. and Lindsay R.P., *Machinery*, July–Nov., 1971.)

2. For low force values, plowing occurs, where the grains plow a furrow in the workpiece, with the extruded material broken off along the sides of the groove, giving a low metal removal rate.

3. For high force values, cutting occurs where the active grains produce chips (in the manner described in section 10.3), giving high metal removal rates.

Clearly for rough grinding and for rapid stock removal, it is desirable to operate under conditions where cutting occurs. The plowing region is important for obtaining a good surface finish during the final passes of the operation.

A more useful way of presenting the results of grinding experiments is shown in Figure 10.9, where the metal removal rate Z_w and the wheel removal rate Z_t are plotted against the thrust force F_t. It can be seen that within the cutting region, the relationships for Z_w and Z_t are both linear. If the threshold thrust force at the beginning of the cutting region is designated F_{t0}, then the following expressions can be written for Z_w and Z_t, respectively:

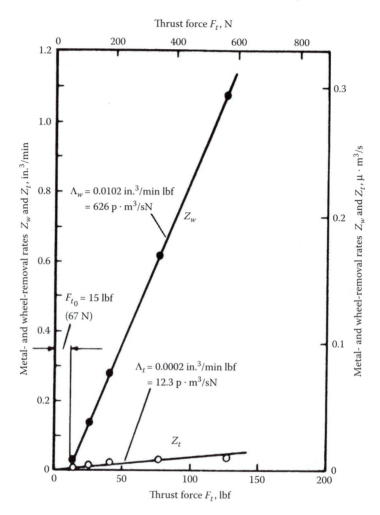

FIGURE 10.9 Effect of trust force F_t on the metal and wheel removal rates Z_w and Z_t, respectively, during internal, plunge, climb grinding of an easy-to-grind material, AISI 52100 at Rockwell C 60-62, using an A70K8V wheel with an equivalent wheel diameter of 29 in. (727 mm), where the wheel surface speed is 12,000 ft/min (61 m/s) and the workpiece surface speed is 250 ft/min (1.27 m/s) and F_{t_0} = threshold force for grinding by "cutting", Λ_w = metal removal parameter, and Λ_t = wheel removal parameter. (After Hahn, R.S. and Lindsay R.P., *Machinery*, July–Nov., 1971.)

$$Z_w = \Lambda_w \left(F_t - F_{t_0} \right) \tag{10.11}$$

$$Z_t = \Lambda_t \left(F_t - F_{t_0} \right) \tag{10.12}$$

where Λ_w and Λ_t are the workpiece and wheel removal parameters [9], respectively, and can be regarded as the volume of workpiece or wheel removed per unit time per unit thrust force, respectively.

The metal removal rate Z_w can also be expressed by

$$Z_w = \pi d_w a_p v_f \qquad (10.13)$$

where d_w is the diameter of the workpiece, a_p is the back engagement (width of cut), and v_f is the feed speed. Also, Z_t can be expressed by

$$Z_t = \frac{\pi d_t a_p \dot{d}_t}{2} \qquad (10.14)$$

Thus, if Equation 10.11, Equation 10.12, Equation 10.13, and Equation 10.14 are combined and substituted in Equation 10.10,

$$v_f = \frac{v_i}{1 + \left(\Lambda_t d_w / \Lambda_w d_t\right)} \qquad (10.15)$$

For an easy-to-grind material the ratio Λ_w / Λ_t, referred to as the grinding ratio, would be on the order of 50; therefore, under steady-state conditions, feed speed v_f approaches the machine infeed speed v_i quite closely. However, for a difficult-to-grind material the grinding ratio can be on the order of 3.0, and Equation 10.15 must be used to estimate the actual feed speed.

10.7.1 Specific Cutting Energy for Grinding Processes

The specific cutting energy for grinding operations is considerably higher than for conventional cutting processes [10], as shown in Table 10.4. The total specific energy during cutting can be divided into three elements as follows:

1. Energy used in chip formation
2. Energy used in plastic deformation without cutting that is plowing
3. Energy expended in sliding friction, in particular on worn areas on the active grains

The specific cutting energy for grinding is higher than for single-point cutting for several reasons, including the following:

1. The size effect: the chips produced by the active cutting edges are very small, and it is well known that during deformation small pieces of material have proportionately greater strength than larger ones. Hence the specific cutting energy for chip formation will be larger.

TABLE 10.4
Specific Cutting Energy for Grinding and Conventional Machining

| Material | Specific Energy (Ws/mm³) | |
	Conventional Machining	Grinding
Aluminum alloys	0.44–1.09	6.8–27.3
Cast irons	1.64–5.46	11.29–60.0
Titanium alloys	3.0–5.0	16.4–54.6
Carbon steels	2.73–9.28	13.6–68.2

Source: Adapted from Kalpakjian, S. and Schmid, S.R., Manufacturing Processes for Engineering Materials, 4th ed., Prentice Hall, NJ, 2003.

2. Effective cutting edge geometry: the active grains have large negative rake angles and the chip formation involves large shear strains. The energy of plastic deformation is higher than for other cutting processes.
3. Wear lands: the size of the worn areas on the active grains is much larger proportionately than for single-point cutting. Consequently, the specific energy expended in sliding friction and plowing is greater.

10.7.2 CYLINDRICAL GRINDING CYCLES

Cylindrical grinding processes are used to improve the tolerance, surface finish, and roundness of workpieces previously machined on lathes. In a typical cylindrical grinding cycle, the first contact with the workpiece is made at the high spots on the work. Consequently, during the first few revolutions or passes of the workpiece, wheel-work contact is intermittent. Each subsequent revolution or pass further improves the workpiece roundness geometry, but this can require several revolutions or passes and significant stock removal. This process is referred to as rounding up the workpiece. Computer-based analysis procedures, which take into account the dynamic behavior of the grinding machine, combined with grinding process equations, have been developed [11] that allow estimates to be made of the amount of stock removal and the time required to round up the workpiece to a required accuracy. The allowance left on workpieces prior to finish grinding must be large enough for the rounding-up process to be complete before the required finished size of the part is reached.

In a cylindrical plunge-grinding operation, the wheel is advanced to the position that, if wheel wear were negligible and the system rigid, would result in the removal of the desired stock from the workpiece. Thus, in the real situation the actual reduction in the workpiece radius when the infeed is stopped is less than that required because of the deflections of the structure and the wheel wear. Continuation of the grinding process without further application of feed results

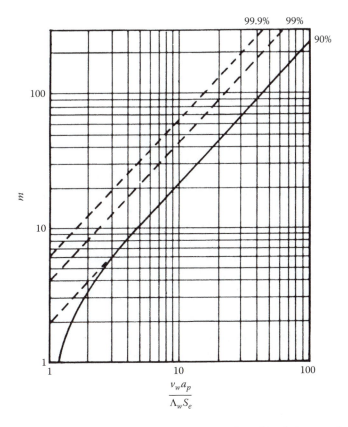

FIGURE 10.10 Effect of grinding conditions on the number of workpiece revolutions, or reciprocations, for spark-out and for various percentages of the material removal, where m = number of workpiece revolutions or reciprocations, v_w = workpiece surface speed, a_p = back engagement (width of cut), Λ_w = workpiece removal parameter, and S_e = system stiffness.

in the removal of the material left as a result of structural deflections; this procedure is known as *spark-out*. The results of an analysis of the sparking-out process [12] are presented in Figure 10.10. This analysis applies to cylindrical grinding in which the contact force between the wheel and workpiece is gradually reduced as the elastic deflections of the machine and workpiece relax.

The solid line in Figure 10.10 shows how grinding conditions affect the number of revolutions of the workpiece to remove 90% of the stock that remains after the machine infeed is disengaged. The linear portion of this relationship can be expressed by

$$m = \frac{2v_w a_p}{\Lambda_w S_e} \tag{10.16}$$

where v_w is the workpiece surface speed, and S_e is the effective stiffness of the grinding-machine system given by $(S_w + S_i + S_t)$. The time taken for the spark-out operation t_s can therefore be obtained from

$$t_s = \frac{m}{n_w} \qquad (10.17)$$

where n_w is the rotational frequency of the workpiece. The value of the constant in Equation 10.16 depends on the percentage reduction in workpiece diameter; for the solid line in Figure 10.10 this is 90%. The remaining lines shown dashed in the figure are for 99 and 99.9% removal of the metal remaining when the infeed is stopped.

It is interesting to note that the spark-out time given by Equation 10.16 is independent of the magnitude of the infeed employed as well as the velocity of the workpiece. From this equation it is possible to reason that a decrease in back engagement (width of cut) a_p decreases the spark-out time and increases in the machine-workpiece-tool stiffness S_e or the metal removal parameter Λ_w (easier-to-grind material) decrease the spark-out time.

It should be pointed out that wheel wear has not been taken into account in this analysis. For difficult-to-grind materials where wheel wear is significant compared to material removal, spark-out time is reduced and an oversize workpiece results.

In addition, toward the end of the spark-out process, the thrust force F_t will fall to values below the plowing-cutting transition (Figure 10.8). After this point the decay rate of the force becomes slower and the metal removal rate is reduced considerably. This final phase of spark-out produces the smooth surface finish required on the workpiece, but it is clearly necessary for the rounding up of the workpiece to be completed before the final phase of spark-out is reached. The spark-out process follows an exponential decay with different time constants corresponding to the cutting and plowing regions [11].

10.7.3 SURFACE GRINDING CYCLES

Similar effects occur during surface grinding, which is used to improve the flatness and surface finish of workpieces previously machined by milling or shaping machines. The finish allowance must again be sufficient for surface flatness to be improved before the required finish size is reached. During plunge grinding on a surface-grinding machine, a spark-out phase similar to that occurring in cylindrical grinding occurs. However, the analysis must be modified to account for the intermittent contact between the wheel and workpiece, which occurs at the end of each stroke of the machine.

10.7.4 EQUIVALENT DIAMETERS OF GRINDING WHEELS

Before considering the material and wheel removal parameters further, it is necessary to explain the effect of conformity of the wheel and workpiece [13].

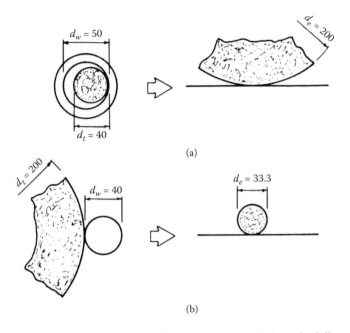

FIGURE 10.11 Equivalent diameter of wheel, where d_e = equivalent wheel diameter, d_t = wheel diameter, and d_w = workpiece diameter. (a) Internal grinding; (b) external cylindrical grinding.

Figure 10.11 shows two situations: internal grinding and external cylindrical grinding. It can be seen that the contact area between the wheel and work is significantly greater during the internal-grinding operation. This amount of contact or conformity affects the grinding process and can be allowed for by estimating the equivalent diameter d_e of the grinding wheel [14].

The equivalent wheel diameter is the diameter of a grinding wheel used in plunge surface grinding that would give the same length of work-wheel contact for the same feed as that obtained in the grinding operation considered. It was shown earlier that the length of work-wheel contact in surface grinding would be given by $(fd_e)^{0.5}$. Figure 10.12 shows the geometry in external cylindrical grinding. It can be seen that the feed is given by

$$f = x_1 + x_2 \qquad (10.18)$$

and the length of work-wheel contact a_b is approximately given by both $(x_1 d_w)^{0.5}$ and $(x_2 d_t)^{0.5}$ where d_w and d_t are the diameters of the workpiece and wheel, respectively. Therefore,

$$fd_e = \left(x_1 d_w\right)^{0.5} = \left(x_2 d_t\right)^{0.5} \qquad (10.19)$$

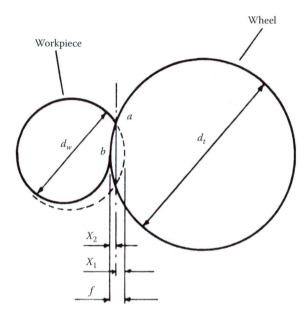

FIGURE 10.12 Geometry in external cylindrical grinding, where d_w = diameter of workpiece, d_t = diameter of wheel, and f = feed.

Combining Equation 10.18 and Equation 10.19 and rearranging gives

$$d_e = \frac{d_t}{1 + d_t / d_w} \qquad (10.20)$$

Similarly, it can be shown that for internal grinding the equivalent wheel diameter is

$$d_e = \frac{d_t}{1 - d_t / d_w} \qquad (10.21)$$

Figure 10.11 gives the equivalent diameter calculated from Equation 10.20 and Equation 10.21 for both external and internal cylindrical grinding. Use of the equivalent diameter of the wheel allows the conditions for a variety of grinding operations to be related and for measurements for one kind of operation to be employed in predicting the conditions for another.

10.7.5 METAL REMOVAL PARAMETER FOR EASY-TO-GRIND MATERIALS

It has been found from previous empirical work [9] that the metal removal parameter Λ_w for easy-to-grind materials is dependent on three main factors: wheel speed, workpiece hardness, and wheel-dressing conditions.

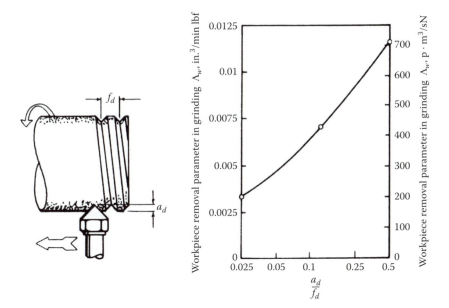

FIGURE 10.13 Geometry during wheel dressing and its effect on the metal removal parameter for AISI 4620 steel at Rockwell C 60-62, where the wheel surface speed is 12,000 ft/min (61 m/s), the workpiece surface speed is 800 ft/min (4.1 m/s), and the feed during dressing is 0.004 in. (0.1 mm) and a_d = depth of dress, f_d = feed during dressing, and Λ_w = metal removal parameter. (After Hahn, R.S. and Lindsay, R.P., *Machinery,* July–Nov., 1971.)

Previous studies [9] have shown that higher wheel speeds tend to increase the magnitude of the metal removal parameter, and higher workpiece hardness tends to decrease it. Also, it has been found [15] that wheel-dressing conditions can influence Λ_w. Figure 10.13 shows the geometry during wheel dressing. The important parameter is found to be the ratio of the depth of the dress a_d and the feed f_d during dressing. It is found [9] that an increased value of a_d/f_d increases the magnitude of Λ_w as illustrated in Figure 10.12.

Lindsay [14] has produced a semi-empirical analysis resulting in the following expression for the metal removal parameter for easy-to-grind metals:

$$\Lambda_w = \frac{7.93 \times 10^5 \times \left(v_w/v_t\right)^{0.158} \left[1 + \left(4a_p/3f_d\right)\right] f_d^{0.58} v_t}{d_e^{0.14} V_b^{0.47} d_g^{0.13} R_c^{1.42}} \qquad (10.22)$$

where

v_w = workpiece surface speed, m/s

v_t = wheel surface speed, m/s

a_d = depth of dress, m

f_d = feed during dressing, m

d_e = equivalent wheel diameter, m

R_c = Rockwell hardness number of the work material (C scale)

Also, d_g is the grain diameter given approximately by

$$d_g = \frac{0.0254}{grainsize} \text{ m} \tag{10.23}$$

and V_b is the percentage volume of bond material in the wheel given by

$$V_b = 1.33 H_n + 2.2 S_n - 8 \tag{10.23}$$

where S_n is the wheel structure number and H_n the wheel hardness number, is obtained from the following scale:

Wheel grade   H I J K L M ...
Wheel hardness number, H_n 0 1 2 3 4 5 ...

Equation 10.21 has been found to give estimates of the metal removal parameter to within about ±20% [9].

10.7.6 EXAMPLE

It is required to estimate the metal removal parameter Λ_w for the internal grinding of an R_c of 60, 52100 steel ring of internal diameter 95.25 mm with a grinding wheel 82.55 mm in diameter. The surface speeds of the wheel and workpiece are to be 63.5 m/s and 4 m/s, respectively. The wheel designation is A8OM4V dressed with a feed of 0.1 mm at a depth of 0.0125 mm.

The wheel structure number S_n is 4, and the wheel hardness number H_n is 5. Therefore, from Equation 10.23

$$V_b = \left(1.33 \times 5\right) + \left(2.2 \times 4\right) - 8 + 7.45$$

The grain size is 80, and therefore from Equation 10.22

$$d_g = \frac{0.0254}{80} = 0.0003175$$

The equivalent wheel diameter for internal grinding is given by Equation 10.20. Thus,

$$d_e = \frac{82.55}{1 - 982.55 / 95.25} = 619 \text{ mm}$$

Substitution of these, and the remaining figures, in Equation 10.21 gives

$$\Lambda_w = \frac{7.93 \times 10^5 \times (4/63.5)^{0.158} \left[1 + (4 \times 0.0125)/(3 \times 0.1)\right] \times 0.0001^{0.58} \times 63.5}{0.619^{0.14} \times 7.45^{0.47} \times 0.0003175^{0.13} \times 60^{1.42}}$$

$$= \frac{7.93 \times 10^5 \times 0.646 \times 1.167 \times 0.00479 \times 63.5}{0.935 \times 2.57 \times 0.35 \times 335}$$

10.7.7 METAL REMOVAL PARAMETER FOR DIFFICULT-TO-GRIND MATERIALS

Tool steels, high-strength alloys, carbides, and so on, are classed as difficult to grind. Figure 10.14 shows how the thrust force affects the metal and wheel removal rates for a difficult-to-grind material. Comparison of these data with those presented in Figure 10.10 for an easy-to-grind material shows that for a difficult-to-grind material, the threshold thrust force F_{t_o} is higher (311 N compared with 67 N), the metal removal parameter Λ_w is lower (13.5 compared with 626 p m³/sN), and the grinding ratio, Λ_w/Λ_t is lower (3.1 compared with 51). This means that higher forces and hence greater power are required and greater wheel wear occurs with the difficult-to-grind materials.

It has not yet been found possible to obtain an empirical equation for the metal removal parameter for difficult-to-grind materials, but as a guide Table 10.5 presents values of Λ_w, for the rough grinding of some of these materials. The low figures are for low values of the thrust force per unit back engagement (width of cut) F_t/a_p, and the high figures are for high values of F_t/a_p.

10.8 THERMAL EFFECTS IN GRINDING

The high cutting speeds in grinding result in temperatures at the grain tip that may be as high as 1700°C. As a result some of the chips may melt during formation. As mentioned already, the specific cutting energy for grinding is higher than for other cutting operations. A high proportion of this energy appears as heat that flows into the work surface. It has been stated that 60 to 90% of the total energy consumed in grinding flows instantaneously into the workpiece, causing rapid local increases in temperature [17]. Increasing the specific metal removal rate reduces the proportion of total energy flowing into the workpiece, and dulling of the wheel increases this value. The proportion of the total energy flowing into the workpiece E has been expressed as [18]:

$$E = \left(0.6 + 0.05 A_g\right) \times 100 \text{ percent} \tag{10.25}$$

where A_g is the relative contact area of the grains expressed as a percentage, ranging from 0 to 5%. The value of A_g increases as the wheel becomes less sharp.

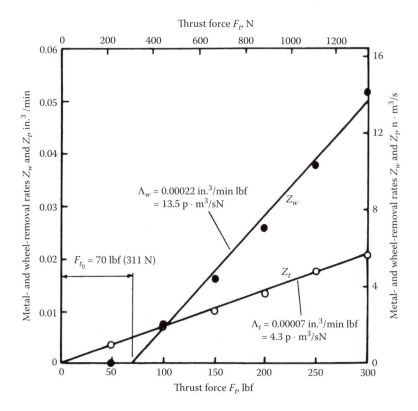

FIGURE 10.14 Effect of trust force F_t on the metal and wheel removal rates Z_w and Z_t, respectively, during internal, plunge, climb grinding of an difficult-to-grind material, M-4 at Rockwell C 65, using an A80K4V wheel with an equivalent wheel diameter of 27.6 in. (701 mm), where the wheel surface speed is 12,000 ft/min (61 m/s), and the workpiece surface speed is 250 ft/min (1.27 m/s) and F_{t_o} = threshold force for grinding by "cutting," Λ_w = metal removal parameter, and Λ_t = wheel removal parameter. (After Hahn, R.S. and Lindsay, R.P., *Machinery,* July–Nov., 1971.)

Various forms of thermal damage may occur in grinding, as shown schematically in Figure 10.15 [19], and the main effects are as follows:

1. Dimensional errors: The heat generated causes expansion of the workpiece, resulting in dimensional errors on cooling after the grinding process.
2. Material structural changes: Metallurgical changes may occur in the surface layers of the workpiece. Softening and tempering of the surface layers of hardened steel components may occur. Regions of untempered martensite may be formed because of rapid cooling after grinding in other steel components.
3. Burning and surface cracking: Excessive temperatures may cause surface oxidation and burning. Thermal cracking of the work surface may result

TABLE 10.5
Approximate Values of the Metal Removal Parameter Λ_w for Some Difficult-to-Grind Materials

Material	Description	Metal Removal Parameter Λ_w	
		in.³/min lbf	p.m³/sN
T-15	High speed steel, tungsten type	0.00015–0.0015	9–90
M-2	High speed steel, molybdenum type	0.0004–0.0011	25–70
Inconel X	Nickel alloy	0.0006–0.0025	37–150
M-4	High speed steel, molybdenum type	0.0002–0.002	12–120
M-50	High speed steel, molybdenum type	0.0019–0.0045	120–280

Source: After Hahn, R.S., *Adv. Mach. Tool Des. Res.*, Pergamon Press, Oxford, 129, 1962.

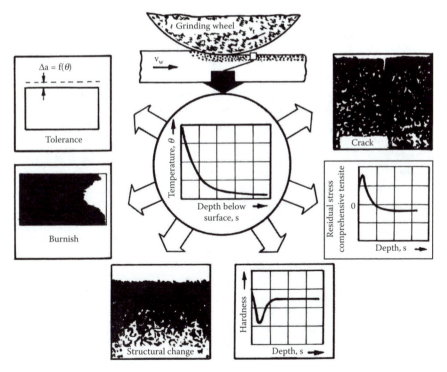

FIGURE 10.15 Thermal effects of grinding at the workpiece surface (Adapted from Konig, W., Honscheid, W., and Lowin, R., *Forschungsb. Landes Nordrh. Nr.* 2648, Westdeutsches Verlag, 1977.)

for some conditions. In severe cases such surface cracks may be readily visible, but for less critical conditions microscopic cracks may occur that reduce the fatigue and stress corrosion resistance of the workpiece.

4. Residual stresses: The temperature changes and gradients in the work surface layers, together with the plastic deformation that takes place, results in residual stresses being set up in the workpiece. Such residual stresses, particularly if they are tensile at the surface, reduce the fatigue strength of the workpiece. These effects can be reduced by using softer grade wheels and lower wheel speeds.

Overall there is a threshold temperature above which significant thermal damage occurs. The principal factors that influence this temperature are wheel speed, work speed, force intensity, and wheel sharpness. In general, higher work speeds and sharp wheels allow higher force intensities and stock removal rates to be used without significant thermal damage occurring. Adaptive control methods for grinding have been developed in which the wheel sharpness and induced force intensity are monitored [11], with cutting conditions adjusted appropriately during the grinding cycle.

10.9 CUTTING FLUIDS IN GRINDING

10.9.1 General Discussion

Cutting fluids in grinding perform the same functions as in other cutting operations. These fluids lubricate the chip-grit and grit-workpiece interfaces, thereby reducing the power consumed and the heat generated. In addition, the workpiece is cooled, reducing the extent of the possible surface damage. A comprehensive survey of friction, lubrication, and cooling in grinding can be found in [20].

Neat oils provide the greatest lubrication, but only limited cooling. Consequently, neat oil use is restricted mainly to difficult-to-grind materials and to form grinding, partly because of higher costs, potential fire hazards, and operator health risks due to the release of toxic fumes. However, there is a recent trend towards increased use of oil-based fluids in some grinding applications [20]. For general-purpose grinding, oil-and-water emulsions are used, with the lubricating properties regulated by adjusting the relative proportions of oil and water in the mixture.

10.9.2 Application of Grinding Fluids

For the grinding fluid to provide the necessary lubrication, it must penetrate into the contact zone between the wheel and workpiece. This is difficult because of the high wheel speeds used in grinding. It has been generally accepted that a thin air film barrier is formed around the rotating wheel that is difficult to penetrate and prevents the fluid from entering the grinding zone (e.g., [20,21]). However, some workers [22] have cast doubt on the validity of this air cushion effect and claim that if the stream of fluid has a velocity significantly different from the surface velocity of the wheel, the fluid forms turbulence and bounces off. If the velocities of the fluid and wheel are similar, then a laminar stream follows the contours of the wheel for some distance without forming turbulence, before finally leaving the wheel surface through centrifugal forces.

Several different methods for applying grinding fluids are used. Internal grinding is the most difficult type of grinding operation in which to obtain an adequate supply of fluid, mainly due to the restricted access to the grinding zone and the greater work-wheel conformity.

1. Flooding is the most common fluid application technique used. A large volume of fluid is applied at low pressure, and this generally gives acceptable results for conventional surface grinding.

2. In jet application, high-pressure jets of fluid are applied through nozzles close to the wheel and workpiece in the direction of wheel rotation. It is generally assumed that the high-pressure jets are required to penetrate the air cushion on the wheel surface. However, the fluid velocity is also high and approaches the wheel speed, so that formation of a laminar stream on the wheel thus dragging fluid into the grinding zone is also more likely to occur.

3. Grinding wheels, particularly those with vitrified bonds, are porous and fluid can be applied through the wheel into the grinding zone. The fluid is normally applied through the spindle or by means of cupped flanges on the sides of the wheel. In some cases radial passages can be formed inside the wheel for fluid transfer [20]. The method can be used in combination with flooding, with neat oil applied through the wheel, to reduce friction in the grinding zone, and an oil-water emulsion applied externally, mainly to provide cooling.

4. Chamber-type or air-deflector flood nozzles are being used increasingly for the low-pressure application of grinding fluids on surface grinding machines, high-speed grinding machines, and other types of cylindrical grinding machines. Figure 10.16 shows a typical configuration of this type of nozzle. Two alternative mechanisms for the effectiveness of chamber-type nozzles have been proposed. One explanation is that the air deflector removes the air cushion from the wheel and causes a low-pressure zone ahead of the grinding zone, pulling fluid from the nozzle into the wheel surface, which carries the fluid into the grinding zone. A second explanation [22] is that the fluid in the enlarged nozzle chamber is accelerated by the wheel itself, forming a laminar stream of coolant in contact with the wheel; this stream is carried into the grinding zone.

5. Some benefits have been found from having radial or spiral grooves in the side faces of the grinding wheel, with the fluid applied to the side faces of the wheel [21]. The grinding fluid then flows radially along these grooves into the grinding zone.

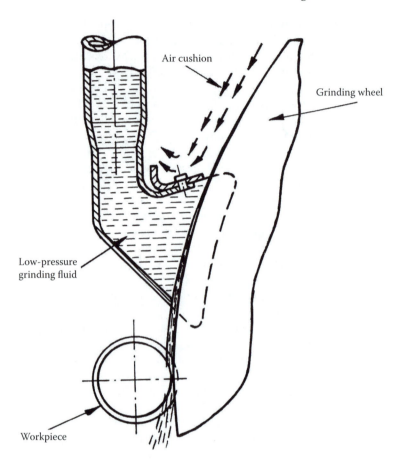

FIGURE 10.16 Air deflector nozzle or chamber-type nozzle used to increase grinding-fluid penetration into the grinding zone. (From Werner and Lauer-Schmaltz [22].)

10.10 GRINDING-WHEEL WEAR

The grinding ratio G_r is an important factor when the performance of grinding wheels is considered; it is given by

$$G_r = \frac{\Lambda_w}{\Lambda_t} \tag{10.26}$$

or, in other words, it is the ratio of the volume of metal removed to the volume of wheel worn away. Curves relating the volume of wheel wear to the volume of metal removed (Figure 10.17) are found to be similar to wear curves for a single-point cutting tool and exhibit an initial breakdown period followed by a region

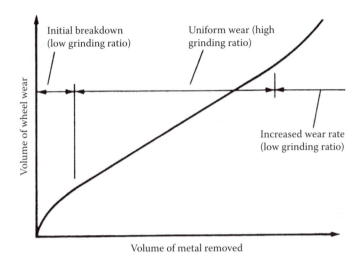

FIGURE 10.17 Wear of grinding wheels.

of uniform wear rate and finally a period of rapid breakdown. Ideally, the grinding wheel should be dressed at the end of the middle region of uniform wear.

As grinding proceeds and the active grains make repeated contact with the workpiece, the sharp edges are worn away, producing flat areas or wear scars on the grains; this type of wear is known as *attritious wear*. As this process continues, the forces on the active grains become sufficiently large either to fracture the grain or to tear it from the wheel bond; this type of wear is known as *fracture wear*. When a sufficiently large proportion of the active grains is fractured or torn from the wheel bond, the wear curve enters the final stage of rapid breakdown, and grinding becomes inefficient. A high grinding ratio is desirable, and Figure 10.18 shows that an optimum grinding ratio will occur within the middle region of uniform wear rate, the grinding ratio being lower in the initial breakdown stage and in the final breakdown period.

The thrust force also affects the grinding ratio, as shown in Figure 10.18. This result shows that the wheel removal rate becomes large, and hence the grinding ratio becomes small, when the thrust force is excessive. It is generally considered efficient to employ conditions where the wheel removal rate has just begun to rise steeply, that is, where F_t is between 150 and 200 lbf in Figure 10.18.

10.11 NONCONVENTIONAL GRINDING OPERATIONS

10.11.1 High-Speed Grinding

It has been demonstrated that increasing grinding speeds well above those used conventionally, up to 180 m/s, can result in increased metal removal rates without significant changes in grinding forces, specific cutting energies, and surface finish [23]. During early investigations of high-speed grinding, it was suggested that

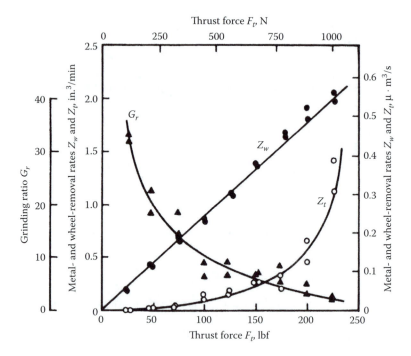

FIGURE 10.18 Effect of trust force F_t on the metal and wheel removal rates Z_w and Z_t, respectively, and the grinding ratio G, during surface grinding of an AISI 52100 steel ring at Rockwell C 60, with an equivalent wheel diameter of 3.25 in. (83 mm), where the wheel surface speed is 12,000 ft/min (61 m/s), and the workpiece surface speed is 500 ft/min (2.54 m/s). (After Hahn, R.S. and Lindsay, R.P., *Machinery*, July–Nov., 1971.)

the process might be developed to compete with other machining operations for large stock removal applications.

However, production applications of high-speed grinding have not increased significantly and use of the process is largely restricted to groove grinding and the grinding of the more difficult-to-machine materials.

Considerable safety problems exist with high-speed grinding because of the increased potential for bursting of the highly stressed wheels. Machines specifically designed to contain broken wheel fragments are required, together with special wheel designs, often of segmental construction.

10.11.2 Creep Feed Grinding

The production utilization of creep feed grinding has increased significantly, largely because of reduced machining times for large stock removal applications, which may be as low as 50% of the times for conventional grinding.

In creep feed grinding, the operation is performed in a single pass with a large depth of cut (Figure 10.19), but with low values of the infeed per revolution

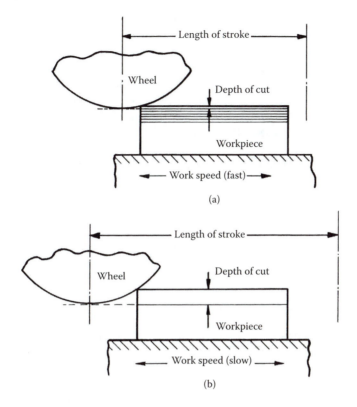

FIGURE 10.19 Feed, length of stroke, and depth of cut. (a) Conventional surface grinding; (b) creep-feed grinding. (From Werner [24].)

[3,24]. Compared to conventional grinding, creep feed grinding has the following characteristics:

1. Increased total grinding forces
2. Reduced average forces on the individual grains
3. Increased temperatures in the wheel-work contact zone
4. Reduced temperatures of the newly generated work surface
5. Large arc of contact between the wheel and work

As a result, creep feed grinding machines must be more rigid than conventional grinding machines, and high-efficiency grinding-fluid application systems are required. The wheel profile stability in creep feed grinding is good and typical applications are in grinding deep slots or the grinding of profiles, particularly those having high depth-to-width ratios or using difficult-to-grind materials. A typical application is the form grinding of turbine blade root profiles of high-temperature superalloys. Adaptation of the process to cylindrical grinding is known as *deep grinding*, and it involves initial plunge grinding into the stationary workpiece, before the workpiece is rotated slowly relative to the wheel. The complete stock removal is then completed in one revolution of the workpiece.

In creep feed grinding a large amount of material is removed in one pass, but it is usual practice to use a final finishing pass to improve flatness and accuracy of the surface produced. This finishing pass is usually taken after the wheel has been redressed for the next part, and hence the wheel is very sharp. Thus a typical machining sequence is as follows [3]:

1. A freshly dressed wheel is used for a skim cut of 0.05 mm (0.002 in.) on the previously cut part.
2. The wheel is then used for the creep feed rough grinding of the next part.
3. The wheel is redressed and makes the final skim cut as in 1.

The dressing of the wheel is usually done by diamond roll dressing or conventional crush dressing is also used [3].

10.11.3 Low-Stress Grinding

By careful selection of grinding conditions, the occurrence of tensile residual stresses near the surface of the workpiece can be largely eliminated [25]. Low-stress grinding is used on workpieces that will be subjected to high stress or stress corrosion environments.

For low-stress grinding, soft and open wheels are used at lower cutting speeds than for conventional grinding, typically 18 m/s rather than 30 m/s. Oil-based grinding fluids are also used, with low infeed rates (0.005 mm per pass), to minimize frictional effects at the work surface. Figure 10.20 shows residual stresses for low-stress, conventional, and high-stress grinding conditions, the latter corresponding to dry grinding with a relatively hard wheel. For low-stress grinding conditions a small compressive residual stress is induced near the work surface, which can be beneficial to the fatigue resistance of the workpiece.

PROBLEMS

1. It is required to slit a 500-mm-square section block of tool steel on a large horizontal surface grinder using a grinding wheel 5-mm-wide and 2 m in diameter. The workpiece removal parameter is 50 p m^3/sN, and the wheel removal parameter is 20 p m^3/sN. Estimate the time taken to complete the operation if the downfeed is set at 2 mm per cutting stroke and the machine operates at 0.2 cutting stroke per second. Also, estimate the thrust force if the threshold thrust force is 900 N and if the workpiece traverse speed is 150 mm/s.

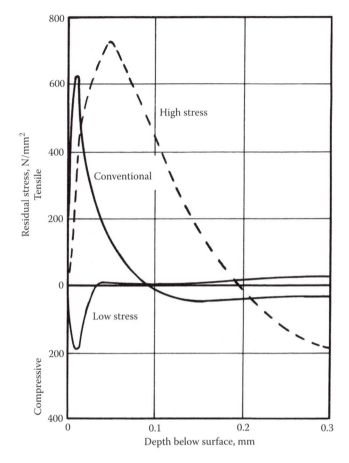

FIGURE 10.20 Residual stresses induced in the workpiece by various grinding conditions. (From Werner, P.C., *SME Technical Paper*, MR 79-3 19, 1979.)

2. A 20-mm-wide groove in a 50-mm diameter steel shaft is to be plunge-ground on a cylindrical grinder. The depth of the groove is to be 10 mm, and the material is easy to grind with a hardness of Rockwell C 60. The rotational frequencies of the workpiece and wheel are 10 and 200 s⁻¹, respectively. The wheel is designated A7OD8V, has a diameter of 300 mm, and is dressed with a feed of 0.1 mm at a depth of 0.01 mm. If the machine infeed speed is set at 0.2 mm/s, estimate the total time for the operation, including spark-out, where 90% of the material remaining after the infeed is stopped is removed. Assume that the effective stiffness of the grinding system in the thrust force direction is 1.6 MN/m.

REFERENCES

1. Tönshoff, H.K., Arendt, C., and Ben Amor, R., Cutting of Hardened Steel, *Ann. CIRP,* Vol. 49/2, 547–566, 2000.
2. Subrumanian, K., Hitchiner, M., and Redington, P., Grinding is Alive and Well, *Manuf. Eng.*, May, 91–100, 2002.
3. Salmon, S.C., *Modern Grinding Process Technology*, McGraw-Hill, NY, 1992.
4. Shaw, M.C., The Grinding of Metals, Proc. Conf. Tech. Eng. Manuf., Inst. of Mech. Eng., London, 433, 1958.
5. Backer, W.R., Marshall, F.R., and Shaw, M.C., The Size Effect in Metal Cutting, *Trans. ASME,* Vol. 74, 1, 61, 1952.
6. Peklenik, J. and Opitz, H., Testing of Grinding Wheels, *Adv. Mach. Tool Des. Res.*, Pergamon Press, Oxford, 163, 1962.
7. Grishrook, H., Cutting Points on the Surface of a Grinding Wheel and Chips Produced, *Adv. Mach. Tool Des. Res.*, Pergamon Press, Oxford, 155, 1962.
8. Hahn, R.S., On the Nature of the Grinding Process, *Adv. Mach. Tool Des. Res.*, Pergamon Press, Oxford, 129, 1962.
9. Hahn, R.S. and Lindsay, R.P., Principles of Grinding, *Machinery,* July–Nov., 1971.
10. Kalpakjian, S. and Schmid, S.R., Manufacturing Processes for Engineering Materials, 4th ed., Prentice Hall, NJ, 2003.
11. Hahn, R.S., A Survey on Precision Grinding for Improved Product Quality, Proc. 25th Int. MTDR Conf., Birmingham, 3, 1985.
12. Boothroyd, G., *Fundamentals of Metal Machining and Machine Tools,* McGraw-Hill, NY, 1975.
13. Hahn, R.S., The Effect of Wheel-Work Conformity in Precision Grinding, *ASME Paper* 54-A-178, 1954.
14. Lindsay, R.P., The Effect of Conformity, Wheel Speed, Dressing, and Wheel Composition on the Grinding Performance of Conventional and High Temperature Alloys, *SME Paper* MR7O-803, 1970.
15. Lindsay, R.P., Dressing and Its Effect on Grinding Performance, *SME Paper* MR69-568, 1969.
16. Lindsay, R.P., On Metal Removal and Wheel Removal Parameters, Surface Finish, Geometry, and Thermal Damage in Precision Grinding, doctoral dissertation, Department of Mechanical Engineering, Worcester Polytechnic Institute, Worcester, MS, 1971.
17. Snoeys, R., Mans, M., and Peters, J., Thermally Induced Damage in Grinding, *Ann. CIRP.* Vol. 27/2, 571–582, 1978.
18. Malkin, S., Thermal Aspects of Grinding, Part I, Energy Partition, *Trans. ASME,* no. 73-WA/Prod., 1973.
19. Konig, W., Honscheid, W., and Lowin, R., Untersuchung der bein Schleifprozess entstehenden Temperaturen und ihre Auswirkungen aufdas Arheitsergebnis, *Forschungsb. Landes Nordrh. Nr.* 2648, Westdeutsches Verlag, 1977.
20. Brinksmeier, E., Heinzel, C., and Wittmann, M., Friction, Cooling, and Lubrication in Grinding, *Ann. CIRP,* Vol. 48/2, 581–598, 1999.
21. Radhakrishnan, V. and Achyutha, B.T., A Method for Reducing the Corner Wear in Plunge Grinding, *Proc. Inst. Mech. Eng.* Vol. 200, B1, 19, 1986.
22. Werner, P.C. and Lauer-Schmaltz, H., Advanced Application of Coolant and Prevention of Wheel Loading in Grinding, Proc. Int. Symp. on Metal Working Lubrication, *ASME,* 228, 1980.

23. Brown, F.L., Schierloh, L.H., and McMillan, A.R., High Speed O.D. Plunge Grinding, *SME Technical Paper* MR 79-952, 1979.

24. Werner, P.C., Application and Technological Fundamentals of Deep and Creep Feed Grinding, *SME Technical Paper,* MR 79-3 19, 1979.

25. Kohls, J.B. and Bellows, G., Low Stress Grinding: Its Parameters and Potential, *Manufacturing Engineering,* 38, October, 1976.

11 Manufacturing Systems and Automation

11.1 INTRODUCTION

The efficiency of a manufacturing facility can be measured in three basic ways: manufacturing cost, productivity, and profit.

Manufacturing cost involves not only the costs of machining, assembly, and so on, but also the costs of moving the components from machine to machine and temporary storage. It can be affected by the efficiency of the various manufacturing processes used and the way these processes are integrated or arranged in the factory. It is also affected by labor costs and the quality of the product design.

Productivity, in contrast, refers only to the efficiency with which labor is utilized. It is generally taken to mean the average output per man-hour. Thus, if factory A produces 300 refrigerators per week and employs 300 workers and factory B can produce only 150 refrigerators per week with the same number of workers, the workers in factory A are twice as productive as those in factory B. This higher labor efficiency of factory A does not necessarily mean, however, that the refrigerators from factory A will be less expensive than those from factory B. Perhaps factory A is using much more expensive automated manufacturing equipment that offsets the savings in labor.

Finally, the profits made by the company will be affected by changes in manufacturing costs or productivity. Manufacturing cost, productivity, and profit cannot be optimized simultaneously, as shown by the example in Chapter 6, where the conditions for minimum cost, minimum production time (maximum productivity), and maximum profit rate were all different.

The purpose of this chapter is to discuss the means of improving productivity in manufacturing by improved manufacturing systems and increased automation. Automation is defined here as any means of helping workers perform their tasks more efficiently.

Before discussing particulars, it is necessary to explain that the approach to automation depends heavily on the volume of production.

11.2 TYPES OF PRODUCTION

Usually, three categories are used in describing the volume of production: mass production, large-batch production, and small-batch production. It may be assumed that mass production involves a volume of more than 1 million components per year and small-batch production less than a few hundred in each production run. Estimates of the proportion of total production in the United States in the batch categories vary from 70 to 85%. Because of the large proportion of manufacturing in the batch categories, much attention is paid to automation in batch manufacturing. The general characteristics of these types of production are as follows.

11.2.1 MASS OR CONTINUOUS PRODUCTION

Mass production is characterized by large numbers of components of the same configuration for which the demand is relatively constant and predictable. The quantities required are sufficient to keep manufacturing equipment loaded continuously or for long periods of time. Consequently, special-purpose machines and tools are used, which closely match the features of the components to be produced. High levels of automation can be justified economically and high productivity can be achieved.

11.2.2 LARGE-BATCH PRODUCTION

In large-batch production, large quantities of similar components are produced in batches. Some flexibility of operation of the machines is required, but many special-purpose machines and tooling, with associated long set-up times, can be justified.

11.2.3 SMALL-BATCH PRODUCTION

In small-batch production a wide variety of components for which the demand fluctuates must be produced in small batches. General-purpose machines, not specifically designed for particular components, must be used. These machines must be frequently reset for batches of different components, resulting in considerable loss of productive time. Automation for this section of industry must be flexible in operation and readily reprogrammed for a wide variety of components.

11.3 TYPES OF FACILITIES LAYOUT

Several different layouts of machine tools and equipment are used in manufacturing, and the appropriate type depends mainly on the quantities and type of product being processed.

11.3.1 FUNCTIONAL OR PROCESS LAYOUT

The functional or process form of layout has been used traditionally for batch manufacture. The plant is organized into sections with similar machines together, that is, a turning section, a milling section, and so on (Figure 11.1). Since most components require a variety of operations, they must be transported frequently from one process-specialized section to another. This results in poor work flow

Functional layout, machines grouped by process

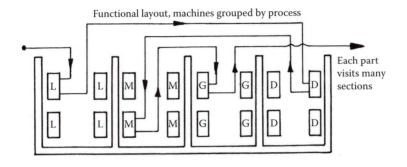

Each part visits many sections

Group layout, machines grouped by part family

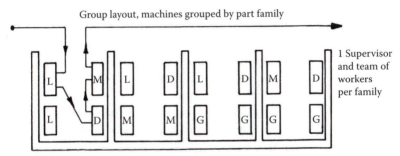

1 Supervisor and team of workers per family

Machines used in a variety of sequences in each cell

Line layout, 1 line per component of family

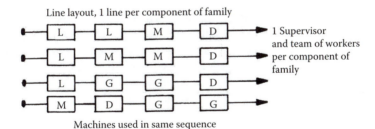

1 Supervisor and team of workers per component of family

Machines used in same sequence

KEY: L = lathe M = miller G = grinder D = drill

FIGURE 11.1 Types of machine tool layout.

and long and uncertain throughput times, giving rise to considerable problems with the scheduling and control of production. High levels of work in progress are also associated with this plant organization. Responsibility for manufacture is in the hands of several different supervisors.

The functional or process layout is very flexible for situations of fluctuating demands, and some benefits are derived from grouping similar skills together in the individual sections.

11.3.2 LINE LAYOUT

In the line layout, machines are arranged in the sequence of operations, so that parts flow from one machine to the next during processing. Large numbers of components with the same processing sequence are required, and consequently flow lines are usually restricted to mass and large quantity production. Such flow lines are usually associated with high levels of dedicated automation.

Short, predictable throughput times are typical for line layouts, together with low levels of work in progress. As a result, production control is considerably simplified. Automatic flow lines require high capital expenditure and considerable design and development costs. They are relatively inflexible and generally unsuitable for situations with variable product demand or frequent product design changes.

11.3.3 GROUP OR CELLULAR LAYOUT

The group or cellular form of layout can be used for small-hatch production when a relatively stable distribution of component types exists. This is usually the case for companies involved in the manufacture of a particular type of product such as machine tools, hydraulic pumps, and so on. Cells or groups of machines are formed to process families of related parts, requiring similar processes in a variety of sequences (Figure 11.1). If possible, components should be completely processed in the particular cell. Compared with the functional layout, the cellular layout results in improved workflow and lower throughput times, together with improved labor and machine utilization. Production control is generally easier and levels of work in progress are reduced.

The successful operation of a cellular layout requires increased labor mobility and flexibility, as there are usually more machines than operators in each cell. The key machines are fully loaded and each requires an operator, but several secondary machines may be used by fewer operators who move from machine to machine as required. A full description of the design and operation of cellular layouts is given by Gallagher and Knight [1]. As will be seen later, an extension of this basic approach leads to modern flexible manufacturing systems (FMS).

Multi-product or group flow lines may be used for some families of components, where each member of the family has a similar sequence of operations so that unidirectional flow lines can be utilized. General-purpose machine tools are used, usually with work handling performed by mechanical conveyors. Some resetting of machines is required for a change of product but advantage can be taken of the similarities between components to keep this to a low level.

Table 11.1 summarizes the general characteristics of the various types of facilities layouts used.

11.4 TYPES OF AUTOMATION

Automation used in manufacturing can be divided into two basic types: fixed automation and programmable automation.

TABLE 11.1
General Characteristics of Different Plant Layout Types

	Type of Layout		
Characteristics	Functional	Group	Line
Specialization	By process	By component type	By component type
Material flow between machines	In batches	Nearly continuous	Continuous
Material throughput times	Long	Short	Shorter
Stocks of work in progress	High	Low	Lower
Responsibility for quality	Many foremen per part	One foreman per part	One foreman per part
Responsibility for delivery by due date	Many foremen per part	One foreman per part	One foreman per part
Investment in special tooling	High: one set per operation per part	Low: one set per tooling family	High: one set per line; one line per part
Investment in buildings	High	Lower	Lower
Control of material flow	Complex	Simpler	Simple

11.4.1 FIXED AUTOMATION

This type of automation is dedicated to a specific task or a limited range of tasks. Cams and other mechanisms are often used to control motions. This form of automation requires high initial investment and long set-up times. It is usually restricted to large-volume production for which high productivity can be achieved. The equipment is relatively inflexible and cannot be readily adapted to new tasks or products. Transfer lines and automatic lathes are examples of fixed or dedicated automation.

11.4.2 PROGRAMMABLE AUTOMATION

This form of automation can be used for lower-volume production since the equipment can be readily reprogrammed for new tasks and products. Two basic forms of programmable automation exist: program sequence control and numerical control.

11.4.2.1 Program Sequence Control

A sequence of motions or operations is controlled by timers, relays, or programmable controllers. The actual motions are limited by preset mechanical stops and limit switches. Many capstan and turret lathes use sequence control, programmed by microprocessor-based sequence controllers.

11.4.2.2 Numerical Control

The most important form of programmable automation for manufacturing is numerical control (NC). The machine controller is fed with a program of instructions in a format specific to the controller of the machine. The position, velocity, and acceleration of the machine motions can be controlled, together with auxiliary functions such as tool changing, turning the coolant supply on-off, and so on. As NC for machine tools was being developed, programs initially consisted of numerical data only, hence the term numerical control. However, in modern controllers a mixture of alphabetic codes and numerical information makes up the program. Some details on the programming of NC machine tools are given in Chapter 12.

Numerically controlled machine tools are very flexible in operation and are highly suited to small-batch manufacturing. When the job changes, a new program of instructions is prepared and the changeover to the new task can be achieved quickly. Fixturing requirements are usually simple, which also facilitates the changeover from one component to another. Programming is carried out away from the machine tool and does not significantly interfere with production time.

Numerical control has been applied to a wide range of manufacturing equipment, including metal-cutting machine tools, welding, sheet metal working, tube bending, flame cutting, and many others. This discussion is limited to metal-cutting machine tools, the most widespread application of NC. Currently, most machine shops have at least one modern machine tool that utilizes this form of control.

Some different forms of automation in machine tools will now be discussed.

11.5 TRANSFER MACHINES

Transfer machines are often the most suitable method for a continuous flow of identical or very similar components in the mass production of consumer goods. They are basically special-purpose systems in which the components are automatically transferred from one machining head to another, often called *unit head machines*. Each machining head carries out one operation until, when the component has negotiated the complete transfer system, all the necessary operations have been completed.

With transfer machines, the component is automatically transferred from one machining operation to the next either by a circular indexing table (the rotary-transfer system shown in Figure 11.2a) or along a conveyor (the in-line transfer system shown in Figure 11.2b). In the in-line system the component is held on a special pallet or fixture, and arrangements have to be made to return empty pallets to the beginning of the line. Often a rapidly moving return conveyor parallel to the main transfer line is used to return the empty pallets.

Transfer machines are often constructed in modular fashion; the machining heads for drilling, boring, and so on, are standard items and are added to the system at the appropriate stations. Rotary-transfer machines can usually accommodate

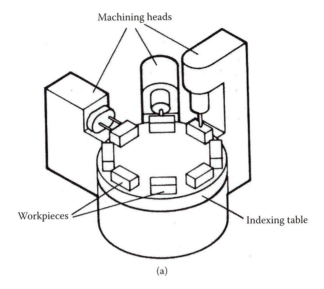

Machining heads

Workpieces

Indexing table

(a)

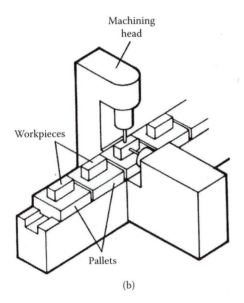

Machining
head

Workpieces

Pallets

(b)

FIGURE 11.2 Transfer machines. (a) Rotary; (b) in-line.

only six or fewer stations because of restrictions on space. The size of an in-line machine can be almost unlimited, and in-line machines have been employed, for example, to perform all the machining operations on an automobile engine cylinder block. In addition to the machining operations, these systems also incorporate stations for inspection and sometimes assembly operations.

11.5.1 ECONOMICS OF TRANSFER MACHINES

When the optimum machining conditions for transfer machines operating on the indexing principle are considered, it is necessary to account for the condition that when one machining station is stopped in order to change tools, the whole indexing transfer machine must stop. It must also be remembered that no advantage exists in having the machining time vary from station to station. If one station has a short machining time, the cutting speed at that station can be reduced without affecting the cycle time for the machine with the advantage that tool costs will be reduced at the station and down time reduced for the whole transfer machine.

For analysis, assume that on a transfer machine having N_s stations the same cutting-tool material is used at each station. For any station s, therefore, the tool-life is

$$\frac{v_s}{v_{r_s}} = \left(\frac{t_{r_s}}{t_s}\right)^n \tag{11.1}$$

where

v_s = cutting speed
t_s = tool life
v_{r_s} = cutting speed for a tool life of t_{r_s}

The machining time t_m for each operation is given by

$$t_m = \frac{K_s}{v_s} \tag{11.2}$$

where K_s is the distance moved by the tool corner relative to the workpiece during the machining time. If Equation 11.1 and Equation 11.2 are combined,

$$t_s = t_{r_s}\left(\frac{v_{r_s} t_m}{K_s}\right)^{1/n} \tag{11.3}$$

The total time to produce N_b components is now given by

$$N_b\left(t_t + t_m\right) + N_b\sum\left(t_m t_{ct_s}/t_s\right) \tag{11.4}$$

where

t_t = time taken to index the machine and advance and withdraw the tools
t_{ct_s} = tool changing time at any station, s
Σ = sum of the terms for all the stations on the machine

Thus, the average production time per component is

$$t_{pr} = t_t + t_m + \sum \frac{t_m t_{ct_s}}{t_s} \tag{11.5}$$

If the rate (cost per unit time) for each station (including the section of the transfer machine associated with that station) is M_s, the total rate for the transfer machine will be M_s, and the production cost per component C_{pr} will be given by

$$C_{pr} = \sum \left(M_s t_{pr} \right) + \sum \frac{t_m C_{r_s}}{t_s} \tag{11.6}$$

where C_{r_s} is the cost of providing a new tool at station s.

Substitution of Equation 11.3 and Equation 11.5 in Equation 11.6 and rearrangement gives

$$C_{pr} = \sum M_s \left(t_m + t_t \right) + t_m^{1-1/n}$$
$$\times \left\{ \sum M_s \sum \left[\left(\frac{K_s}{v_{r_s}} \right)^{1/n} \frac{t_{ct_s}}{t_{r_s}} \right] + \sum \left[\left(\frac{K_s}{v_{r_s}} \right) \frac{C_{t_s}}{t_{r_s}} \right] \right\} \tag{11.7}$$

Equation 11.7 can now be differentiated with respect to t_m to find the machining time t_{m_c} giving minimum production costs. Thus,

$$t_{m_c} = \left(\left(\frac{1}{n} - 1 \right) \left\{ \sum \left[\left(\frac{K_s}{v_{r_s}} \right)^{1/n} \frac{t_{ct_s}}{t_{r_s}} \right] + \frac{\sum \left[\left(K_s / v_{r_s} \right)^{1/n} \left(C_{r_s} / t_{r_s} \right) \right]}{\sum M_s} \right\} \right)^n \tag{11.8}$$

The minimum production costs per component can now be found by substitution of t_{m_c} for t_m in Equation 11.7, and the optimum tool life at each station can be obtained from Equation 11.3.

To illustrate the effect of combining a series of machining processes on one indexing transfer machine, it will now be assumed that a series of similar operations are to be performed at each station and that the parameters at each station are identical. In this case Equation 11.8 becomes

$$t_{m_c} = \left\{ \left(\frac{1}{n} - 1 \right) \left[\left(N_s \frac{t_{c_t}}{t_r} \right) + \frac{C_t}{M t_r} \right] \right\}^n \frac{K}{v_r} \tag{11.9}$$

Substitution of Equation 11.9 in Equation 11.3 now gives the optimum tool life at each station:

$$t_c = \left(\frac{1}{n} - 1 \right) \left(N_s t_{c_t} + \frac{C_t}{M} \right) \tag{11.10}$$

Comparison of this result with the optimum tool life for a single operation (obtainable by writing $N_s = 1$ in Equation 11.10) indicates that as the number of stations on a transfer machine increases, so does the optimum tool life. Thus, the optimum cutting speeds would be lower on a transfer machine than those on a single machine tool.

11.5.2 EXAMPLE

Assume that high-speed steel tools are used ($n = 0.125$) on a five-station indexing transfer machine and that the values of K_s, t_{ct_s}, C_{t_s}, v_{r_s}, and t_{r_s} for each station are as shown in Table 11.2. From these figures the following values are obtained:

$$\sum M_s = \$0.0155/s$$

$$\sum \left(\frac{K_s}{v_{r_s}} \right)^{1/n} \frac{t_{ct_s}}{t_{r_s}} = 1.49 \times 10^{19} s^8$$

$$\sum \left(\frac{K_s}{v_{r_s}} \right)^{1/n} \frac{C_{t_s}}{t_{r_s}} = \$6.0 \times 10^{17} s^7$$

From Equation 11.8, the optimum machining time is

$$t_{m_c} = \left[7 \left(1.49 \times 10^{19} \right) + \frac{6 \times 10^{17}}{0.0155} \right]^{0.125} = 373 \ s$$

TABLE 11.2
Machining Parameters for a Five-Station Indexing Machine

Station	K_s, m	t_{ct_s}, s	C_{t_s}, \$	M_s, \$/s	t_{r_s}, s (for $v_{r_s} = 1$ m/s)
1	183	180	2	0.003	65
2	243	60	3	0.0027	65
3	61	120	2	0.0038	65
4	91	120	2	0.003	65
5	122	240	1	0.003	65

From Equation 11.7 if the transfer time t_t is 12 s, the minimum production cost per component is

$$C_{\min} = 0.0155\left(12+373\right) + \frac{\left(0.0155 \times 1.49 \times 10^{19}\right) + \left(6 \times 10^{17}\right)}{10^{18}}$$

$$= 5.97 + 0.83$$

$$= \$6.80$$

where $C_{\min}$ is the minimum cost of production, that is, the minimum value of C_{pr}. Finally, the optimum cutting speed at each station can be obtained from Equation 11.2; thus, for stations 1, 2, 3, 4, and 5 the cutting speeds are 0.49, 0.65, 0.16, 0.24, and 0.33 m/s, respectively.

11.6 AUTOMATIC MACHINES

Where transfer machines are used exclusively for mass production, the class of machine tool referred to as automatic is used in both mass and large-batch production. The most common of these machines is the automatic lathe, sometimes referred to simply as an "automatic."

Those single-spindle automatic lathes that are designed to produce components from bar stock will continually produce these components without needing attention, except for the replacement of the bar stock and the changing of worn tools. Feeding of the bar stock through the collet is carried out intermittently after each machined component has been cut off, or, with the swiss-type automatic, the bar is rotated and fed continuously while the machining operations are carried out. Where feeding is intermittent, the machine is basically an automatic turret lathe; where feeding is continuous, the workpiece is fed past the various tools mounted radially around the working region.

Movement of the tool slides is often controlled by flat disk cams, which must be produced for the particular job. In general, in a single-spindle automatic, three cams are required, each cam being milled with an end mill of the same diameter as the cam follower. Clearly, the expense of producing cams for an automatic lathe can only be justified for large production runs.

Single-spindle automatic lathes with chucks are used for those components that cannot be machined from bar stock. In these machines the individual workpieces are loaded manually with one operator able to service a group of machines.

A final class of automatic lathe is the multi-spindle automatic. These lathes are basically flexible, rotary-transfer machines where each spindle holding a collet or chuck is indexed around the various tooling positions so that during each indexing cycle a completed component is produced. Drum cams instead of disk cams are used in these machines.

Another class of machine tool that can be mechanized is the internal or external cylindrical grinder. When fitted with mechanisms for loading workpieces automatically and with automatic wheel-dressing devices, these machines can continually produce finished components. The workpieces are often stacked in a magazine, and the operator simply ensures that the magazine is replenished with workpieces at suitable intervals. Because the components are small and the machining forces light, magnetic chucks are often employed to facilitate automatic work holding.

11.6.1 PLANNING FOR MULTI-SPINDLE AUTOMATIC LATHES

On multi-spindle automatic lathes parts can be produced from bar stock or from separate workpieces that are gripped in a chuck or fixture. Figure 11.3 shows the construction features of a typical six spindle bar machine [2]. A bar machine uses the work material in bar form, with one bar for each spindle of the machine. When one workpiece has been fully machined, it is cut off from the bar, which is then fed forward and the same sequence of operations repeated.

The spindles are arranged radially about the end working tool slide (also known as tool slide, end slide, or end working slide), as shown in Figure 11.4. The number of spindles is usually four, five, six, or eight. Each spindle is rotated at the same speed and has one cross slide and one end working position on the end slide. For bar machines, one cross slide must be used for cutting off, whereas for chucking machines, one position must be used for loading and unloading. Figure 11.4 also shows the tooling area for a typical six-spindle bar machine. All of the cross slide tools are fixed to the face of the machine. Their only motion is towards and away from the side of the bar, with the motion governed by a series of cams. Each cross slide is independently controlled and thus the feed rates for each cross slide tool can be different. The end working tools are usually mounted on the tool slide such that they all move simultaneously towards and away from the end of the bars. Thus only one feed rate for all the tool slide tools is possible.

On completion of one machining cycle, the spindle carrier indexes, and one of the bars arrives at the cutoff station. The productive output of these machines is related to the number of spindles employed. Multi-spindle automatics require long setup times, and so they are usually employed when batch sizes are large (above 50,000 components).

To obtain a cycle time estimate for a component, it is necessary to process plan the component [3]. This involves identifying each operation to be performed on the workpiece, selecting the appropriate tooling for the operations, and then optimizing the tool layout to achieve a minimum cycle time. For each tool, the number of revolutions of the workpiece required to complete its operation (*revs*) must be determined. The recommended cutting speed for each tool determines the required rotational speed for the corresponding spindle. The lowest required spindle speed is the speed at which all the spindles must rotate, and is known as the machine speed setting, n_s.

The cycle time is obtained by selecting the tool requiring the maximum number of *revs*, and dividing by n_s. Optimization of the tooling layout to decrease

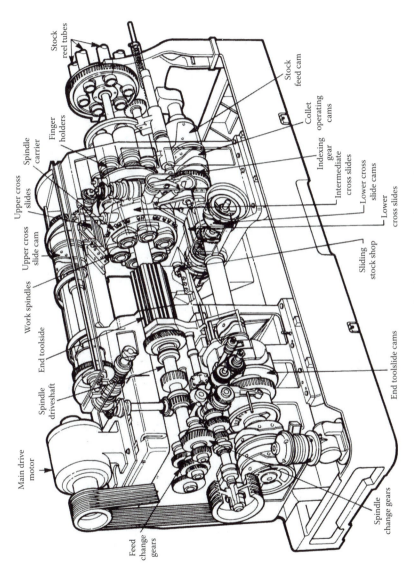

FIGURE 11.3 Typical six-spindle automatic lathe with bar feed. (National Acme Co.)

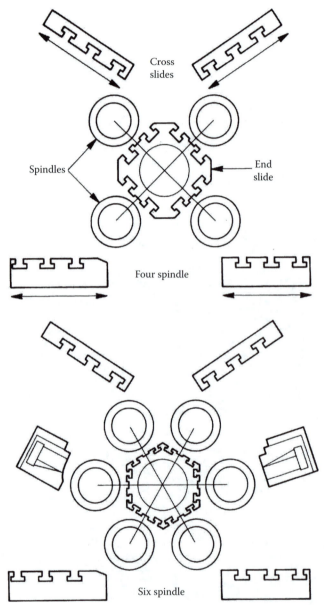

FIGURE 11.4 Layout of cross slides and end tooling on a multi-spindle automatic lathe. (National Acme Co.)

cycle time is an iterative procedure which involves the redistribution of operations between the stations. This is achieved by increasing the amount of tooling or by altering the type of tooling used to perform the required operations, and thus is

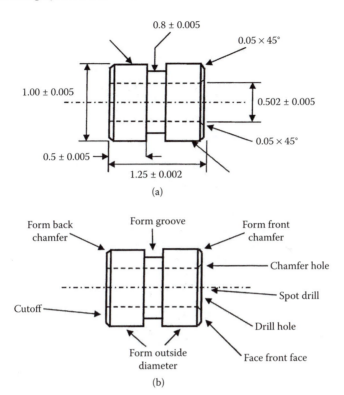

FIGURE 11.5 Sample part to illustrate planning of operations for a multi-spindle automatic lathe. (a) Part dimensions; (b) operations required for the part.

restricted by component features, and the limitations of the machine. The longest machining operation determines the cycle time.

In order to illustrate the planning process, Figure 11.5a shows a simple part that will be called *sample part*. The stock diameter is obtained from the next available increment above the maximum part diameter. For the *sample part*, a bar stock diameter of 1 1/16 in. (27 mm) is suitable, and thus a 1 3/8 in. six-spindle New Britain Model 52 would be a suitable machine for this part.

A list of the operations that are required to produce the component features must be prepared. For bar machines, a cutoff operation is usually required, however it may be combined with a forming operation (using a form tool) in special situations. All operations are first randomly listed, so that bias is not introduced into the selection of appropriate tooling. The list of operations required for the *sample part* is as shown in Figure 11.5b. The list must be examined and divided into those operations that require using standard tooling (i.e., standard operations) and those requiring nonstandard tooling (i.e., nonstandard operations). Standard tooling is defined as that which is commonly available from commercial suppliers.

Having identified the initial tooling requirements, the amount of tooling should be reduced as much as possible by combining operations. If a nonstandard tool is required, then as many operations as possible should be combined with it. The standard operations should be examined with the aim of combining operations, provided standard tooling can still be used. This process should reduce the number of tools used and provide greater control of accuracy between features.

The tooling layout sheet is shown in Table 11.3; this is used to obtain an initial estimate of the cycle time. With the exception of the cutoff tool, the tools should be placed, in sequence order, in the first technically feasible spindle position on the machine. The cutoff tool should be placed in the last position. For the *sample part* machined using a New Britain six-spindle Model 52 lathe, the initial tooling layout would be:

Tool Slide   Position 1. Spot & chamfer
                Position 2. Drill
Cross Slides   Position 1. OD form
                Position 6. Cutoff

The controlling diameter of any cut is the maximum diameter of the workpiece which is in contact with the tool. This diameter is used to determine the rated surface cutting speed of the workpiece for the particular tool and the recommended feed, f. The depth of cut is defined as the distance that the tool must travel to complete the cut, in the direction parallel to the slide upon which it is mounted. The width of cut, w, is defined as the length of the contact between the tool and the workpiece, Figure 11.6 and is restricted by the limiting recommended width to cut diameter ratios shown. Similarly drill diameters are restricted by limiting depth to diameter ratios for different materials as shown in Figure 11.6.

The motion of the tools is automatically controlled with the use of cams. The distance that the tool travels whilst under feed motion is used to find the number of revolutions required to complete the cut (*revs*), once a recommended feed is known. This distance is known as the rise or throw of the cam, and it differs from the depth of cut in that additional allowances must be made for machine backlash errors and also for required tool overtravel. The tool approach allowances for machine backlash errors are constant for all tools in a particular machine and typical values of which are shown in Table 11.4. The amounts of overtravel required to complete various operations are also given in Table 11.4. The overtravel for the cross slide operations ensures that the tool travels past the center of the workpiece. The overtravel for the threading tools accounts for the backlash in the reversing clutch in the special attachments.

Recommended cutting speeds are usually quoted as the surface speed of the workpiece. The corresponding spindle speeds must be determined by dividing by the maximum diameter to be cut. For the *sample part* these values are given in Table 11.3. The tool which requires the minimum machine speed setting, n_s, is called the *speed control tool*. For the *sample part*, the OD form tool is the speed control tool with a maximum rated machine speed of 449 rpm.

TABLE 11.3
Initial Tool Layout Sheet for *Sample Part*

Part Name: Sample part								
Material: 1018 cold drawn steel								
Stock Size: 1.063								

Workplan No. 1.1
Machine: NB Model 52 - 6 spindle 1.375 in.
Index Time (s): 1.00

Station	Operation	Actual Cut Dimensions (in.)			Required Rise (in.)	Recommended		Rated rpm	Required revs
		Dia	Depth	Width		Cutting Speed (ft/min)	Feed (in./rev)		
[1]	Spot & chamfer	0.522	0.261		0.339	82	0.0043	600	79
1#	OD form	1.063	0.281	1.250	0.328	125	0.0015	449	218
[2]*	Drill	0.500	1.250		1.428	82	1.250	626	332
2									
[3]									
3									
[4]									
4									
[5]									
5									
[6]	Cut off	0.980	0.239	0.125	0.316	136	0.0019	530	166
6									

Machine speed setting (rpm) = 449
Maximum number of revs = 332

Cycle time (s) = 45.40

[-] Tool slide mounted tool
* Revs control tool
# Speed control tool
Tool slide feed setting (in./rev) = 0.0043

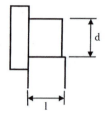

Form tool ratios	
d (in)	Limiting l/d ratio
0.125–0.156	1.0
0.156–0.818	1.5
>0.818	2.5

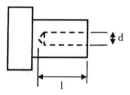

Operation	Drilling ratios (1/d)	
	Steel	Brass and aluminum
First	3.5	5.0
Second	2.0	3.5
Subsequent	1.0	2.0

FIGURE 11.6 Limiting dimensions for form tools and drills.

TABLE 11.4
Allowances for Tool Approach and Overtravel

Tool Type	Approach Allowance
Tool slide mounted tools	0.063
Cross slide mounted tools	0.032
Threading tools	2 times thread pitch

Operation Type	Overtravel Allowance
Facing	0.015
Cutoff	0.015
Shaving	0.015
Threading tools	2 times thread pitch

The number of revolutions required to complete each operation (*revs*) is calculated using the required rise, C_R (in or mm) and the recommended feed

$$revs = C_R/f \tag{11.11}$$

Thread cutting tools present a different situation where certain threading ratios for the required threading attachments must be known. The threading on ratio, R_{tn}, is defined as the effective speed of the tool as it cuts the thread, i.e., tool rpm/spindle rpm, a typical value of which is 0.25. The threading off ratio, R_{to}, is the effective speed of the tool as it retracts from the thread, with a typical value of 0.500. For self opening die heads, the required *revs* is then found by,

TABLE 11.5
Typical Machine Indexing Times

Machine Bar Size (in.)	Indexing Time (s)
Up to 0.75	0.5
0.875 to 1.375	1.0
1.5 and greater	2.5

$$revs = \frac{C_R}{fR_{tn}} \qquad (11.12)$$

and for solid taps or acorns dies,

$$revs = \frac{C_R}{f}\left(\frac{1}{R_{tn}} + \frac{1}{R_{to}}\right) \qquad (11.13)$$

The operation requiring the maximum number of workpiece revolutions ($revs_{max}$) to complete its operation is named the *revs control tool*, and this governs the cycle time. For the *sample part*, it can be seen in Table 11.3 that the drill is the *revs control tool* requiring 332 revolutions of the workpiece.

The cycle time for the particular process plan may be calculated from,

$$t_c = \frac{60\,revs_{max}}{n_s} + t_{in} , \text{ s} \qquad (11.14)$$

where t_{in} is the indexing time, which is the nonproductive portion of the machine cycle during which the tools retract from the workpieces, the spindle carrier indexes, and the tools rapidly advance to a position where the feed is engaged. The indexing time is dependent upon the particular machine being used, but some typical values for a range of machine sizes are given in Table 11.5. For the *sample part* initial process plan, with a machine indexing time of 1.0 s, the cycle time is estimated to be 45.40 s.

11.6.1.1 Process Plan Optimization

The cycle time obtained is naturally dependent upon the nature of the process plan for which it was calculated. It is unlikely that the initial process plan is the one for which the cycle time is minimized. Reduction of the initial cycle time is termed the *optimization* of the process plan [3]. This requires the use of iterative procedures in which the number of revolutions required to produce the component (*revs*) is reduced and/or the maximum limit for the machine speed (n_s) is increased. Two strategies for reducing cycle time can be employed and these are known as *revs reduction* and *rated rpm increase*.

11.6.1.1.1 Revs Reduction

The two variables which control the number of revolutions required for an operation are cam rise and feed. Thus, a reduction in the number of *revs* can be accomplished in two ways: (1) by dividing the *revs control operation* into two or more separate operations which individually would require a smaller cam rise, and/or; (2) by altering the *revs control tool* to one which could be run at a greater feed. This latter procedure is rarely performed because it may either be a consequence of the former, or it may conflict with the rated rpm optimization described later.

The limiting factors for these optimization procedures include: the availability of positions on the machine for the additional operations; the technical feasibility of performing the additional operations; the availability of tooling which could be run at greater feed rates; and the ability to sub-divide certain operations — for example, thread rolling must be completed in one pass due to the effect of work hardening of the material.

It is important to bear in mind that the division of operations should still enable the use of standard tooling wherever possible. Similarly, the division of operations with nonstandard tooling should result in an increase in the usage of standard tooling wherever possible.

The *revs reduction* procedure can be summarized as follows:

1. Identify the *revs control tool*.
2. Divide the operation into two identical operations.
3. Calculate the new cam rises for both operations.
4. Check the feasibility of the new operations (i.e., the availability of positions on the machine, and the limiting drilling/forming ratios).
5. Determine the new cycle time for the new process plan iteration.
6. Repeat steps 1 to 5 until any of the limiting factors governing the division of operations is encountered.

11.6.1.1.2 Rated RPM Increase

An increase in the required spindle speed for individual operations is achieved by changing the tool material, starting with the *speed control tool*, to one that enables greater surface speeds to be used. Increases in required spindle speeds may also result from the division of operations in the *revs reduction* procedures as this may result in changes in the cut diameters. However, this is independent of the choice of tool materials. Therefore, to maximize the rated spindle speed for a particular tool, the following procedure should be adopted:

1. Identify the *speed control tool.*
2. Change the tool material to one which can be used at a greater surface cutting speed.
3. Calculate of the new rated spindle speed for the modified tool.
4. Determine the new cycle time for the new process plan iteration.
5. Repeat steps 1 to 4 for all tools until the cycle time cannot be reduced further by changing tool materials.

It is important to note that a change in tool material usually entails a change in the recommended feed in addition to the surface cutting speed. One consequence of this is that with each change in tool material (and thus feed), the number of revolutions required for that tool will also change. If the new tool then becomes the new *revs control tool*, then the whole previously described *revs reduction* procedure becomes invalid. This is because optimization of the new *revs control tool* would most likely result in the generation of a different process plan. The interdependence of the variables involved in the optimization procedures suggests that both the *revs* and spindle speed should be optimized simultaneously, using the following procedure:

1. Perform the *revs reduction* procedure for the whole operation sequence.
2. Perform the *rated rpm increase* optimization procedure for the first *speed control tool.*
3. If the change in this tool's recommended feed does not result in this tool becoming the new *revs control tool*, then continue the *rated rpm increase* procedures for the remaining tools until n_s is maximized.
4. If the *speed control tool* does become the new *revs control tool*, then repeat the whole *revs reduction* procedure, altering the operations sequence where necessary; identify the first *speed control tool* again (the previous one may have been eliminated due to the changes in the operations sequence); and repeat steps 1 to 4 for all tools, until the *revs* are minimized and spindle speed is maximized.

In practice, this simultaneous optimization results in the procedures becoming cumbersome. Therefore, for the following discussions, it will be assumed that any changes in tool materials, to increase the recommended surface cutting speed, will not result in any changes in the recommended feed rates. Thus, for a particular tool, the feed remains constant throughout the analyses.

For the *sample part*, the *revs reduction* and *rated rpm increase* optimization procedures are summarized in Table 11.6 and Table 11.7, respectively [3], with Iteration Number 10 representing the process plan at the final stage of the *revs reduction* optimization procedures. It is shown that a change in the tool materials enables the cycle time of the *sample part* to be reduced from 13.28 s to 7.39 s.

TABLE 11.6
Revs Reduction Optimization Summary for *Sample Part*

Iteration Number	Revs Control Tool	Solution	New Cam Rise (in.) Tool 1	New Cam Rise (in.) Tool 2	New Cycle Time (s)
1	—	Initial plan (Table 11.3)	—	—	45.40
2	Drill	Add an extra drill	0.678	0.813	30.19
3	OD form tool	Face then OD form	0.318	0.164	26.28
4	Drill	Add an extra drill	0.438	0.438	23.23
5	Cut off	Add a breakdown tool	0.197	0.182	22.08
6	Drill	Add an extra drill	0.366	0.376	15.57
7	OD form tool	Add a groove tool	0.074	0.164	14.83
8	Cut off	Add a breakdown tool	0.137	0.122	14.62
9	Drill (×2)	Add an extra drill	0.292	0.292	13.77

TABLE 11.7
Rate RPM Increase Optimization Summary for *Sample Part*

Iteration Number	Speed Control Tool	Solution	New Rated rpm	New Cycle Time (s)
10	OD form tool	Brazed carbide OD form Tool	898	13.28
11	Groove	Brazed carbide groove tool	934	11.82
12	Cut off	Carbide insert cut off tool	1059	11.23
13	Face	Solid carbide face tool	1120	10.56
14	Spot and chamfer	Coated HSS spot/chamfer tool	1199	10.16
15	Drill (×5)	Carbide tipped drills	1252	7.39

11.6.2 ECONOMICS OF AUTOMATIC MACHINES

It will be assumed that one operator can service N_a automatic machines, that the operator's rate (including overheads) is W_o', and that the rate for one machine (including overheads) is M_t'. The cost of production C_{pr} per component will then be given by

$$C_{pr} = \frac{C_b}{N_b} + \left(W_0' + M_t'\right)t_l + \left(\frac{W_0'}{N_a} + M_t'\right)t_m + \left[\left(W_0' + M_t'\right)t_{c_t} + C_t\right]\frac{t_m}{t} \quad (11.15)$$

where

    C_b = cost of setting up the machine (including manufacture of cams, etc.)
    N_b = batch size
    t_{c_t} = tool-changing time

C_t = cost of providing one new cutting edge

t_m = machining time

t = tool life

t_l = loading and unloading time

11.6.3 EXAMPLE

Assume the following values apply to a bar-fed single-spindle automatic lathe (where t_l is negligibly small) and one operator is servicing six machines: t_m = 30 s, t = 4000 s, t_{c_t} = 60 s, W_0' = \$2.20 × 103/s, M_t' = \$1.40 x 103/s, C_t = \$0.50, and C_b = \$1000.

Substitution of these figures in Equation 11.11 gives

$$C_{pr} = \frac{1000}{N_b} + \left[\left(\frac{2.2 \times 10^{-3}}{6} + 1.4 \times 10^{-3} \right) 30 \right]$$

$$+ \left[\left(2.2 \times 10^{-3} + 1.4 \times 10^{-3} \right) 60 + 0.5 \right] \frac{30}{4000}$$

$$= \frac{1000}{N_b} + 0.058$$

It can be seen that in this example a batch size of about 200,000 components is required before the set-up costs are reduced to 10% of the production cost per component.

These costs have been estimated for all of the *sample part* process plan iterations described earlier. The cost breakdown per component and the cycle time have plotted for each iteration as shown in Figure 11.7. This demonstrates that the ultimate effects of the process plan optimization procedures were to reduce the cycle time from 0.7657 min to 0.1232 min (i.e., 84%) whilst reducing the total component cost from \$0.4949 to \$0.1808 (i.e., 64%).

11.7 NUMERICALLY CONTROLLED (NC) MACHINE TOOLS

11.7.1 MAIN FEATURES OF NC SYSTEMS

The complete NC machine tool system consists of the three basic elements shown in Figure 11.8: the program of instructions, the controller unit, and the machine tool.

11.7.1.1 Program of Instruction

The program of instructions, or *part program*, as it is often called, is the detailed set of directions telling the machine tool what to do. It is a mixture of alphabetic

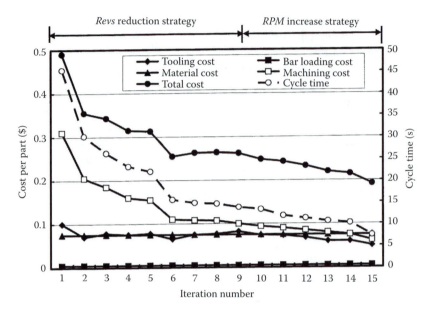

FIGURE 11.7 Cycle time and cost reductions during process plan optimization for the sample part. Adapted from Young, J., Masters thesis, University of Rhode Island.

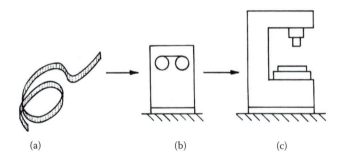

(a)   (b)   (c)

FIGURE 11.8 Basic components of a numerical control machine tool system. (a) Program; (b) controller; (c) machine tool.

codes and numerical data which is interpreted by the NC controller. In the past, programs were fed to the controller using a paper tape reader, this is now done electronically, usually from an on-line programming system remote from the machine. The program is prepared independently of the machine tool by a part programmer, and programming an NC machine tool is essentially a specialized process-planning exercise. Some details of the preparation of NC part programs are discussed in Chapter 12.

The program contains instructions telling the NC controller, for example, how to move the machine slides to generate the required surfaces on the workpiece. The displacements, velocities, and accelerations of the various machine tool components can usually be controlled. In addition, the various auxiliary functions of the machine tool can be controlled, including such items as coolant on or off, automatic tool change or turret index, and so on.

11.7.1.2 Machine Tool Controller

The machine tool controller accepts the part program of instructions and converts these instructions into signals that control the various actions of the machine tool. The controller forms part of a feedback control system for the motions of the machine tool slideways.

The machine tool control system also performs a number of functions (Figure 11.9). Input instructions are usually provided by the program but they can also be provided by the operator who enters information manually from a control panel. The control panel also contains data displays for the operator. Instructions are fed initially into a data input buffer, which stores a block of information. This represents one complete set of instructions for the next step in the machining sequence. Each block is usually terminated by an end-of-block character, equivalent to the carriage return-line feed at the end of an instruction for a computer. Each instruction in the block is then decoded and the information is stored within the controller and acted upon.

Other main functions of the controller are interpolation and feed rate generation. In order to machine complex curved surfaces at a constant feed rate, instructions must be fed to the drives for the various slides and spindles so that

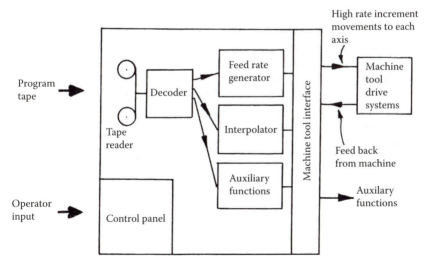

FIGURE 11.9 NC controller functions.

the sum of their individual motions produces the required shape. The controller breaks down (interpolates) these curves into small individual increments for each controlled motion of the machine tool.

Finally, signal input-output channels connect the controller to the drive motors, transducers, and other controls of the machine tool.

11.7.1.3 Machine Tool

The third element of the NC system is the machine tool itself. This is a general-purpose machine tool fitted with drive motors and controls to operate the spindle and slides automatically and to switch on or off the various auxiliary functions of the machine. Transducers are fitted to feed back data on the position and velocity of, for example, the slideways, to the controller.

Numerically controlled machine tools range in complexity from simple drilling machines to complex machines that utilize a number of controlled motions. The majority of NC machines in use are basically similar in configuration to the general purpose manual machines described in Chapter 1. The most versatile general purpose small NC machines are vertical milling machines. The most complex NC machine tools are so-called NC machining centers. These machines incorporate multi-axis control and are usually horizontal- or vertical-spindle milling machines, fitted with high-capacity tool magazines and automatic tool changers. The high capacity tool magazines increase the flexibility of the machines, by enabling complex parts with a wide range of features or a range of different parts to be processed without any change of tools. The objective in using these machines is to carry out a large amount of machining in one workpiece setup, and they are capable of performing a wide variety of operations including milling, drilling, tapping, and boring. Additional motions, using, for example, indexing tables, are usually possible so that several orientations of the workpiece can be presented during the program.

Numerically controlled lathes are widely used and most incorporate turrets of tools to increase the flexibility of operation. The most versatile machines are termed *turning centers* and these are capable of machining complex turned parts. The addition of so called live tools using secondary turret mounted driven spindles enables off axis milling and drilling operations to be carried out in one setup and this increases the versatility of turning machines considerably. Ranges of turning and mill-turning centers have been developed.

The availability of secondary milling spindles on turning machines has led to the application of a hybrid operation called *turn-milling*. In this operation, cutting with a rotating milling tool occurs while the workpiece is rotated slowly with the main spindle such that the spindle rotation effectively provides the feed motion. This process results in high material removal and improved chip control relative to conventional turning, with reduced chip loads. Eccentric diameters and odd shapes can also be readily machined. A number of highly versatile turn-milling centers have been developed that are capable of a wide variety of operations in one setup [4].

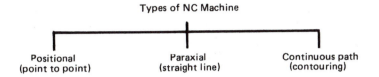

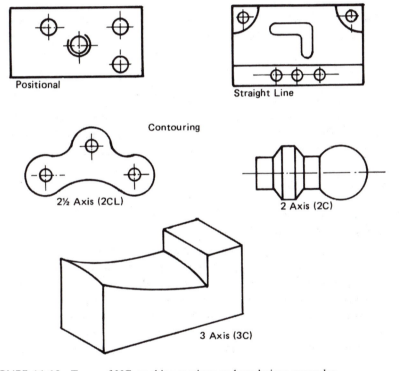

FIGURE 11.10 Types of NC machine motions and workpiece examples.

11.7.2 NC MOTIONS

There are three basic types of NC motions (Figure 11.10), as follows:

11.7.2.1 Point-to-Point or Positional Control

In point-to-point control the machine tool elements (tools, table, etc.) are moved to programmed locations and the machining operations performed after the motions are completed. The path or speed of movement between locations is unimportant; only the coordinates of the end points of the motions are accurately controlled. This type of control is suitable for drill presses and some boring machines, where drilling, tapping, or boring operations must be performed at various locations on the workpiece.

11.7.2.2 Straight-Line or Linear Control

Straight-line control systems are able to move the cutting tool parallel to one of the major axes of the machine tool at a controlled rate suitable for machining. It is normally only possible to move in one direction at a time, so angular cuts on the workpiece are not possible. Consequently, for milling machines, only rectangular configurations can be machined or for lathes only surfaces parallel or perpendicular to the spindle axis can be machined. This type of controlled motion is often referred to as *linear control* or a *half-axis of control* (see below). Machines with this form of control are also capable of point-to-point control.

11.7.2.3 Continuous Path or Contouring Control

In continuous path control, the motions of two or more of the machine axes are controlled simultaneously so that the position and velocity of the tool are changed continuously. In this way curves and surfaces can be machined at a controlled feed rate. It is the function of the interpolator in the controller to determine the increments of the individual controlled axes of the machines necessary to produce the desired motion. This type of control is referred to as *continuous control* or a *full axis of control*. The majority of NC machines now produced have continuous control on all axes.

Some terminology concerning controlled motions for NC machines has been introduced. For example, some machines are referred to as four-or five- or even six-axis machines. For a vertical milling machine, three axes of control are fairly obvious, these being the usual X, Y, Z coordinate directions. A fourth or fifth axis of control would imply some form of rotary table to index the workpiece or possibly to provide angular motion of the workhead. Thus, in NC terminology an axis of control is any controlled motion of the machine elements (spindles, tables, etc). A further complication is the use of the term half-axis of control; for example, many milling machines are referred to as 2½-axis machines. This means that continuous control is possible for two motions (axes) and only linear control is possible for the third axis. Applied to vertical milling machines, 2½-axis control means contouring in the X-Y plane and linear motion only in the Z direction. With these machines, three-dimensional objects have to be machined with water lines around the surface at different heights. With an alternative terminology the same machine could be called a $2CL$ machine (C for continuous, L for linear control). Thus, a milling machine with continuous control in the X, Y, Z directions could be termed a three-axis machine or a $3C$ machine. Similarly, lathes are usually two axis or $2C$ machines. Figure 11.10 illustrates some of this terminology and the corresponding shape-producing capabilities.

11.7.3 COMPUTERS IN NUMERICAL CONTROL

Rapid developments in and increased availability of low cost, high-speed computing systems have influenced the development and operation of NC systems. Early NC controllers were control systems with all the functions handled by hardware, logic circuits, switches, relays, and so on. Adaptation of these controllers to applications where more axes of control are acquired involved the addition of more hardware. The development of microcomputer systems has allowed many of the functions to be provided by software programs in modern NC systems. These systems are referred to as *CNC* systems, meaning computer numerical control systems. Practically all modern NC controllers are of this type.

The external appearance of a CNC machine is very similar to that of older NC machines. The program is stored in memory within the CNC controller. The various internal functions of the controller, interpolation, decoding, and so on, are carried out by software rather than hardware. Thus, CNC systems are sometimes referred to as soft-wired systems. Computer numerical control machine tools offer greater flexibility over older NC machine tools particularly because the user can edit programs at the machine and can use special program macros provided for frequently used cutting sequences. In modern integrated systems, the NC controller is simply a special purpose computer system that can be networked with others for data transfer, etc., in an open architecture configuration.

11.7.4 ECONOMICS OF NUMERICALLY CONTROLLED MACHINES

For NC machines, the cost of the preparations for machining a batch of components C_b will mostly consist of programming and tape preparation costs. There is generally a direct relationship between these costs and the machining time t_m as follows:

$$C_b = K_p t_m \qquad (11.16)$$

where K_p is the cost of programming and tape preparation per unit machining time.

The production cost per component is therefore given by

$$C_{pr} = \frac{K_p t_m}{N_b} + M\left(t_l + t_m\right) + \frac{t_m}{t}\left(M t_{c_t} + C_t\right) \qquad (11.17)$$

where

N_b = batch size
M = total machine and operator rate
C_t = cost of a sharp tool
t = tool life
t_m = loading and unloading and tool advance and tool withdrawal time
t_{c_t} = tool-changing time

11.7.5 EXAMPLE

Consider a large NC turret lathe. The tools are fitted with disposable carbide inserts costing $0.50 per cutting edge and taking an average of 60 s to reset. The cost of tape preparation per unit of machining time would typically be $0.l0/s, and a total cycle time of 300 s is assumed of which 50% is spent on actual chip removal. The lathe initially cost $125,000 with a payback period of 2 years, operator's wages are $4.00/hr, and 100% overheads are applied to the machine and operator. The total machine rate M therefore becomes

$$M = \frac{2 \times 125,000}{2 \times 50 \times 40 \times 3600} + \frac{2 \times 4}{3600} \approx \$0.02/s$$

Finally, the loading and unloading time is assumed to be 150 s and the tool life 300 s. Substitution of these figures in Equation 11.13 gives

$$C_{pr} = \frac{0.1 \times 150}{N_b} + 0.02\left((150 + 150) + \frac{150}{300}\left[(0.02 \times 60) + 0.5\right]\right) = \frac{15}{N_b} + 9.85$$

This result shows clearly that the system is economic for small batches and that the cost of programming and tape preparation ($15/N_b$) falls below 10% of the total production cost per component when the batch size is greater than 15.

11.8 COMPARISON OF THE ECONOMICS OF VARIOUS AUTOMATION SYSTEMS

Selection of the appropriate level of automation for a particular manufacturing situation is determined largely by the batch sizes and variety of parts to be processed. This can be illustrated by considering the economics of the manufacture of representative types of components on equivalent machines with the different types of automation fitted. Figure 11.11 shows a typical result [5] that compares the use of a center (engine) lathe, a turret lathe, and an NC turret lathe for manufacturing a typical component.

The more automated machines are not economic for very small hatches because the preproduction costs for setup and programming must be spread over a small number of components. For larger batches, the reduced cycle times for the automated equipment compensates for the greater preproduction costs. This result is typical and shows that manual machines are most suitable when only a few components are required and that NC machines are economic for medium-batch quantities, with more dedicated automation restricted to large-quantity production.

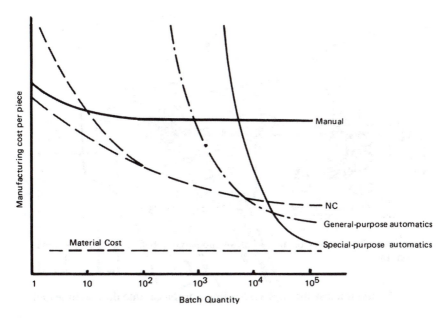

FIGURE 11.11   Variation of manufacturing costs with batch size for different levels of automation.

11.9   HANDLING OF COMPONENTS IN BATCH PRODUCTION

It has been mentioned that batch production can be automated only by using programmable devices that are preferably controlled by a central computer. One of the greatest problems in realizing the objective of complete computer-controlled manufacturing facilities is the development of general-purpose, programmable handling devices. One such device is an industrial robot such as that depicted in Figure 11.12. This robot is capable of six axes of motion and of handling objects weighing as much as 120 kg (300 lb). Such robots are increasingly employed in the loading and unloading of groups of machine tools.

Industrial robots are essentially programmable handling devices, which have several degrees of freedom and which can be fitted with tooling for a variety of tasks, including welding, paint spraying, and so on. A gripper or hand is required to handle workpieces and a wide variety of grippers have been devised for holding various items. In adapting a robot to a new task, the grippers must usually be redesigned to accommodate the new item to be handled, and the robot must be reprogrammed.

A common task for robots is loading workpieces into the machine tool and unloading the machined components. However, most applications of robots to date have been for dedicated or large-batch production tasks, rather than the

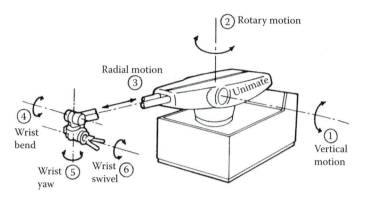

FIGURE 11.12 Six axes of motion of an industrial robot. (Courtesy of Unimation, Inc.)

small- to medium-batch production that their inherent programmability would suggest. This is partly because the grippers and tooling must be adapted to a new task, even though reprogramming the robot may be relatively simple. The use of robots along with NC machines in a cellular layout is attractive provided adaptation of the tooling and grippers to new parts can be made efficient. Engleberger [6] states "if parts are classified into families and machines aggregated into complementary families the handling of the parts throughout the manufacturing process becomes robot work." However, the grippers must be designed to handle the family of component parts without need modification.

Figure 11.13 shows a small cell of machines, serviced by a robot, used in the manufacture of commercial vehicle gearbox components [7]. The group of machines that form the automated manufacturing cell is made up of an automatic straightening press, two CNC cylindrical-grinding machines, and a pallet shuttle system. There are four basic families of components produced in the cell, with over 100 different components in all.

11.10 FLEXIBLE MANUFACTURING SYSTEMS

A logical step from the concepts of group layout and of NC machine tools and robotics are computer-controlled interlinked multi-station machining complexes, or flexible manufacturing systems (FMS) as they have become called. Such systems can be looked upon as highly automated cells manufacturing families of components.

The concept of FMS is not a new one; the first proposals were made in the mid 1960s [8–10]. In recent years, there has been a growth in the number of systems, particularly in Japan, such that it is estimated that in excess of a hundred systems had been installed worldwide by the mid 1980s [11]. A flexible manufacturing system contains a number of features as follows:

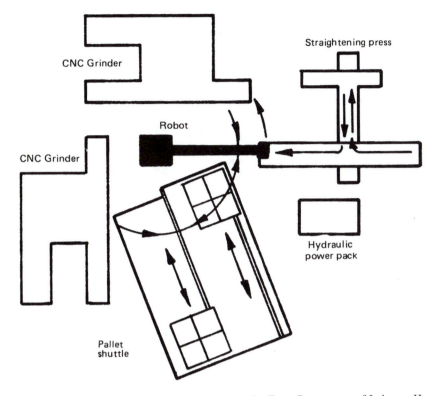

FIGURE 11.13 Robot work handling in small cell. (From Department of Industry, Her Majesty's Stationery Office, United Kingdom, 18, 1980.)

1. Interlinked NC workstations operating on a limited range or family of workpieces. In early proposals the machines were of modular construction, but in more recent systems essentially general-purpose NC machines, in particular machining centers, are most commonly used.
2. Automatic transportation, loading and unloading of workpieces and tools, using automatic guided vehicles (AGVs), robots, etc.
3. Workpieces mounted on pallets for transportation, partly to overcome the problems of new setups at each workstation.
4. Centralized NC or DNC, together with overall computer control of the system.
5. Operation for significant periods of time with little or no manual intervention.

With FMS the term "flexibility" means the ability to process a variety of components without having to adjust machine setups or change tooling. High flexibility implies that a large family of different components can be produced

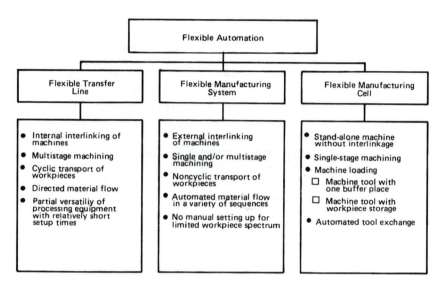

FIGURE 11.14   Variants of the FMS concept. (Adopted from Spur, G. and Mertins, K., Proc. 1st Int. Conf. on Flexible Manuf. Syst., 37, Brighton, U.K., Oct 20–22 1982.)

by the particular system. Figure 11.14 shows that several variants of the basic FMS concept exist [12]. These are:

1. Flexible manufacturing cells (FMCs): These are basically machining centers but with the addition of a pallet pool or magazine (Figure 11.15). The aim is to machine the workpiece with one setup. This type of machine can be operated unmanned for long periods of time, with the palletized workpieces transferred automatically to and from the machine. Flexible manufacturing cells of this type must be served by machines or operators engaged in blank preparation and palletization of workpieces. These cells are highly flexible in operation, having the ability to deal with a wide range of parts (40 to 800), in small batches of from 15 to 500.

2. Flexible transfer lines (FTLs). These systems consist of a number of NC or head-changeable machine tools connected by automatic material transfer systems. The system can machine different components but without flexible routing of the workpieces. The family of components is relatively small (<20) and the components must be quite similar to one another, as the overall flexibility of the system is too low for a larger variety to be accommodated. In consequence, the work cycles at each station must be quite well balanced. Figure 11.16 shows a typical system of this type [13]. Production quantities must be quite large for economic use of these systems (1,500 to 15,000 per annum for each component).

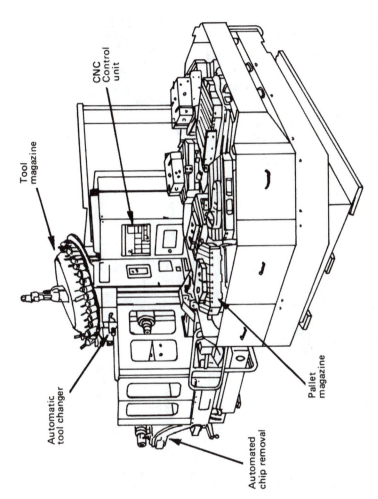

FIGURE 11.15 Typical machining center designed for unmanned operation with pallet and tool magazines and CNC controller able to communicate with host computers.

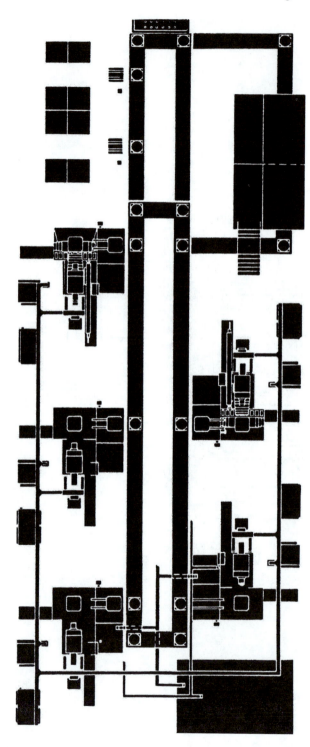

FIGURE 11.16 Flexible transfer line. (From Popplewell, F. and Schmoll. P., Proc. 1st Int. Conf. on Flexible Manuf. Syst., 501, Brighton, U.K., Oct 20–22 1982.)

3. Flexible manufacturing systems (FMSs). In these systems NC work-stations are linked by automatic workpiece transfer and handling, with flexible routing and automatic workpiece loading and unloading. Machining times at each station can differ considerably. The number of different components that can be processed by these systems is from 10 to 150 in general and moderate quantities can be produced (15 to 500 components per annum for example).

11.10.1 MACHINE TOOLS IN FMSs

The majority of FMSs developed have been for the processing of prismatic components [11], with a smaller number for rotational components. The most common machine tool type included in these systems is the horizontal-spindle machining center, but head-changing or head-indexing machines are also frequently used [14]. The latter machines are used for the drilling, tapping, and so on, of multiple holes, and operate using multi-spindle heads capable of processing several holes simultaneously. Separate heads are required for each workpiece processed.

The choice of machine tools for an FMS is influenced by a number of factors, but above all should be based on a detailed analysis of the family of components to be processed. Some of these factors are illustrated in Figure 11.17 [14]. Using horizontal machining centers with high-capacity tool magazines increases the flexibility of the system as these are able to process a wide variety of components without modification. Having several identical machine tools in the system also increases flexibility as this increases the availability of machine tools on which to process a particular component at any given time. Machining centers are relatively inefficient for the machining of large numbers of holes, since only one tool can be used at a time. Consequently, head-changing or head-indexing machines may be introduced to improve productivity if numerous holes must be machined. However, the overall flexibility of the system is thereby reduced, as these machines are suitable for only a relatively small variety of components. A requirement for high precision would also dictate the inclusion of a precision boring machine in the system.

11.10.2 WORK HANDLING FOR FMSs

Workpieces are usually mounted on standard pallets for processing in FMSs and these pallets locate automatically at each workstation in the system. A variety of work-handling devices are used to transport parts, pallets, and tools around the system. Some of these are as follows:

1. Tow carts: These are the most common devices used; they consist of a simple platform on castors and are towed around the system by engagement with under floor, continuously moving chains. Carts stop at workstations by means of a mechanism that releases the tow pin at the appropriate time. Branches and loops are controlled in a similar

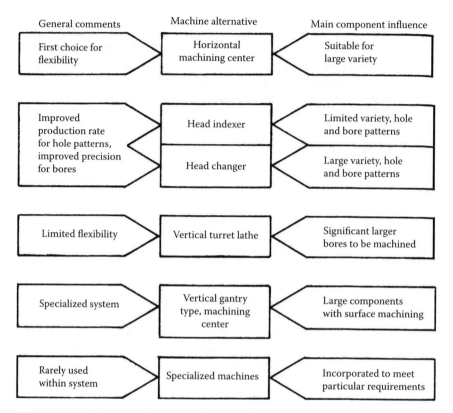

FIGURE 11.17   Factors affecting machine tool selection for FMS. (From Hannam [14].)

manner to railway systems. The main advantage of tow carts is their simplicity and low cost, since no on-board power is required for their movement or control. Facilities must be available at each workstation to load and unload pallets from the carts. Also, the circulation of carts must be unidirectional.

2. Automatic guided vehicles (AGVs): These devices are usually designed to follow wires buried in the floor of the plant or lines painted on the floor. On-board power and control is required for both movement and steering and for the handling of pallets. Automatic guided vehicles are more expensive than tow carts and are both larger and heavier. The main advantage of AGVs is their greater flexibility of operation. These devices may move in either direction, but for ease of control, circulation is usually restricted to one direction only in practice.

3. Rail carts: These carts move on rails and are generally restricted to backward and forward motion along straight tracks. Overhead conductors or extra rails transfer power and control instructions. Rail carts often accommodate two pallets to allow for pallet exchange at the system workstations.

4. Roller conveyors: Most of the early FMS developments utilized powered-roller conveyors for moving workpieces from station to station. The use of these conveyors in modern systems is less common. Roller conveyors are expensive to install and occupy valuable floor space. In addition, these conveyors are relatively inflexible in operation and difficult to alter if the overall system is expanded.

5. Industrial robots: Robots are used in FMSs but not extensively unless the cell consists of only a few machines. They may be used as secondary handling devices, particularly for turned workpieces, which may be transported around the system in hatches on pallets by other handling devices and then transferred to the machine tool by robots at each workstation. Gripper designs suitable for handling a wide variety of components are important in this case.

11.10.3 LAYOUTS FOR FMSs

A variety of different layouts for the machine tools in FMSs have been adopted. The choice depends on the scope of the system and the type of handling devices used for transporting workpieces from workstation to workstation. The use of rail carts means that a straight track must be used, with machines located at the side of the track. Figure 11.18 shows the layout of a typical rail-cart-based system. Early systems using roller conveyors usually employed a simple loop configuration, with branches to the workstations.

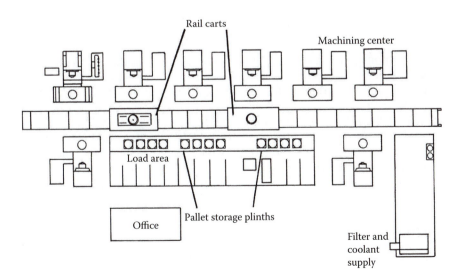

FIGURE 11.18 In-line rail-cart layout with eight horizontal machining centers and a load/unload area alongside the track. (From Hannam, R.G., *Proc. Inst. Mech. Eng.*, Vol. 198, 82, 111, 1985.)

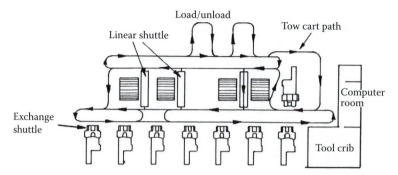

FIGURE 11.19 Typical tow-cart layout with eight horizontal machining centers and four head changers. (From Hannam, R.G., *Proc. Inst. Mech. Eng.,* Vol. 198, 82, 111, 1985.)

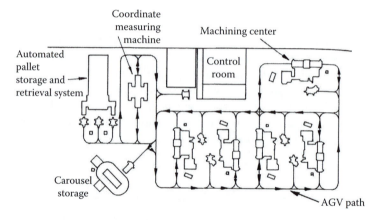

FIGURE 11.20 A typical automatic guided-vehicle layout with five horizontal machining centers, a coordinate measuring machine, automatic pallet storage and retrieval system, and carousel storage. (From Hannam, R.G., *Proc. Inst. Mech. Eng.,* Vol. 198, 82, 111, 1985.)

The increased use of tow carts and AGVs has resulted in more complex multiloop or tree-type layouts being used. The latter type is most suitable for AGVs and is particularly useful if expansion of the system with additional workstations is anticipated. Figure 11.19 shows a typical multiloop layout using tow carts, and Figure 11.20 shows a typical layout where AGVs are used for work handling.

11.10.4 TOOLING IN FMSS

The need for large numbers of tools in an FMS tends to restrict flexibility. Sufficient tools must be available to process all the component types. As explained earlier, overall flexibility can be increased by installing several machine tools of the same type, but if any component can be processed on any of these machines, this means that all the required tools must be available on all the machines. This leads to the requirement for very large capacity tool magazines capable of accommodating up to 250 tools or more on each machine. Without such high tool capacity other solutions must be adopted that tend to reduce the overall flexibility of the system. These solutions include:

1. Dedication of some machines to a restricted range of components
2. Restricted times for processing subgroups of components, with the tooling in the system changed within the allowed time

Long-term solutions may require a means of automatically changing the tools available in the system, but this results in considerable additional capital expenditure and development costs. Approaches being adopted include:

1. Interchangeable tool magazine carousels
2. Automated tool transfer from fixed to moving carousels at each workstation
3. Automatic handling of tools and components individually, together with supplying these together at each workstation

11.10.5 PALLETS AND FIXTURES FOR FMSs

The pallets used to transport workpieces around FMSs are precision items and represent a significant capital investment. The pallets usually require some form of fixturing for locating and holding the workpiece. This leads to a requirement for multiple identical fixtures if several workpieces of the same configuration are to be processed at the same time. In conventional batch production only one fixture item is generally required for each machine tool. This leads to the need for forms of flexible fixturing that can be readily rebuilt or adapted to the range of workpieces being processed by FMS.

PROBLEMS

1. A rotary-transfer machine is designed to drill six 25-mm-diameter holes in a workpiece. At the feed employed it requires 1000 revolutions of a drill to complete the operation. Each drill costs \$6.00 and can be reground on average 10 times at a cost of \$2.00. The time taken to remove and replace each drill is 100 s. If the total machine rate including the operator and overheads is \$0.012/s, the indexing time is 3 s, and the time for advance and withdrawal of the tools is 5 s, estimate the minimum cost of production for each component and the corresponding production time. Assume that the Tayor tool-life relation for the high-speed steel drill is

$$vt^{0.125} = 2.65$$

where v is the cutting speed, in meters per second (m/s), and t is the tool life, in seconds (s).

2. A numerically controlled milling machine has a linear interpolating control system for continuous contouring that will accept information from a punched tape prepared by the programmer. The programming time for this system is estimated to be 600 s per change point, and the hourly charges of the programmer are $0.00 139/s, including over-heads. The job under consideration has 20 change points and the total distance of cutter travel is 1.02 m. The appropriate feed speed for the particular cutter and work material is 1 mm/s and the total set-up time for each component is 300 s. If the total machine rate is $0.0033/s, including operator costs and overheads, what minimum batch size is required for economic use of the numerically controlled machine tool? (Assume that on a hand-operated milling machine the set-up time and machine time per component are 300 s and 2.7 ks, respectively, and that the total machine rate is $0.00278/s.)

REFERENCES

1. Gallagher, C.C. and Knight, W.A., *Group Technology Production Methods in Manufacture*, Ellis Horwood, Ltd., Chichester, England, 1986.
2. Drozda, T.J. and Wick, C. (eds), Tool and Manufacturing Engineers Handbook, Vol. 1, Machining, 4th Ed., Soc. of Manufacturing Engineers, Dearborn, 1983.
3. Young, J., Masters thesis, University of Rhode Island, Kingston, 1992.
4. Byne, G., Dornfeld, D., and Denkena, B., Advancing Cutting Technology, *Ann. CIRP*, Vol. 53(2), 1–25, 2003.
5. Koenigsberger, F. and De Barr, A.E., *The Selection of Machine Tools*, Institution of Production Engineers, London, 1979.
6. Engleberger, J.F., *Robotics in Practice*, Kogen-Page, London, 1980.
7. Department of Industry, *A Human Guide to Robots*, Her Majesty's Stationery Office, United Kingdom, 18, 1980.
8. Ohmi, T., Ito, Y., and Yoshida, Y., Flexible Manufacturing Systems in Japan-Present Status, Proc. 1st Int. Conf. on Flexible Manuf. Syst., 23, Brighton, U.K., Oct 20–22 1982.
9. Ranky, P.C., *Computer Integrated Manufacturing*, Prentice-Hall, Englewood Cliffs, NJ, 1986.
10. Hitomi, K., *Manufacturing Systems Engineering*, Taylor & Francis, London, 1979.
11. Steinhilper, R., Step-by-Step Access to Flexible Automation in Manufacturing, Proc. 2nd European Conf. on Automated Manuf., Birmingham, 299, 1983.
12. Spur, G. and Mertins, K., Flexible Manufacturing Systems in Germany; Conditions and Development Trends, Proc. 1st Int. Conf. on Flexible Manuf. Syst., 37, Brighton, U.K., Oct 20–22 1982.
13. Popplewell, F. and Schmoll, P., The Road to FMS, Proc. 1st Int. Conf. on Flexible Manuf. Syst., 501, Brighton, U.K., Oct 20–22 1982.
14. Hannam, R.G., Alternatives in the Design of Flexible Manufacturing Systems for Prismatic Parts, *Proc. Inst. Mech. Eng.*, Vol. 198, 82, 111, 1985.

12 Computer-Aided Manufacturing

12.1 INTRODUCTION

Computers are used extensively in the scientific and commercial fields. Computer systems are now available that combine high computing power with high reliability, small size, and low cost, making their use viable in increasing ranges of applications. These computer developments have made possible the rapid performance of tasks that are complicated and time consuming to do manually. Computers can be programmed to carry out complex calculations and to process and store large amounts of data with high speed and efficiency. As a result, computers have become extensively used in both design and manufacture, leading to the concepts of computer-aided design (CAD) and computer-aided manufacture (CAM) or CAD/CAM for short. The objective is to increase the efficiency of all aspects of design and manufacture and to integrate these activities by means of shared databases, together with automatic data transfer between the various design and planning activities to be carried out. In this chapter only those aspects of CAD/CAM concerned with the technological aspects of machine tool application are considered in detail; in particular, process planning and numerical control (NC) programming applications are discussed. Details of other CAD/CAM applications can be found in a variety of publications [1–4]. Before proceeding with details of computer-aided process planning and NC processing systems, an outline of the scope of CAD/CAM applications will be given.

12.2 SCOPE OF CAD/CAM

Computer-aided design is the use of computer systems to facilitate the creation, modification, analysis, and optimization of a design [1]. In this context the term computer system means a combination of hardware and software. Computer-aided manufacturing is the use of a computer system to plan, manage, and control the operation of a manufacturing plant. An appreciation of the scope of CAD/CAM can be obtained by considering the stages that must be completed in the design and manufacture of a product, as illustrated by the product cycle shown in Figure 12.1. The inner loop of this figure includes the various steps in the product cycle, and the outer loop shows some of the functions of CAD/CAM superimposed on the product cycle.

Based on market and customer requirements, a product is conceived, which may well be a modification of previous products. This product is then designed

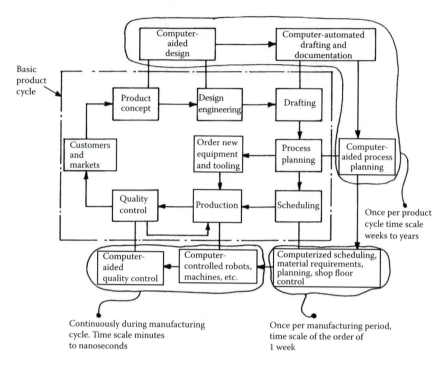

FIGURE 12.1 Product cycle with computer-aided procedures superimposed. (Adapted from Groover, M.P. and Zimmers, E.W., Pearson Educational, Englewood Cliffs, NJ, 1997.)

in detail, including any required design analysis, and geometric models and parts lists are prepared. Subsequently, the various components and assemblies are planned for production, which involves the selection of sequences of processes and machine tools and the estimation of cycle times, together with the determination of process parameters, such as feeds and speeds for machining operations. When the product is in production, scheduling and control of manufacture take place, and the order and timing of each manufacturing step for each component and assembly is determined to meet an overall manufacturing schedule. The actual manufacturing and control of product quality then takes place according to the schedule and the final products are delivered to the customers.

Computer-based procedures have been developed to facilitate each of these stages in the product cycle, and these are shown in the outer loop of Figure 12.1. Computer-aided design and drafting techniques have been developed. These allow a geometric model of the product and its components to be created in the computer. This model can then be analyzed using more specialized software packages such as those for finite element stress analysis, mechanisms design, and so on. Subsequently, drawings, if required, and parts lists can be produced with computer-aided drafting software. Computer-aided process-planning systems, including the preparation of NC programs, are available that produce work plans, estimates, and manufacturing instructions automatically from geometric descriptions of the components and assemblies.

For scheduling and production control, large amounts of data and numerous relatively simple calculations must be carried out. One example is the determination of order quantities by subtracting stock levels from forecasts of the number of items required during a particular manufacturing period. Many commercial software packages are available for scheduling, inventory control, and shop floor control, including materials requirements planning (MRP) systems [2]. At the shop-floor level computers are used extensively for the control and monitoring of individual machines (see the description of CNC machines in Chapter 11).

There is a difference in the time scale required for processing data and the issuing of instructions for these various applications of computers in the product cycle. For example, design and process-planning functions are carried out once for each new product and the time scale required is on the order of weeks to years for the completion of the whole task. Scheduling and production control tasks will be repeated once every production period (usually one week) throughout the year. At the machine-control level instructions must be issued continually with a time scale of micro- or nanoseconds in many cases.

One of the major objectives of CAM is the integration of the various activities in the product cycle into one unified system, in which data is transferred from one function to another automatically. This leads to the concept of computer-integrated manufacture (CIM), with the final objective being the "paperless" factory. Since the design and process-planning functions are carried out once in the product cycle, these are the most suitable functions for integration. This integration is particularly desirable because the geometric data generated during the design process is one of the basic inputs used by process planning when determining appropriate manufacturing sequences and work plans. Consequently, various activities in design and process planning can share a common design and manufacturing database, as illustrated in Figure 12.2 [4]. With such a system, geometric models of the products and components are created during the design process. This data is then accessed by various downstream activities, including NC programming, process planning, and robot programming. The programs and work plans generated by these activities are also added to the database. Production control and inventory control programs can then access the work plans, time estimates, and parts lists (bill of materials file), in preparing the manufacturing schedules, for example.

12.3 PROCESS-PLANNING TASKS

Process planning has been defined by the Society of Manufacturing Engineers as "the systematic determination of the methods by which a product is to be manufactured, economically and competitively." Process plans are used to specify the sequence of production operations and the required tools and facilities that transform

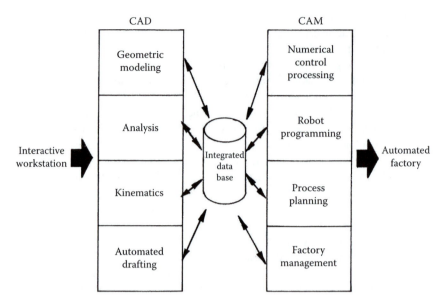

FIGURE 12.2 Integrated design and manufacturing database. (From Knox, P., Marcel Dekker, NY, 1983.)

raw stock into finished parts. Process planning requires detailed knowledge of both the manufacturing processes available (e.g., machining, forming) and the production capabilities of the specific plant in which the parts are to be manufactured.

Figure 12.3 shows the sequence of steps in the design and planning for manufacture of a product and its component parts. The first three steps are concerned with the design of the items to be produced, and CAD systems are aimed at increasing the efficiency of these tasks. In manual systems the output of these stages is in the form of component and assembly drawings, specifications, and parts lists. In integrated computer-based systems, this information is stored in a common database which includes geometric modeling information and bill of materials (BOM) files.

Steps 4 to 9 in Figure 12.3 describe the main steps required in process planning. The process planner converts the geometric information (drawings) about the items to be manufactured into appropriate sequences of manufacturing operations and determines the processing parameters (e.g., feeds and speeds) for each operation, taking into consideration the part geometry, material, and quantities required. The first stage is to determine the operation sequence in general terms, followed by the selection of the machines for each operation. It is at this stage that the level of automation of the machines will be determined and, as explained in Chapter 11, this will be largely based on estimates of the quantities required for the product.

Steps 6 to 9 are concerned with the detailed planning of each operation on each machine and include the selection of work-holding methods, sequences of

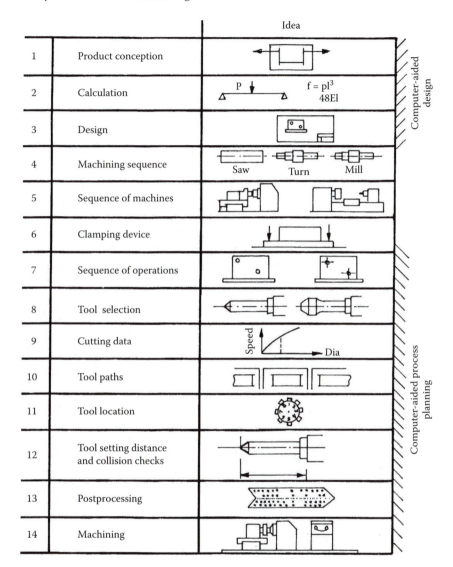

FIGURE 12.3 Stages in design and process planning.

operations and tools, and finally the determination of cutting data such as feeds and speeds, together with estimates of cycle times. At this stage the machinability of the material and economics of the process must be considered in determining the most appropriate cutting parameters. The output at step 9 is in the form of detailed work plans, which list all operations to be carried out in sequence, together with the processing parameters and operation times. Computer-aided process-planning systems have been developed to deal with most of the steps 4 to 9 in Figure 12.3.

For NC machines further steps are necessary to determine the tool paths necessary to machine the objects and to prepare the programs that drive the machines. Many computer-aided programming systems for NC machines have been developed. Initially these systems dealt mainly with the geometric calculations required to determine tool paths to machine the workpiece, but increasingly the automatic selection of tools and cutting conditions are determined by these systems, so that steps 7 to 9 in Figure 12.3 are also incorporated into the programs.

For manual process planning, engineering part drawings are examined to develop manufacturing process plans and instructions. The process relies to a great extent on experienced planners to develop economical and feasible plans. The process planner must manage and retrieve a great deal of data and many documents, including established standards, machinability data, machine specifications, tooling inventories, stock availability, and existing process plans. Thus, process planning is an information-processing task to which the computer is well suited. In addition, because process planning involves many interrelated activities, determining the most appropriate plan for a particular case may require several iterations. The use of the computer enables alternatives to be readily explored. Another advantage of computer-aided process planning is greater consistency in the plans determined. This is a particularly important feature because the process-planning task involves the interpretation of the design features of a component to meet its manufacturing requirements. It is well known that in most companies a variety of different sequences exist for basically similar parts [5,6] and these have been generated by different process planners at different times. Similarly, giving the same component to several different planners will usually result in several different work plans being determined, dependent on the experience and choice of the individual planners [6].

Process planning is not an easy task to computerize because it is not an activity that can easily be flow-charted. The planner must consider many parameters simultaneously and make involved and detailed decisions that are difficult to structure in an ordered manner. Ideally the computer-aided process-planning system should follow the planning logic of an experienced planner.

12.4 COMPUTER-AIDED PROCESS PLANNING

Computer-aided process planning (CAPP) systems are aimed at carrying out the planning tasks outlined above and shown in Figure 12.3. Comprehensive surveys of CAPP developments can be found in [7–9]. By far the greatest proportion of CAPP developments have been devoted to machined parts, but some developments in the area of sequence design for metal forming operations, assembly, and fabrication have also occurred [9]. The following discussion will be mainly concerned with planning for machining operations.

Plans for machining processes can be developed in two basic ways: forward planning and backwards planning [2]. In forward planning the operation sequence is determined by working forward from the raw material geometry to the final shape, such that the initial decision is the basic workpiece shape and size. The sequence of operations is then developed by removing material progressively in the successive operations until the final part geometry is achieved. In backwards planning the starting point is the final part shape. The operation sequence is developed by adding the material removed by each operation in the reverse sense until the geometry of the starting workpiece is achieved.

Two basic approaches to computer-aided systems for work planning have evolved: (1) the variant or retrieval approach, and (2) the generative approach [2,4,7–10]. The variant approach utilizes the principles of Group Technology [5,10], and work plans are based on families of similar parts. In this manner the plan for a new part can be obtained more or less automatically from relatively small amounts of input data and with little interaction by the user of the system. Such systems are generally very efficient, but it may not be possible to develop systems using this approach for the whole range of parts to be processed.

The variant approach is an extension of manual planning. Plans for new parts are developed through recalling, identifying, and retrieving existing plans for similar parts and making necessary modifications. In some variant systems, standard plans for each family of parts are developed and filed using code numbers based on group technology principles. Upon entering the code number for a proposed new part, the standard plan for the family is retrieved to be either used or modified. A great deal of preliminary work is necessary to establish the standard plans, categorize the family of parts groupings, and establish the classification and coding systems. The computer, acting much like a word processor, then helps to assemble and edit appropriate plans for a new part. During the planning stage, various subroutines or canned programs can be called up to help the planner with the many decisions that must be made, generally in an interactive, conversational mode at a computer terminal. A number of variant or retrieval planning systems have been developed, including that prepared for CAM-I [11], as illustrated in Figure 12.4. A similar system, based on the storage, retrieval, and editing of work plans for similar parts, identified by a classification system, has been developed by the Organization for Industrial Research and is shown in Figure 12.5 [12].

Generative process-planning systems attempt to synthesize a unique plan for each individual part from general-purpose procedures, algorithms, and stored manufacturing knowledge that define the various technological decisions that must be made. For truly generative systems, the sequences of operations as well as the process parameters are established without reference to previous work plans stored in the system. The system synthesizes or generates a specific process plan for a specific part, and relies on sophisticated decision algorithms, part data, and the specific manufacturing capabilities available. From an input of the component geometry, decision algorithms identify geometric features and the specific processes required to produce them.

Ideally this approach would allow a planner to present any part to the system and receive a complete, automatically determined plan. Processes would be selected, machine tools specified, processing sequences established, and process parameters determined. However, the decision logic for such systems is complex; it varies from industry to industry and company to company, and often process

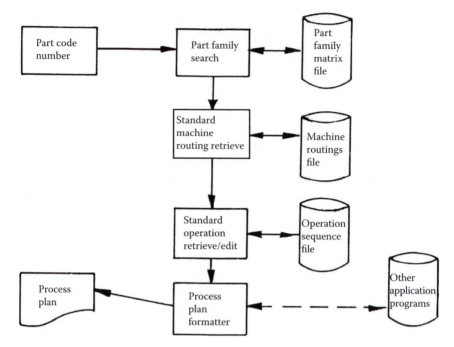

FIGURE 12.4 Retrieval process-planning system. (From Link, K.H., *SME Technical Paper*, MS 77-3 14, 1977.)

parameters need to be adjusted interactively to suit particular cases. Although considerable progress has been made in the development of generative process-planning systems, there are still considerable problems to be solved for the widespread use of comprehensive generative planning systems to be possible. Current systems tend to be restricted to specific operations or a particular type of manufacturing process (e.g., for rotational machined components) and still require considerable interaction by the user in selecting the appropriate sequences. This will probably be the case for some time to come, although considerable development work is in progress.

In a generative planning system, a geometric description of the item to be machined is processed through some decision logic to result in a process plans for the part. Thus the initial problem to be approached is how to input the part geometry to the CAPP system. This can be approached in several ways [2,9]: (1) geometric classification codes; (2) a descriptive language; (3) extraction of relevant data from a CAD geometry model of the part.

In a way similar to variant CAPP systems, description of the part to be processed can be described in the form of a classification code, and a number of systems have been developed with this form of input [2,9]. However, the descriptive codes must be much more detailed than for the variant approach and in general must be supplemented by specific parameter values, particularly for processing details associated with, for example, surface finish and tolerances. A

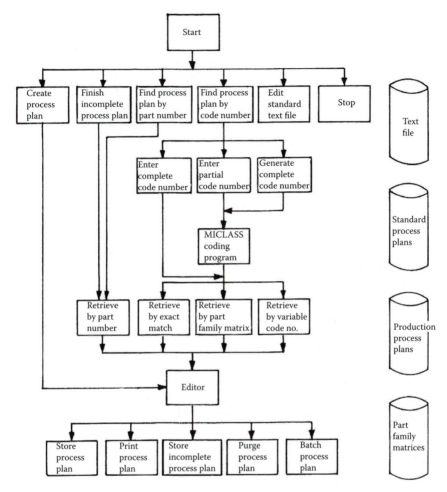

FIGURE 12.5 Retrieval or variant process-planning system. (From Houtzeel, A., *American Mach.*, Vol. 123, 115, 1979.)

restriction on this form of input is that detailed coded descriptions are difficult to generate automatically and hence a well designed user interface is required to operate the system.

Description of the item geometry through a specifically designed descriptive language has been the preferred input method for a number of successful CAPP systems [2,3,7–9]. This approach is similar to that used in computer aided NC processing systems, as outlined in section 12.7 and a language is defined to provide all the information needed for the process planning system. The process planner must become proficient in the descriptive language to operate the CAPP system.

Most of the information necessary for the generation of process plans for machining is contained in the geometric models used in modern CAD systems.

Additional information on the part material and surface conditions may be required. However, the extraction of the required information from CAD models is by no means straight forward and algorithms for machining feature extraction from the underlying CAD data are difficult to implement in a generalized way. At present with this approach it is usually necessary to rely on significant interaction from the user of the system or to narrow down the problem by considering only certain classes of parts such as rotational parts or 2½ axis milling, for example.

Manufacturing process knowledge can be transformed into sets of rules for process planning decision making. This decision logic can be applied in several ways [2], including:

1. Decision trees
2. Decision table
3. Artificial intelligence, e.g., rule-based expert systems

Decision trees [2] are a form of flow diagram in which conditions (IF, etc.) are set on branches of a tree and actions are placed at the junction of each branch. The decision tree can be implemented as computer code or represented as data. In the former case, the tree translates directly into a program flow chart. The same information can be represented by decision tables [2]. Decision tables can be readily implemented on a computer but usually require a preprocessor or a specific language to implement the tables and control operation.

A number of generative process-planning systems have been developed but these are generally restricted to limited ranges of component types or processes [2,9]. Many of the developed systems are able to handle only the machining of rotational parts [13]. As an example, the AUTAP system, developed at the University of Aachen, is able to process operation plans only for rotational parts and simple sheet metal parts. The general structure of the AUTAP system is shown in Figure 12.6. The workpiece geometry is defined in terms of individual shape elements, which have associated manufacturing operations and sequences. This is achieved through the use of a specifically designed descriptive language as illustrated in Figure 12.7 [14]. On the basis of the individual elements present in the workpiece, the overall machining sequence is determined, and this is followed by the calculation of machining and set-up times for the individual operations. The final stage is to determine the quality control and inspection processes, followed by issuing the operation plan.

This basic approach has been extended to prismatic parts in a separate system [15]. The overall structure of this latter system is shown in Figure 12.8. Again the basic description of the component is in terms of detailed shape elements, which have an associated manufacturing sequence. The relationship between the shape features and associated processing sequences is illustrated in Figure 12.9. The geometric description must follow a prescribed sequence, in which the overall shape is defined first, followed by the secondary shape features. Combination of the processes associated with each shape element can result in several alternative overall sequences from which the most economic is selected.

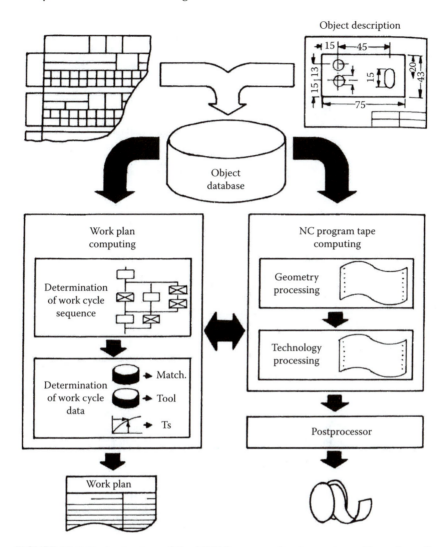

FIGURE 12.6 Basic structure of the AUTAP process-planning systems. (From Spur, G. and Krause, F.L., *Computer-Aided Design and Manufacture*, Rembold, U. and Diliman, R., eds., Springer Verlag, NY, 1986, 65.)

As explained previously, the development of computer-aided process-planning systems for widespread application is difficult, mainly because the planning decision logic varies from industry to industry, company to company, and so on. In addition, heavy reliance on experienced planners in determining suitable process plans is frequently required. For these reasons, recent concentration has been on applying expert systems programming methods to process planning and to developing procedures for feature extraction from geometric models [9]. This approach is attractive because in expert systems the manufacturing decision rules

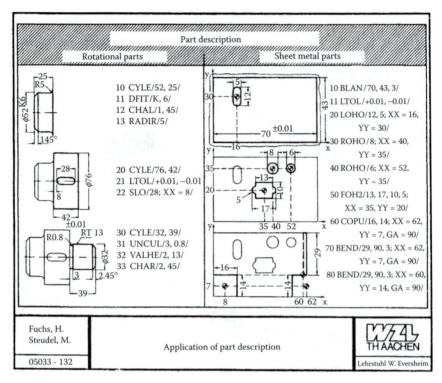

FIGURE 12.7 Part of the geometric description language used for the AUTAP process-planning systems. (From Eversheim, W.H., Fuchs, H., and Zon, K.H., Proc. 12th CIRP, Int. Sem. on Manuf. Sys., Belgrade, 779–800, 1980. Courtesy of the Laboratory for Machine Tools and Production Engineering, Technical University of Aachen.)

can be isolated in a knowledge base, rather than incorporated into complex algorithms. It is then possible to alter and add to the rule base, without altering the inference logic of the system. Thus, the overall system should be readily adaptable to different manufacturing environments.

12.5 PROCESSING OF NC PROGRAMS

Numerical control programming is a specialized work planning process, as the programmer must make similar decisions and use similar information to that of the process planner but with the additional complication that such work as the detailed geometric calculation of tool paths must be done. Numerical control machines may be programmed by manual methods, using the basic instruction set of the machine controller or with a computer-aided NC processing system or as an output of an integrated CAD/CAM system. All methods require a good deal of skill and experience on the part of the programmer.

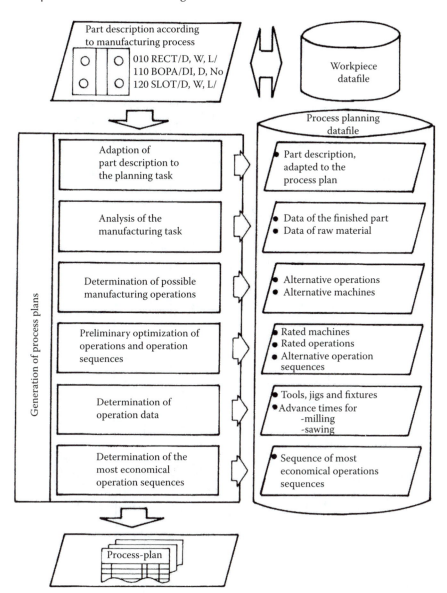

FIGURE 12.8 Basic structure of process-planning system for prismatic parts. (From Eversheim, W., Automated Generation of Process Plans for Prismatic Parts, *Ann. CIRP*, Vol. 32, 361, 1983.)

Many computer NC processing systems have been developed over the years, and a high proportion of these have been based around the language structure of the automatically programmed tools (APT) system [1,2,16]. With such systems the geometry of the part and tool motions is described by the programmer in the language of the system, and then the tool paths and similar calculations, are

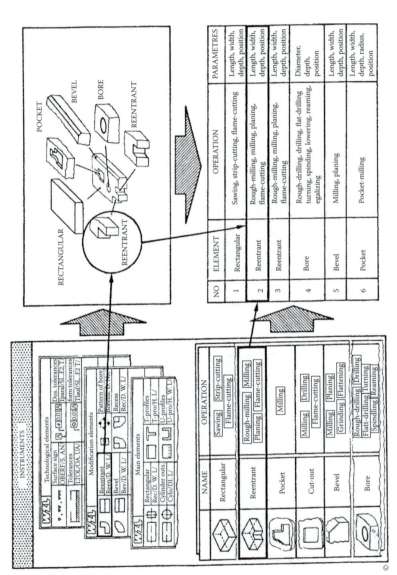

FIGURE 12.9 Relationship between shape features and operation sequences for process planning. (From Eversheim, W., Automated Generation of Process Plans for Prismatic Parts, *Ann. CIRP*, Vol. 32, 361, 1983.)

determined by the processing system. Although simpler than manual programming for all but simple parts, NC processing systems still require significant experience and effort to use effectively. The underlying procedures of these NC processing systems have been incorporated into most of the modern graphics-based CAM systems and solid model-based CAD/CAM systems.

The programming of NC machines involves the following tasks (Figure 12.3):

1. Determination of the sequence of operations required to machine the workpiece. This involves examining the shape features of the component to be produced and selecting the sequence of machining operations to produce these features.
2. Choice of cutting tools and order of use, corresponding to the sequence of operations.
3. Determination of feeds and cutting speeds for each operation. The appropriate values of feed and speed are dependent on the materials of the workpiece and of the selected cutting tools and can be selected using the economic models described in Chapter 8.
4. Determination of tool path coordinates, including both cutting and rapid traverse motions, taking into account possible collisions as with clamps. Tool path coordinates are offset from the surface of the workpiece, so that when the paths are followed, the correct size and shape of the workpiece is produced. This is necessary because the program point of the tool does not correspond to the point of contact between the tool and workpiece. Figure 12.10 shows the program point, that is, the point that traces out the tool path, for end milling cutters and lathe tool. For all but simple components, determination of the tool path coordinates involves time-consuming, repetitive, and often difficult geometric calculations.
5. Setting of program auxiliary functions, including tool changes, spindle starts and stops, coolant flow on and off, and so on.

To accomplish these tasks the following information sources are used:

1. Drawings or geometric models of the component to be machined
2. Data on machine tool capabilities
3. Catalogs of machinability data for materials, recommended feeds, and cutting speeds
4. Catalogs of available cutting tools, which should be standardized as much as possible
5. Standard geometric formulas

12.6 MANUAL PROGRAMMING OF NC MACHINES

Numerical control machines respond to a series of coded instructions stored in the controller. These coded instructions were developed for feeding the NC program into the controller by means of punched paper tape. In modern applications,

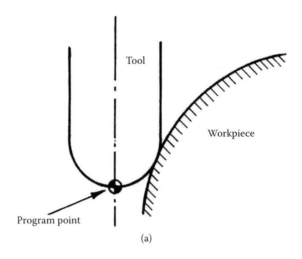

(a)

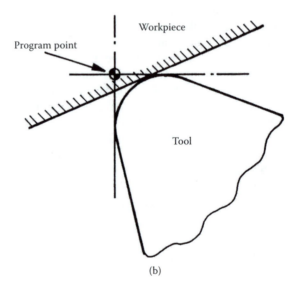

(b)

FIGURE 12.10 Tool program points for tool path determination. (a) Ball-ended milling cutter; (b) lathe tool. The program point sweeps out the offset tool path.

NC programs are input to the machine controllers electronically. However, the basic instruction set is still maintained for modern machines. The format of the basic instructions is now largely standardized and modern controllers utilize what is known as a *word address format* [17]. Older machines may use other tape code formats but these will not be described here.

Each set of instruction codes that the machine acts upon is called a *NC block* and is terminated by means of an end of block character (eob), which corresponds to the carriage return, line feed character. The NC machine acts upon each block of coded instructions in turn. A typical block of instruction is of the following form:

```
n001 g80 x 75.0 y 65.0 z 10.0 f60.0 s2000 t001 m03 eob
```

The individual instructions are conventionally entered in this order, but with word address format any order can be used, except for the first item (n001). This is because earlier controllers required the codes to be entered in a specified order. Each instruction consists of a letter code, which tells the controller the type of data to follow and indicates the address to which this data is to go, hence the term word address format. The explanations of each word or code type are as follows:

1. Sequence number (n-codes): This is an identification number for each block of instructions and increases sequentially through the program.
2. Preparatory function (g-codes): These are the main motion commands of the machine and are identified by the prefix g. For this reason the programming of NC machines at this level is often colloquially referred to as g-code programming. These instructions tell the controller what type of motion or action is to be carried out. These codes are largely standardized and Table 12.1 shows a typical list. An example of a preparatory function is g02, which indicates that the next motion will be circular interpolation in a clockwise direction. Many of the preparatory functions can indicate what are called canned cycles, as explained in more detail later.
3. Coordinate data (x, y, z-codes, etc.): These indicate the coordinates for the tool to move to (the end of the motion), in the manner indicated by the preparatory function currently in force. The x, y, z-codes indicate the conventional Cartesian coordinates. In four- and five- axis machines, other codes (a, b, c) may be used to indicate angular positions. For circular interpolation i, j, k-codes may be used to indicate the center of the circular arc to be followed.
4. Feed rate (f-code): This specifies the feed rate for the operation. The units are usually inches or millimeters per minute.
5. Spindle speed (s-code): This specifies the spindle speed to be used for the operation and maybe an actual value (in rpm) or a code for a fixed spindle speed.
6. Tool number (t-code): This indicates to the controller which tool is being used for the operation. In machines with manual tool-changing facilities, changing the t-word may simply tell the controller which previously stored tool length compensation data to use.
7. Miscellaneous functions (m-codes): These codes program various auxiliary functions on the machine tool. For example, m03 starts the spindle with clockwise rotation. Miscellaneous functions are again

TABLE 12.1
Preparatory Functions (EIA Standard RS-273)

Code	Function
g00	Point to point, positioning
g01	Linear interpolation (straight line motion)
g02	Circular interpolation arc CW
g03	Circular interpolation arc CCW
g04	Dwell
g05	Hold
g06 & g07	Unassigned
g08	Acceleration
g09	Deceleration
g10	Linear interpolation (long dimensions)
g11	Linear interpolation (short dimensions)
g12	Unassigned
g13–g16	Axis selection
g17	xy plane selection
g18	zx plane selection
g19	yz plane selection
g20	Circular interpolation arc CW (long dimensions)
g21	Circular interpolation arc CW (short dimensions)
g22–g24	Unassigned
g25–g29	Permanently unassigned
g30	Circular interpolation arc CCW (long dimensions)
g31	Circular interpolation arc CCW (short dimensions)
g32	Unassigned
g33	Thread cutting, constant lead
g34	Thread cutting, increasing lead
g35	Thread cutting, decreasing lead
g36–g39	Reserved for control use only
g40	Cutter compensation cancel
g41	Cutter compensation — left
g42	Cutter compensation — right
g43–g49	Cutter compensation if used; otherwise unassigned
g50–g59	Unassigned
g60–g79	Reserved for positioning only
g80	Fixed cycle cancel
g81	Fixed cycle 1
g82	Fixed cycle 2
g83	Fixed cycle 3
g84	Fixed cycle 4
g85	Fixed cycle 5
g86	Fixed cycle 6
g87	Fixed cycle 7
g88	Fixed cycle 8
g89	Fixed cycle 9
g90–g99	Unassigned

TABLE 12.2
Miscallaneous Functions (EIA Standard RS-273)

Code	Function
m00	Program stop
m01	Optional (planned) stop
m02	End of program
m03	Spindle on CW
m04	Spindle on CCW
m05	Spindle off
m06	Tool change
m07	Coolant No.2 on
m08	Coolant No.1 on
m09	Coolant off
m10	Clamp
m11	Unclamp
m12	Unassigned
m13	Spindle on CW & coolant on
m14	Spindle on CCW & coolant on
m15	Motion +
m16	Motion -
m17–m24	Unassigned
m25–m29	Permanently unassigned
m30	End of tape (file)
m31	Interlock bypass
m32–m35	Constant cutting speed
m36–m39	Unassigned
m40–m45	Gear changes if used; otherwise unassigned
m46–m49	Reserved for control use only
m50–m99	Unassigned

largely standardized and Table 12.2 shows a typical list. The miscellaneous functions may be acted upon at the start or end of the motion described by a block of instructions. For example, m03, spindle rotation, is carried out before the motion takes place and m05, spindle off, occurs after the motion takes place.

With the word address format, information need not be repeated in successive blocks if it is to remain the same for subsequent blocks. For example,

```
n001 g01 x75.0 y65.5 z 10.0 f60.0 s2000 m03 eob
n002 x87.5 y75.0 eob
```

In this case the g-, z-, f-, s-, and m-codes remain the same for the second block. For safety reasons some preparatory functions are canceled after the block is acted upon and must be repeated in a subsequent block if required. This is generally the case for circular interpolation (g02, g03) and codes for thread cutting.

12.6.1 Fixed or Canned Cycles

Many NC machines can be programmed using what are called fixed or canned cycles. Preparatory codes from g80 upwards are usually assigned to these cycles. On modern CNC machines, the user can define his own fixed cycles of commonly used operations. Many drilling and 2½ axis milling machines are programmed largely with fixed cycles. Figure 12.11 shows two typical fixed cycles for a drilling-milling machine. The first, g81, is a drilling cycle and the second, g84, is a tapping cycle. Each fixed cycle causes a sequence of motions to take place.

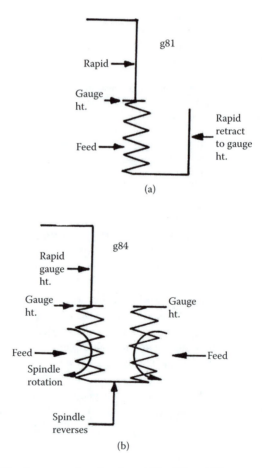

FIGURE 12.11 Fixed or canned cycles for a NC milling-drilling machine. (a) Drilling cycle; (b) tapping cycle.

Using the preparatory function g81 will cause a drill to move at rapid traverse above the work surface to a specified x, y position, then to drill down to the specified z depth at the specified feed rate, followed by rapid retraction to just above the work surface. The g84 cycle is similar except the spindle is reversed and the tool feeds out at the appropriate tapping feed.

12.6.2 TOOL PATH COORDINATES

The NC programmer must carry out the tasks outlined in section 12.4. Determination of the tool path coordinates is an important part of NC programming and is a task that involves numerous repetitive calculations, many of which may be quite complex. As explained above, the tool paths are offset from the surface of workpiece, as it is the program point of the tool that is driven along the tool path to machine the workpiece. Each path will be made up generally of linear or circular arc (circular interpolation) line segments. Thus, the manual programmer must determine the coordinates of the change points between each of these line segments. This can be illustrated by Figure 12.12, which shows a milling operation in which an end mill is required to machine the outside profile of the workpiece. The change point coordinates that must be determined by the programmer are shown as the centers of the dotted circles which represent the tool positions. In addition, the coordinates of the centers of the circular arcs in the profile must be determined. It can be seen that even in this fairly simple case, some relatively complicated calculations must be carried out to determine the required tool path coordinates. Determination of the required tool paths from complex three-dimensional surfaces is obviously difficult to do manually.

In many cases the programmer must carry out large numbers of relatively simple repetitive calculations in determining the relevant tool paths. Three such

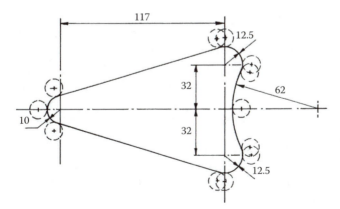

FIGURE 12.12 Tool path segment change points for a milling operation in which an end mill machines the outside profile of the workpiece.

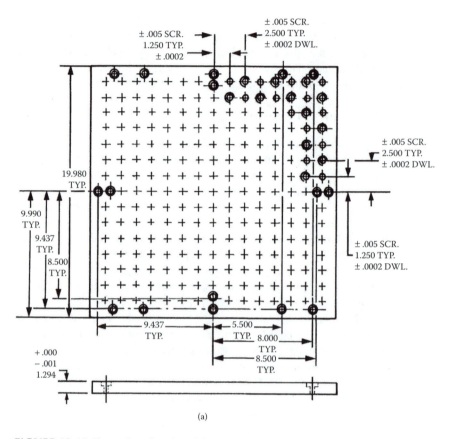

(a)

FIGURE 12.13 Examples of tool positions and paths for NC machining operations. (a) Drilled hole patterns; (b) layers of cuts for rough-turning operation; (c) area clearance for milling operation.

cases are illustrated in Figure 12.13. In drilling operations (Figure 12.13a) with complex patterns of holes, many coordinate positions must be determined. For roughing cuts in both milling and turning, many change points must be determined. Figure 12.13b shows the layers of cuts necessary for a rough turning operation, and Figure 12.13c illustrates the clearance of an area using an end mill. In both cases many tool path coordinates must be determined, using relatively simple repetitive calculations.

Relatively little direct programming of NC machines in this underlying instruction set is now done and the programming of NC machines is now largely achieved with computer-based processing systems. However, the final program delivered to the NC controller is still in this long established format and this is often referred to as a tape-image file, since the program is still a list of instructions in essentially the same format that in the past used to be fed to the controller through a paper tape reader.

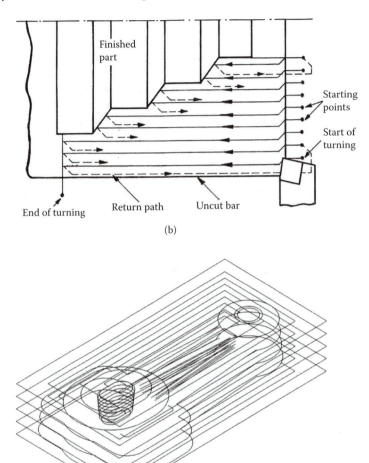

(b)

(c)

FIGURE 12.13 (continued).

12.7 COMPUTER-AIDED NC PROCESSING

Computer-aided NC processing systems have been developed to facilitate the programming of NC machines. The objective of these systems is to carry out some or all of the programming tasks listed in section 12.6, using stored data on tools, materials, machines, and geometric algorithms, from an input of the geometric description of the component to be machined. Computer-aided NC processing systems are economically justified for companies having a number of NC machine tools installed. Additionally, such systems may be the only practical way of preparing programs for the machining of complex workpieces, involving three-dimensional surfaces.

Computer-aided NC processing is carried out in two or sometimes three stages (Figure 12.14). Input description of the part geometry is in a certain format called a NC language or derived from a geometric model of the part. The program that carries out the main geometric calculations is called a processor. The NC processor may have two stages. First, a geometric processing stage is required, which determines the offset tool path positions to machine the workpiece. This may be followed by a technological processing stage, which selects tools and determines appropriate feeds and speeds. For this latter purpose, files of standard tool data and materials data must be accessed. The main output from the processor is a file, called a cutter location (CL) file, containing mainly the coordinates of the tool positions in a standard format.

The NC processor program is largely independent of the machine tool that will be used to machine the workpiece. Consequently, a further program, called

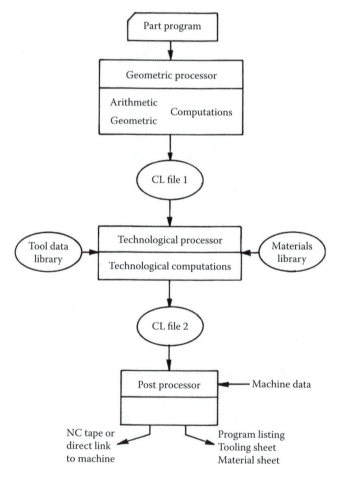

FIGURE 12.14 Processing stages for computer-aided NC program preparation.

a postprocessor, must be used to convert the general CL data into the specific input instructions for the particular NC machine to be used. A different postprocessor is required for each of the various machine tool models for which the main NC processor may be used. The basic output of the post processor is a text file that corresponds to the punched paper tape that used to be input medium for NC machines. This file is essentially a tape image file.

The basic functions of a postprocessor program are:

1. Read in CL data.
2. Convert data to the machine tool coordinate system.
3. Convert data to absolute or increment coordinates as required.
4. Check for machine limitations (slide travel, interference, allowable feeds and speeds, etc.).
5. Develop appropriate feeds and speeds for the machine.
6. Develop motion commands for the specific machine and controller.
7. Allow for interpolator type.
8. Output the machine tool program type in appropriate block format and code set (g-codes).
9. Print out listing of program for checking.
10. Print out error codes and diagnostics.

12.8 NUMERICAL CONTROL PROCESSING LANGUAGES

The development of NC programming languages began in the 1950s and before the introduction of modern graphics-based CAM systems, became the most common way to generate NC programs. Their use today is much less widespread. However, because of its ability to deal with complex parts and surfaces, the language APT is still used in some industries, particularly some aerospace companies. It is estimated that about 10 to 15% of NC programs generated are still generated using APT-based processors. Some knowledge of the APT system particularly is useful to NC programmers because many of the routines incorporated into modern graphics-based CAD/CAM systems have their origins in APT. In addition, the standard CL format and post processing procedures used in most CAM systems are APT based.

A large number of NC processing languages have been developed over the years [2,16–19] and these may be classified in various ways, based on the input information required and the internal similarities of these languages.

12.8.1 INPUT INFORMATION LEVEL

12.8.1.1 Group I

With this group of languages, the basic input is the finished part geometry and the cutter location data is determined from this. The programmer specifies all motions and selects feeds, speeds, tools, and so on. Most of the early-developed

NC languages fall within this group, including ADAPT and early versions of the APT processor.

12.8.1.2 Group II

With this group of languages, the basic input is the same as for group I, but with the additional specification of details of the initial workpiece geometry. The program contains routines for the automatic processing of tool paths for area clearance, pocket milling, machining patterns of holes, and so on. Feeds, speeds, and tools are selected by the programmer. This group includes languages such as APT, NELAPT, IFAPT, and COMPACT II.

12.8.1.3 Group III

With this group of languages, technological decisions such as tool selection and feed and speed selection are made by the program, based on files of tool and material data. The basic input is the same as in group II, with the addition of details of the workpiece material. Sequences of tools and operations are determined automatically by the processor. For example, specification of the tapping of a hole at a certain location in a rough surface will lead to the selection of a sequence of operations, including spot facing, center drilling, core drilling and tapping, together with the tools, feeds, and speeds corresponding to each operation. Languages such as the EXAPT processors [18,19] and MITURN are in this group of languages.

12.8.2 SIMILARITY OF LANGUAGE STRUCTURE

Available NC processing languages can also be classified according to the similarity of their structures, in particular, in free- and fixed-format languages.

12.8.2.1 Free-Format Languages

Free-format languages are similar in concept to high-level computer programming languages such as BASIC or FORTRAN. The free-format system incorporates a vocabulary of words that can be used in a variety of ways to describe both part geometry and tool motions. For example, there will be a variety of definitions for points, lines, and circles and these can be used in many different ways to define a particular part geometry. Consequently, free-format NC languages are generally flexible to use, but require considerable programming skill and training to be used effectively. It is relatively easy to add more facilities to free-format languages and to use the system for a variety of different machine tool types.

12.8.2.2 Fixed-Format Languages

With fixed-format languages, programming is simplified by restricting the input to a series of numbers and codes that must be entered in a predetermined sequence.

Generally these languages are for a particular machine tool type such as lathes. The aim is to produce an easy-to-use language so that machines can be programmed with very little training required. These languages are much less flexible to use than free-format languages, and it is difficult to expand their facilities. Less computer power is required for fixed-format than for similar free-format languages. However, the majority of languages in common use are free-format. NC programming languages can also be divided into:

1. APT-based languages (free-format) based on the APT language structure, which is described in more detail later. Included in this group of languages are APT, ADAPT, NELAPT, IFAPT, EXAPT, and others.
2. Non-APT languages, including COMPACT H, AUTOSPOT, SPLIT, and others.

The APT-based languages were very commonly used, as was the COMPACT II system. The APT system is important because many of the associated procedures have found their way into modern CAM systems. Also the full APT system is very powerful and enables the generation of NC programs for very complex parts, using multi-axis machine tools.

12.9 NC PROGRAMMING USING APT-BASED LANGUAGES

Development of the APT (automatically programmed tools) [16] processor began at MIT in the early 1950s and has continued gradually since then. The full APT processor can deal with the programming of machine tools having five or more axes of control, but several systems of more limited capabilities with the same language structure have been developed. For example, ADAPT is restricted to programming these machines having 2½ axes of control.

12.9.1 LANGUAGE STRUCTURE

The APT language contains more than 300 English-like words, which can be used in a variety of ways to describe the component geometry and tool motions.

12.9.1.1 Geometric Statements

The APT language contains a large number of ways of defining the geometry of the component or the workpiece in terms of the points, lines, circles, surfaces, and so on, which make up this geometry. Each geometric statement is of the form:

```
SYMBOL (NAME) = WORD/MODIFIER, MODIFIER,
```

For example, a point can be defined by

```
P = POINT/b, 20, 40
```

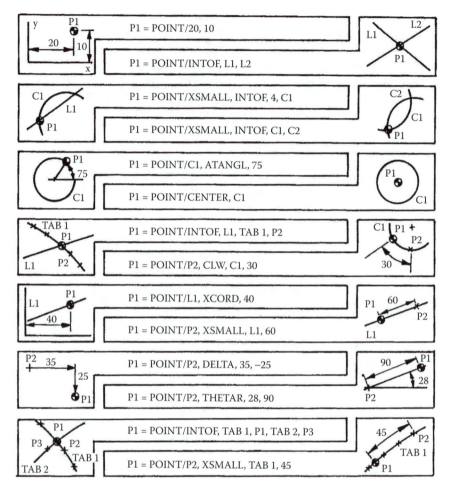

FIGURE 12.15 Some POINT definitions for the APT system.

or a circle as:

```
CIRC1 = CIRCLE/CENTER, P, RADIUS, 57.75
```

There are a large number of ways of defining each type of geometry, with many of the definitions being in terms of previously defined geometry. For example, Figure 12.15 shows some of the available ways of defining points. Further details of the variety of geometry definitions in APT can be found elsewhere [16].

12.9.2 MOTION STATEMENTS

These commands describe how previously defined geometry is to be machined. Each statement is of the form:

```
motion command/descriptive data (modifier)
```

A simple point-to-point motion statement is

```
GOTO/P1
```

and this would cause the tool to move from the current position to the previously defined coordinates of point P1. Incremental motions are specified using the statement GODLTA. For example, GODLTA/2.O, 3.0, 4.0 moves the tool 2.0, 3.0, and 4.0 increments in the x, y, z directions, respectively.

For contouring operations, the general method for specifying motions is in terms of three intersecting surfaces, and the procedure can be illustrated by considering the orthogonal surfaces shown in Figure 12.16. The three surfaces are as follows:

1. Drive surface: This is the surface that the axis of the tool moves parallel to during cutting so that the side of the cutter remains in contact with this surface.
2. Part surface: This is the surface that the end of the tool remains in contact with during cutting.
3. Check surface: This is the surface the tool is brought up to in order to end the cuts.

For the more general case, the drive, part, and check surfaces may be three-dimensional curved surfaces but the basic principle is the same as for orthogonal surfaces. The APT processor determines the appropriate tool path, which is offset from the drive and part surfaces so that the side of the tool remains in contact with the drive surface and the end of the tool remains in contact with the part surface.

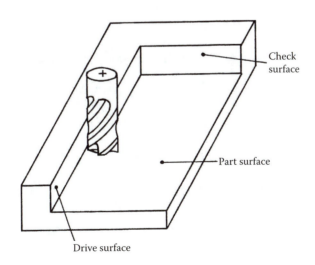

FIGURE 12.16 Surfaces for APT motion statements for contouring operations.

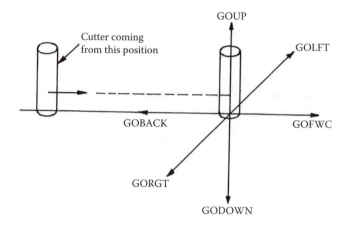

FIGURE 12.17 Motion commands for the APT system.

In general, the tool follows the drive surface, leaving the part surface cut until it reaches the check surface. Once a cut has been initiated, subsequent cuts can be specified by the following motion commands (Figure 12.17):

    GOLFT, GOFWD, GOUP, GORGHT, GOBACK, GODOWN

In using these commands, the programmer must know the direction of approach of the tool after the previous tool path segment.

Motion commands may have several modifiers. In particular, these designate the finishing position at the check surface. The main modifiers are: TO, ON, PAST, and TANTO. The effects of these commands are illustrated in Figure 12.18.

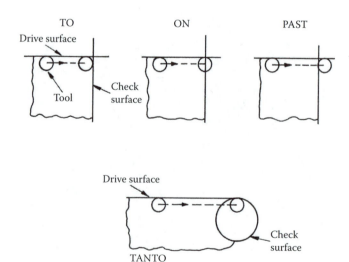

FIGURE 12.18 Motion command modifiers for the APT system.

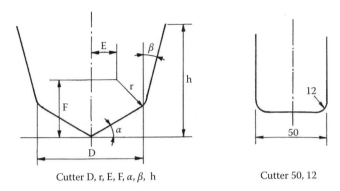

Cutter D, r, E, F, α, β, h   Cutter 50, 12

FIGURE 12.19 Cutter definitions in the APT system.

12.9.3 AUXILIARY STATEMENTS

Auxiliary statements include:

    PARTNO, INTOL/, OUTTOL/, CUTTER/, etc.

The cutter specification is important since one of the main functions of the APT processor is to determine the offset tool paths, and for this the tool geometry must be known. For milling and drilling operations the general CUTTER specification is shown in Figure 12.19. The default value of many of the parameters is zero and consequently the statement:

    CUTTER/.500

defines a 0.5-in-diameter end-milling cutter.

Programs in the APT language contain other statements which are passed directly to the postprocessor, including COOLNT/, RAPID, SPINDL/, FEDRAT/, TURRET/, MACHIN/, and so on. The words FEDRAT and SPINDLE specify the feed rate and spindle speed, respectively. The values for these are chosen by the programmer for most of the APT-based systems.

12.9.4 PROGRAMMING EXAMPLE IN APT

The preceding discussion has given a brief overview of the APT language, which can be further illustrated by considering the example shown in Figure 12.12. The outside profile of this component is to be machined with a 25-mm-diameter end-milling cutter. The APT program to generate the NC program to machine the part is shown in Table 12.3. Statements 1 to 9 are header and postprocessor statements. The tool, feed, and speed are selected by the programmer and specified by these commands.

Statements 10 to 21 are the geometry statements the programmer has decided to use to define the component geometry. Figure 12.20 shows the geometry specified by the geometry statements. The geometry defined at this stage is "unbounded," that is, the planes and lines extend to infinity and complete circles

TABLE 12.3
APT Program for Part Shown in Figure 12.11

SEQ. No.	INPUT STATEMENT	
1	PART No. STOP PLATE (DRG. No. AJ001X7)	
2	REMARK HEADING STATEMENTS FOLLOW	
3	NOPOST	
4	FULIST/ON	
5	CLPRNT	Header
6	CUTTER/25	
7	SPINDL/350	
8	TOLER/0.03	
9	FEDRAT/15	
10	REMARK GEOMETRIC STATEMENTS FOLLOW	
11	P1 = POINT/0, 0, 6	
12	P2 = POINT/117, 32, 6	
13	P3 = POINT/117, -32, 6	
14	C1 = CIRCLE/CENTER, P1, RADIUS, 10	
15	C2 = CIRCLE/CENTER, P2, RADIUS, 12.5	Geometry
16	C3 = CIRCLE/CENTER, P3, RADIUS, 12.5	statements
17	L1 = LINE/RIGHT, TANTO, C1, RIGHT, TANTO, C3	
18	L2 = LINE/LEFT, TANTO, C1, LEFT, TANTO, C2	
19	C4 = CIRCLE/XLARGE, OUT, C2, OUT, C3, RADIUS, 62	
20	SP = POINT/-25, -25, 25	
21	PL1 = PLANE/P1, P2, P3	
22	REMARK MOTION STATEMENTS FOLLOW	
23	FROM/SP	
24	GO/TO, L1, TO, PL1	
25	GORGT/L1, TANTO, C3	
26	GOFWD/C3, TANTO, C4	
27	GOFWD/C4, TANTO, C2	
28	GOFWD/C2, TANTO, L2	Motion
29	GOFWD/L2, TANTO, C1	statements
30	GOFWD/C1, TANTO, L1	
31	GOTO/SP	
32	STOP	
33	PRINT/3, ALL	
34	FINI	

are defined. The geometry is converted to the profile required by the tool motions specified by the motion commands.

Statements 22 to 31 are the motion commands. The programmer specifies that the tool start from the point SP, be brought to line L1, and then follow the profile in a counterclockwise direction. These motion commands convert the unbounded geometry into the required profile. This can be illustrated by considering

12-19 Remark geometric statements follow
P1 = POINT/0, 0, 6
P2 = POINT/117, 32, 6
P3 = POINT/117, -32, 6
C1 = CIRCLE/CENTER, P1, RADIUS, 10
C2 = CIRCLE/CENTER, P2, RADIUS, 12.5
C3 = CIRCLE/CENTER, P3, RADIUS, 12.5
L1 = LINE/RIGHT, TANTO, C1, RIGHT, TANTO, C3
L2 = LINE/LEFT, TANTO, C1, LEFT, TANTO', C2
C4 = CIRCLE, X LARGE, OUT, C2, OUT, C3, RADIUS, 62
SP = POINT/ -25, -25, 25
PL1 = PLANE/ P1, P2, P3

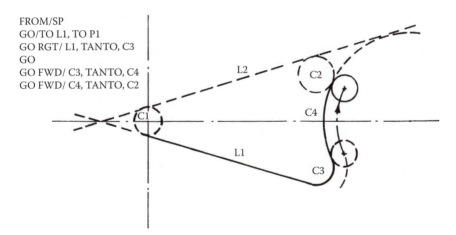

FIGURE 12.20 Geometric statements in APT for the milling operation in Figure 12.12.

FROM/SP
GO/TO L1, TO P1
GO RGT/ L1, TANTO, C3
GO
GO FWD/ C3, TANTO, C4
GO FWD/ C4, TANTO, C2

FIGURE 12.21 Motion statements convert the unbounded geometry into the required part profile.

the effect of statement 27 (Figure 12.21). This causes the tool to move from the tangency point between circles C3 and C4 to the tangency point between circles C4 and C2. Only the arc between these two points is cut, as required for the profile of the component.

As can be seen, the part geometry is defined and the offset tool positions are determined by the APT processor. The tool path points shown in Figure 12.12 for manual programming do not need to be determined by the programmer in this case. This is a relatively simple example, but the programmer is relieved from some fairly difficult calculations. For more complex parts, the saving in programming time is considerably greater. Many of the processors that utilize the APT language have facilities for automatically determining tool paths for machining processes that require many tool path points to be calculated. For example, area clearance routines are normally included for rough-turning operations, pocket milling, and so on. With these routines the profile to be machined and the original workpiece shape are defined. The various tool path points to clear out the required area are then determined automatically.

With many of the APT-based NC processing systems, the programmer selects the sequences of operations, tools, and cutting processing parameters. The incorporation of procedures for making technological decisions on sequences, cutting tools, and process parameters has also taken place by adding a technological processing stage after the geometric processing. At the technological processing stage [18], operation sequences, tools, and cutting process parameters are determined utilizing files of tools, materials, and operation sequences (Figure 12.22). Additional instructions are included in the language, which use the same basic APT language format. These instructions enable both single operations and sequences of operations to be defined. In response to these instructions, the tools and cutting-process parameters are selected from the files associated with the system. The material of the workpiece and its surface condition may also be specified. For example, the production of a reamed hole in a workpiece will require a sequence of operations, including center drilling, spot facing if the work surface is rough, drilling, predrilling for large-diameter holes, and finally reaming. Figure 12.23 shows a flow diagram for deciding this sequence, and rules for each decision are incorporated into the program [19]. For each operation in the sequence, the appropriate tools will be chosen from the tool file and the feeds and speeds selected to suit the operation and workpiece materials.

12.10 GRAPHICS-BASED NC PROCESSING SYSTEMS

In recent years the language-based NC processing systems have been largely replaced by graphics-based systems, either as part of comprehensive CAD systems that utilize solid modeling of the part geometry or as standalone systems that are able to take the output of available CAD systems for the processing of NC programs. This means that the part geometry no longer has to be described by means of a descriptive language structure such as in APT. However, the underlying geometry processing and logic for NC program development in these graphics-based systems is still essentially the same as developed for APT and similar language-based systems. In particular the post processing stage is carried out on APT compatible CL files developed by the graphics-based systems.

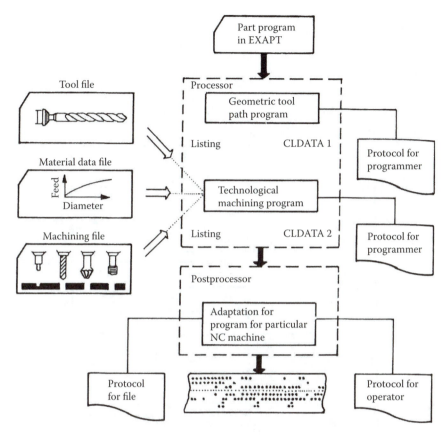

FIGURE 12.22 Sequence of programming a NC machine with EXAPT. (From Rembold, U. and Epple, W., *Computer-Aided Design and Manufacturing*, Rembold, U. and Dillman, R., eds., Springer Verlag, NY, 1986, 279.)

The processing of NC programs is now available as one of the secondary features of modern solid model-based CAD systems. It is generally necessary for part programmers to select the features to be machined interactively and also select the tools to be used from an associated database. Sequences of operations and tool paths are generated automatically using routines that have their basis in the procedures developed for APT and other language-based systems. An associated material database enables the automatic determination of appropriate feeds and speeds for the various machining operations.

When NC processing is integrated with CAD systems, it is possible to associate the tool paths generated with the CAD geometric model. This feature is usually referred to as *associativity*. A consequence of this provision is that if the geometric model is subsequently modified, the associated tool paths can change automatically without the need for reprocessing.

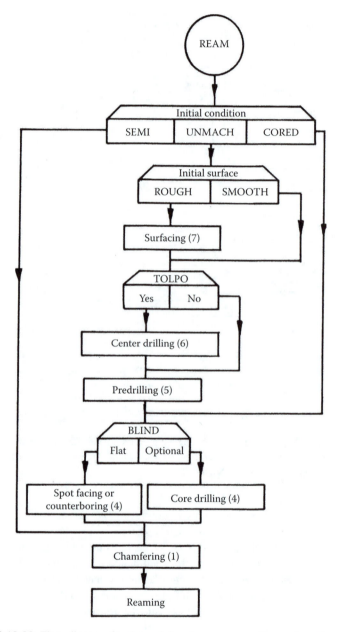

FIGURE 12.23 Flow diagram for determining the sequence of operations for a reamed hole in EXAPT 1. (From Opitz and Simon [19].)

A number of standalone graphics NC processing systems are available commercially that run on graphics workstations or PCs. These systems usually incorporate tool and materials databases that can be maintained by the user. The part geometry can be entered into the system using a special purpose graphics-based user interface and geometry processes or can be read into the system from an available CAD system either directly or through one of the standard graphics exchange formats such as IGES or STEP [2]. The routines for tool path generation similar to those used in APT or similar language processors are incorporated into the software. However, the part programmer generally has to select interactively the various geometric features to be machined, together with the associated cutting tools to be used.

In general APT compatible CL files are produced from these systems which must then be post processed for the specific machines to be used to produce the parts. Most commercial systems now include simple editing routines for creating postprocessors for new machines and controllers, whilst providing a library of post processors for popular machine tools available on the market.

REFERENCES

1. Groover, M.P. and Zimmers, E.W., *CAD/CAM: Computer-Aided Design and Manufacturing*, Pearson Educational, Englewood Cliffs, NJ, 1997.
2. Chan, T-C., Wysk, R.A., and Wang, H-P., *Computer Aided Manufacture*, 2nd ed., Prentice Hall, NJ, 1998.
3. Amirouche, F., *Principles of Computer-Aided Design and Manufacture*, 2nd ed., Pearson Education, Englewood Cliffs, NJ, 2003.
4. Knox, P., *CAD/CAM Systems Planning and Implementation*, Marcel Dekker, NY, 1983.
5. Gallagher, C.C. and Knight, W.A., *Group Technology Production Methods in Manufacture*, Ellis Horwood, Ltd., Chichester, England, 1986.
6. Halevi, G., *The Role of Computers in Manufacturing Processes*, Wiley, NY, 1980.
7. Ham., I. and Lu, S.C-Y., Computer Aided Process Planning: The Present and the Future, *Ann. CIRP*, Vol. 34/2, pp. 591–602, 1988.
8. Tonshoff, H.K. and Anders, N., Survey of Developments and Trends in CAPP Research within CIRP, *Ann. CIRP*, Vol. 39/2, 707–710, 1990.
9. El Maraghy, H., Evolution and Future Perspectives of CAPP, *Ann. CIRP*, Vol.42/2, 739–751, 1993.
10. Tultoff, J., Process Planning: An Historical Review and Future Prospect, Proc. 19th CIRP Int. Sem. on Manuf. Sys., Pennsylvania State University, University Park, June 1–2, 1987, 207.
11. Link, K.H., Computer-Aided Process Planning (CAPP), *SME Technical Paper*, MS 77-3 14, 1977.
12. Houtzeel, A., The Many Faces of Group Technology, *American Mach.*, Vol. 123, 115, 1979.
13. Spur, G. and Krause, F.L., Technological Planning for Manufacture — Methodology of Process Planning, in *Computer-Aided Design and Manufacture*, Rembold, U. and Diliman, R., eds., Springer Verlag, NY, 1986, 65.
14. Eversheim, W.H., Fuchs, H., and Zon, K.H., Automatic Process Planning with Regard to Production by Application of the System AUTAP for Control Problem, Proc 12th CIRP, Int. Sem. on Manuf. Sys., Belgrade, 779–800, 1980.

15. Eversheim, W., Automated Generation of Process Plans for Prismatic Parts, *Ann. CIRP*, Vol. 32, 361, 1983.

16. Hori, S., *APT Part Programming*, ITT Research Institute, McGraw-Hill, NY, 1967.

17. Pressman, R.S. and Williams, J.E., *Numerical Control and Computer-Aided Manufacturing*, Wiley, NY, 1977.

18. Rembold, U. and Epple, W., Present State and Future Trends in the Development of Programming Languages for Manufacturing, in *Computer-Aided Design and Manufacturing*, Rembold, U. and Dillman, R., eds., Springer Verlag, NY, 1986, 279.

19. Opitz, H. and Simon, W., *EXAPT I Part Programmer Reference Manual*, NEL Report No. 293, East Kilbride, June, 1967.

13 Design for Machining

13.1 INTRODUCTION

Many techniques are available to reduce manufacturing costs and increase manufacturing productivity. These measures include using:

1. Improved materials, tools, and processes
2. More effective organization and factory layout, materials handling, and assembly techniques
3. Automation, wherever it contributes to greater efficiency

Unfortunately, designers often consider that their job is to design the product for performance, appearance, and possibly reliability, and that it is the manufacturing engineer's job to produce whatever has been designed. There is often a natural reluctance to change a proven design for the sake of a reduction in manufacturing cost, and although designers generally assert that they take manufacturing problems into account whenever possible, the fact is that as a subject, design for manufacture is hardly recognized compared with design for performance. Sometimes design for manufacture is mentioned briefly in textbooks on manufacturing, but only rarely is it mentioned in textbooks on design.

Many companies ensure that their designers obtain some manufacturing experience, and many university engineering programs include courses in manufacturing. All this is beneficial, but design for manufacture is one of those subjects a person is expected to assimilate through experience rather than learn from a textbook or instructor. This chapter explains one important aspect of this subject: design for machining.

For obvious reasons machining is a wasteful process, and many engineers feel that the main concern should be to design components that do not require machining. Since 80 to 90% of manufacturing machines are designed to machine metal, the view that machining should be avoided must be considered impracticable for the immediate future. However, the trend toward the use of "near net shape" processes that conserve material is clearly increasing, and when large-volume production is involved, this approach should be foremost in the designer's mind.

The following discussion deals first with standardization, then the choice of raw material for the component, the form of the raw material, the ways in which this form can be readily changed to the desired form by machining, and the ways in which the surfaces of the component are finished. Finally, an introduction to early cost estimating for designers is presented.

In the discussion, certain design principles that can help to simplify the machining of components and reduce manufacturing costs are introduced; these are summarized in the chapter.

The terms "workpiece," "part," "piece part," "component," or "component part" are often used interchangeably in describing the machined object. In this chapter, to avoid confusion, the object being machined will be termed a workpiece; after machining is complete, it will be called a component.

13.2 STANDARDIZATION

Provided it has been established that a separate component is necessary, perhaps the first rule in designing for machining is to use standard components as much as possible. Many small components such as nuts, washers, bolts, screws, seals, bearings, gears, and sprockets are manufactured in large quantities and standard sizes that should be employed wherever possible. The cost of these components is much lower than that of similar, nonstandard components. Clearly, the designer will need catalogues of the standard items available; these can be obtained from suppliers. Supplier information is provided in standard trade indexes, where companies are listed under products.

A second rule is to minimize the amount of machining by preshaping the workpiece if possible. Workpieces can sometimes be preshaped by using castings or welded assemblies or metal deformation processes such as extrusion, deep drawing, blanking, or forging. Obviously, the justification for the preforming of workpieces depends on the size of the batch (the number of components to be produced). In general, for small batches the tendency is to produce the desired shapes by machining. Again, standardization can play an important part when workpieces are to be preformed. The designer may be able to use preformed workpieces designed for a previous similar job because the necessary patterns for castings or the tools and dies for metal-forming processes are already available.

Finally, even if standard components or standard preformed workpieces are not available, the designer should attempt to standardize the machined features to be incorporated in the design. Standardizing machined features means that the appropriate tools, jigs, and fixtures will be available, which can reduce manufacturing costs considerably. Examples of standardized machined features might include screw threads, keyways, seatings for bearings, splines, and so on. Information on standard features can be found in various reference books, for example, *Machinery's Handbook* [1].

13.3 CHOICE OF WORK MATERIAL

When choosing the material for a component, the designer must consider applicability, cost, availability, machinability, and the amount of machining required. Each of these factors influences the others, and the final optimum choice will generally be a compromise between conflicting requirements. The applicability of various materials will depend on the component's eventual function and will be decided by such factors as strength, resistance to wear, appearance, corrosion resistance, and so on. These features of the design process are outside the scope of this chapter, but once the choice of material for a component has been narrowed by the application of these criteria, the designer must then consider factors that help to minimize the final cost of the component. It should not be assumed, for example, that the least-expensive work material will automatically result in minimum cost for the component. For example, it might be more economical to choose a material that is less expensive to machine (more machinable) but has a higher purchase cost. In a constant-cutting-speed, rough-machining operation, the production cost C_{pr} per component is given by

$$C_{pr} = Mt_l + MKv^{-1} + Kt_r^{-1}v_r^{-1/n}\left(Mt_{c_t} + C_t\right)v^{(1-n)/n} \tag{13.1}$$

where

M = total machine and operator rate

t_l = nonproductive time

K = constant for the machining operation

v_r = cutting speed for a tool life of t_r

t_{c_t} = tool changing time

C_t = cost of a sharp tool

v = cutting speed

n = constant depending mainly on the tool material

It was shown that the cutting speed v for minimum cost is given by

$$v_c = v_r\left(\frac{t_r}{t_c}\right)^n \tag{13.2}$$

where t_c is the tool life for minimum cost and is given by

$$t_c = \left(\frac{1}{n} - 1\right)\left(t_{c_t} + \frac{C_t}{M}\right) \tag{13.3}$$

Thus, if Equation 13.1, Equation 13.2, and Equation 13.3 are combined, the minimum cost of production C_{min} is given by

$$C_{min} = Mt_l + \frac{MK}{(1-n)v_r}\left(\frac{t_c}{t_r}\right)^n \tag{13.4}$$

The first term in this expression is the cost of the nonproductive time on the machine tool and is not affected by the work material chosen or by the amount of machining carried out on the workpiece. The second term is the cost of the actual machining operation, and for a given machine and tool design depends on the values of n, v_r, t_r, and K. The factor n depends mainly on the tool material (for high-speed steel $n = 0.125$ and for carbide $n = 0.25$ approximately); vt_r^n is a measure of the machinability of the material; K is proportional to the amount of machining to be carried out on the workpiece and can be regarded as the distance traveled by the tool corner relative to the workpiece during the machining operation. For a given operation on a given machine tool and with a given tool material, it is shown in Equation 13.3 that the tool life for minimum cost would be constant and hence (from Equation 13.4) that the machining costs would be inversely proportional to the value of $v_r t_r^n$. Since v_r is the cutting speed giving a tool life of t_r, more readily machined materials have a higher value of $v_r t_r^n$ and hence give a lower machining cost.

Taking, for example, a machining operation using high-speed steel tools ($n = 0.125$) and typical figures of $M = \$0.008/s$, $t_1 = 300$ s, $t_c = 3000$ s, $K = 183$ m (600 ft), and $v_r = 0.76$ m/s (150 ft/mm) when $t_r = 60$ s, then from Equation 13.4 the minimum production cost per component C_{min} is \$6.22. If, however, an aluminum workpiece for which a typical value of v_r is 3.05 m/s (600 ft/mm) when t_r is 60 s could be used, the use of aluminum would reduce the production cost to \$3.43. In other words, an additional amount equal to the difference between these two costs could be spent for each workpiece in order to employ the more machinable material (i.e., as much as \$2.79 additional per workpiece). For a workpiece 25 mm (1 in.) in diameter, a value of K of 183 m (600 ft) would correspond to finish-machining of a component of about 1.64×10^{-5} m (1 in.³) in volume. Since the cost per unit volume of aluminum rod is only a few cents higher than the cost per unit volume of mild-steel rod, the total cost of the machined component would be much less for the aluminum material.

Clearly, the designer should select work materials that will result in minimum component cost.

13.4 SHAPE OF WORK MATERIAL

With the exception of workpieces that are to be partially formed before machining such as forgings, castings, and welded structures, the choice of the shape of the work material depends mainly on availability. Metals are generally sold in plate, sheet, bar, or tube form (Table 13.1) in a wide range of standard sizes. The designer should check on the standard shapes and sizes from the raw material supplier and then design components that require the minimum of machining.

Components manufactured from a circular or hexagonal bar or tube are generally machined on machine tools that apply a rotary primary motion to the workpiece; these types of components are called *rotational components*. The remaining components are manufactured from square or rectangular bar, plate, or sheet and are called *nonrotational components*. Components partially formed

TABLE 13.1
Standard Material Shapes and Range of Sizes

Name	Size	Shape
Plate	6–75 mm (0.25–3 in.)	
Sheet	0.1–5 mm (0.004–0.2 in.)	
Round bar or rod	3–200 mm dia. (0.125–8 in. dia.)	
Hexagonal bar	6–75 mm (0.25–3 in.)	
Square bar	9–100 mm (0.375–4 in.)	
Rectangular bar	3 × 12–100 × 150 mm (0.125 × 0.5–4 × 6 in.)	
Tubing	5 mm dia., 1 mm wall–100 mm dia., 3 mm wall (0.1875 in. dia., 0.035 in. wall–4 in. dia., 0.125 in. wall)	

before machining can also be classified as either rotational or nonrotational components.

13.5 SHAPE OF COMPONENT

1 3.5.1 Classification

As mentioned above, component shapes can be classified as rotational and nonrotational. The rotational components are those whose basic shape can be machined on lathes, boring mills, cylindrical grinders, or any other machine tool that applies a rotary primary motion to the workpiece. In defining a component shape as rotational, the important factor is that in machining its basic shape, the workpiece is rotated. The nonrotational category includes all shapes other than rotational shapes (and, of course, in machining its basic shape, the motion of the workpiece is linear). The basic shape of a rotational component is a plain cylinder with dimensions sufficiently large to enclose the final shape (Figure 13.la); the basic shape of a nonrotational component is a right-rectangular prism with dimensions sufficiently large to enclose the final shape (Figure 13.lb).

In considering design for machining, it is important to know the ways in which the basic shapes can be readily changed by machining processes. Before proceeding, however, it will be helpful to describe briefly a method of classifying machined components developed by Opitz [2]. This classification system allows components to be identified by geometric code numbers for the purposes of analysis of complete production systems. The code number depends on the types of machining processes required to produce the component, and therefore provides a suitable basis for discussion of the ways in which basic shapes can be changed by machining. The main use of the classification system is to be able to categorize and group those components having similar features and requiring similar sequences of machining operations. This information allows manufacturing engineers to efficiently plan the layout of machines in the factory to reduce the handling and transfer of components as much as possible, as described in Chapter 11.

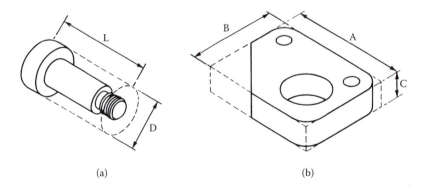

(a)   (b)

FIGURE 13.1 Basic component shapes. (a) Rotational; (b) nonrotational.

However, it will also assist the designer who wishes to standardize components and avoid specifying machined features the company is not equipped to handle.

The basic geometric code number consists of five digits, the first digit describing the component class (either rotational or nonrotational) and the proportions of the overall shape, the remaining digits describing the variations from this shape. Table 13.2 shows the general basis of this coding system and the means of determining the first digit. Thus, for example, if the rotational component shown in Figure 13.la has a length-to-diameter (L/D) ratio of 2, the first digit in the code number is 1; if the nonrotational component shown in Figure 13.lb has ratios A/B of 1.2 and A/C of 4.5, it is classified as a flat component, and the first digit in the code number is 6. (In labeling the basic rectangular prism the largest side is given the length A and the smallest the length C.) Components with first digits of 3 or 4 are those rotational components machined from other than cylindrical shapes (hexagonal bar, for example), and components with first digits of 5 or 9 are those that do not fit within the other categories. For components with first digits of 0, 1, 2, 6, 7, and 8, the determination of the remaining digits is shown in Table 13.3, Table 13.4, and Table 13.5.

A supplementary code can also be employed. This supplementary code consists of four further digits: the first gives the basic dimension of the component (D for a rotational component and A for a nonrotational component), the second specifies the material used, the third specifies the form of the raw material, and the fourth specifies the accuracy of the machined surfaces. For a complete description of the classification system, the reader should see Reference [2].

The rotational component shown in Figure 13.la can be machined from a bar by gripping one end in a lathe chuck, turning to form the steps, and then generating the screw thread. A second digit of 2 describes such a shape (a step to one end with a screw thread), and since the component has no internal shape, no plane-surface machining, and no auxiliary holes or gear teeth, the third, fourth, and fifth digits are all zero. Thus, the code number for this component is 12000. For the nonrotational component shown in Figure 13.lb a second digit of 1 describes a plane component with one angular deviation, a third digit of 1 describes one smooth principal bore, a fourth digit of 0 describes no plane-surface machining, and a fifth digit of 3 describes a component with no gear teeth or forming and two or more holes drilled in one direction and related by a drilling pattern. Thus, the code number would be 61103. It should be noted that the rounding of the corners of the basic rectangular shape is ignored in deciding the geometric code.

From these brief examples it can be seen that the coding system is based on the more common machining techniques used to alter the initial workpiece shape. Some of these machining techniques will now be described and will help to illustrate further design rules for machined components.

13.5.2 ROTATIONAL COMPONENTS [(L/D) ≤ 0.5]

Rotational components where the length-to-diameter ratio is less than or equal to 0.5 have a first digit of 0 and may be classified as disks. For diameters to

TABLE 13.2
Geometric Code Devised by Opitz [2] for Classification of Machined Components

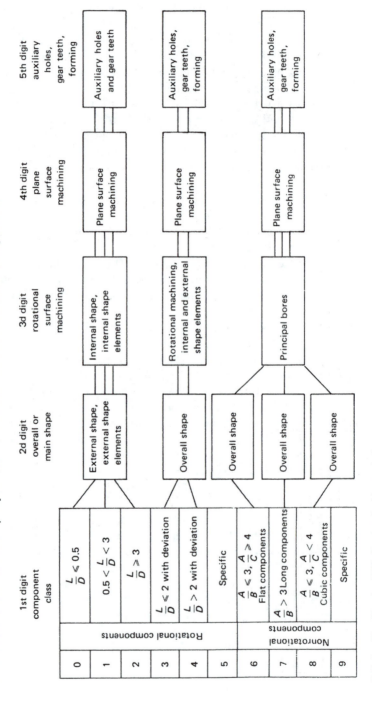

TABLE 13.3
Geometric Code for Rotational Components[a]

	1st digit	2d digit	3d digit	4th digit	5th digit
	Component class	External shape, external shape elements	Internal shape, internal shape elements	Plane surface machining	Auxiliary hole(s) and gear teeth
0	$L/D \leq 0.5$	Smooth, no shape elements	Without through bore blind hole	No surface machining	No auxiliary hole(s)
1	$0.5 < L/D < 3$	Smooth or stepped to one end — No shape elements	Without through bore blind hole — No shape elements	External plane surface and/or surface curved in one direction	Axial hole(s) not related by a drilling pattern
2	$L/D > 3$	With screwthread	With screwthread	External plane surfaces related to one another by graduation around a circle	Axial holes related by a drilling pattern
3		With functional groove	With functional groove	External groove and/or slot	Radial hole(s) not related by a drilling pattern
4		Stepped to both ends (multiple increases) — No shape elements	Stepped to both ends (multiple increases) — No shape elements	External spline and/or polygon	Holes axial and/or radial and/or in other directions not related
5		With screwthread	With screwthread	External plane surface and/or slot and/or groove, spline	Holes axial and/or radial and/or in other directions related by drilling pattern
6		With functional groove	With functional groove	Internal plane surface and/or groove	Spur gear teeth
7		Functional taper	Functional taper	Internal spline and/or polygon	Bevel gear teeth
8		Operating thread	Operating thread	External and internal splines and/or slot and/or groove	Other gear teeth
9		Others (> 10 functional diameters)	Others (> 10 functional diameters)	Others	Others

(Column rows 0–4 of 5th digit: No gear teeth; rows 5–9: With gear teeth. First-digit column grouped under "Rotational components.")

[a] After Opitz [2] © Verlag Wu. Gerardet 1970. Reprinted with permission.

TABLE 13.4
Second Digit[a] of Geometric Code for Flat and Long Nonrotational Components[b]

1st digit Component class	2d digit Overall shape		
6 Flat components $\frac{A}{B} < 3$, $\frac{A}{C} > 4$	Plane	0	Rectangular
		1	Rectangular, with one deviation (right angle or triangular)
		2	Rectangular with angular deviations
		3	Rectangular with circular deviation
		4	Any flat shape other than 0 to 3
		5	Flat components, rectangular or right angled with small deviations due to casting, welding, forming
		6	Flat components, round or any shape other than position 5
		7	Flat components regularly arched or dished
		8	Flat components irregularly arched or dished
		9	Others

1st digit Component class	2d digit Overall shape		
7 Nonrotational components — Long components $\frac{A}{B} > 3$	Uniform cross section / Shape axis (straight)	0	Rectangular
		1	Rectangular with one deviation (right angle or triangular)
		2	Any cross section other than 0 and 1
	Varying cross section	3	Rectangular
		4	Rectangular with one deviation (right angle or triangular)
		5	Any cross section other than 3 and 4
	Shape axis curved (bent)	6	Rectangular, angular, and other cross sections
		7	Formed component
		8	Formed component with deviations in the main axis
		9	Others

a For the third, fourth and fifth digits, see Table 13.5.

b After Opitz [2]. ç Verlag Wu. Girardet 1970. Reprinted with permission.

TABLE 13.5
Geometric Code for Cubic Components[a]

Digit value	1st digit — Component class	2d digit — Overall shape	3d digit — Principal bore, rotational surface machining	4th digit — Plane surface machining	5th digit — Auxiliary hole(s), forming, gear teeth
0		Rectangular prism	No rotational machining or bore(s)	No surface machining	No auxiliary holes, gear teeth, and forming
1		Rectangular with deviations (right angle or triangular)	One principal bore, smooth	Functional chamfers (e.g., welding prep.)	Holes drilled in one direction only
2		Compounded of rectangular prisms	One principal bore stepped to one or both ends	One plane surface	Holes drilled in more than one direction
3		Components with a mounting or locating surface and principal bore	One principal bore with shape elements	Stepped plane surfaces	Holes drilled in one direction only (related by a drilling pattern)
4		Components with a mounting or locating surface, principal bore with dividing surface	Two principal bores, parallel	Stepped plane surfaces at right angles, inclined and/or opposite	Holes drilled in more than one direction (related by a drilling pattern)
5		Components other than 0 to 4	Several principal bores, parallel	Groove and/or slot	Formed, no auxiliary holes
6		Approximate or compounded of rectangular prisms	Several principal bores, other than parallel	Groove and/or slot and 4	Formed, with auxiliary holes
7	Nonrotational components	Components other than 6	Machined annular surfaces, annular grooves	Curved surface	Gear teeth, no auxiliary hole(s)
8	Cubic components $A < 3$, $\frac{A}{B} < 4$, $\frac{A}{C} < 4$	Approximate or compounded of rectangular prisms	7 + principal bore(s)	Guide surfaces	Gear teeth, with auxiliary hole(s)
9		Components other than 8	Others	Others	Others

Overall shape groupings: values 0–5 = Block and blocklike components (0–4 Not split, 5 Split); values 6–9 = Box and boxlike components (6–7 Not split, 8–9 Split).

5th digit groupings: values 0–4 = No gear teeth, no forming; values 5–6 = Forming, no gear teeth; values 7–8 = Gear teeth.

[a] After Opitz [2]. © Verlag Wu. Girardet 1970. Reprinted with permission.

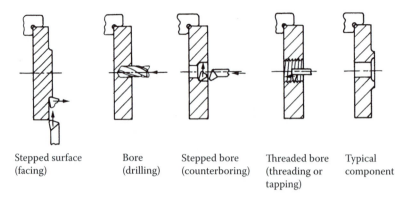

Stepped surface	Bore	Stepped bore	Threaded bore	Typical
(facing)	(drilling)	(counterboring)	(threading or tapping)	component

FIGURE 13.2 Some ways of machining a disk shaped workpiece.

approximately 300 mm (12 in.) the workpiece is generally gripped in a lathe chuck; for larger diameters it is necessary to clamp the workpiece on the table of a vertical borer. The simplest operations that could be performed would be machining of the exposed face and drilling, boring, and threading a concentric hole. All these operations could be performed on one machine without having to grip the workpiece more than once (Figure 13.2). The realization that neither the unexposed face nor a portion of the outer cylindrical surface can be machined leads to a further general guideline for design: if possible, design the component so that machining is not necessary on the unexposed surfaces of the workpiece when it is gripped in the work-holding device. Also, the diameters of the external features should gradually increase, and the diameters of the internal features should gradually decrease from the exposed face.

Of course, with the examples shown in Figure 13.2 it would probably be necessary to reverse the workpiece in the chuck to machine the opposite face. However, if the diameter of the workpiece were less than about 50 mm (2 in.), the desired surfaces could probably be machined on the end of a piece of bar stock and the component then separated from the bar by a parting or cut-off operation (Figure 13.3). It should be remembered that when a workpiece must be reversed in the chuck, the concentricity of features will be difficult to maintain (Figure 13.4).

When machined surfaces intersect to form an edge, the edge is square; when surfaces intersect to form an internal corner, however, the edge is rounded to the shape of the tool corner. Thus, the designer should always specify radii for internal corners. When the two intersecting faces are to form seating for another component in the final assembly, the matching corner on the second component should be chamfered to provide clearance (Figure 13.5). Chamfering ensures proper seating of the two parts and presents little difficulty in the machining operations. These corners or chamfers do not affect the geometric code of the components.

On rotational components some features may be necessary that can only be produced by machine tools other than those that rotate the workpiece. Consequently, the batch of workpieces requiring these features will have to be stacked

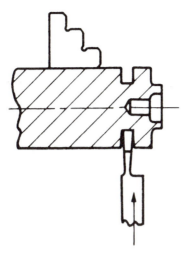

FIGURE 13.3 Part finished components from bar stock.

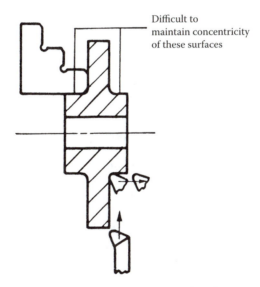

FIGURE 13.4 Machining of components stepped to both ends.

temporarily and then transferred to another machine tool that may be in another part of the factory. This storage and transfer of workpieces around a factory presents a major organizational problem and adds considerably to manufacturing costs. Thus, if possible, the components should be designed to be machined on one machine tool only.

Plane-machining operations may also be required on a rotational component. Such operations might be carried out on a shaper (or slotting machine) or milling

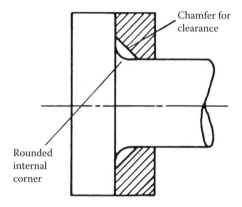

FIGURE 13.5 Rounded corners and chamfers.

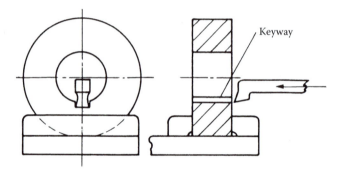

FIGURE 13.6 Machining a keyway on a shaper.

machine. A typical example is a keyway in the bore of one of the disk-shaped parts discussed earlier. Such a keyway could be readily machined in a shaper, as shown in Figure 13.6. Finally, auxiliary holes (those not concentric with the component axis) and gear teeth may be required. Auxiliary holes would be machined on a drill press and would generally form a pattern as shown in Figure 13.7. It should be noted that axial auxiliary holes are usually the easiest to machine because one of the flat surfaces on the workpiece can be used to orient it on the work-holding surface. Thus, the designer should avoid auxiliary holes inclined to the workpiece axis. Gear teeth would be generated on a special gear-cutting machine, and this process is generally slow and expensive.

13.5.3 ROTATIONAL COMPONENTS [0.5 < (L/D) ≤ 3]

The class of rotational components having a length-to-diameter ratio of between 0.5 and 3 are short, cylindrical components. They have a first digit of 1 and are the most suitable for machining on engine and turret lathes. The workpieces from which these components are produced would often be in the form of bar stock,

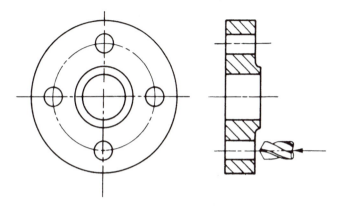

FIGURE 13.7 Drilling a pattern of auxiliary holes.

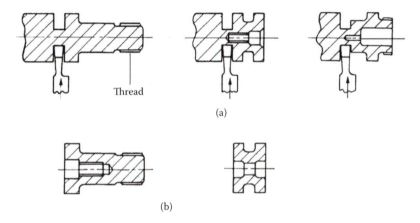

FIGURE 13.8 Machining components from bar stock. (a) Components that can be parted off complete; (b) components that cannot be parted off complete.

and the machined component would be separated from the workpiece by parting or cutoff, as was shown in Figure 13.3. The whole of the outer surface of this type of component can be machined without interference by the jaws of the chuck. However, it is important with this type of component for the designer to ensure (if possible) that the diameters of a stepped internal bore are gradually decreasing from the exposed end of the workpiece (Figure 13.8) and that no recesses or grooves are required on the surface produced in the parting or cutoff operation.

13.5.4 Rotational Components [(L/D > 3]

The class of rotational components having a length-to-diameter ratio greater than or equal to 3 are long, cylindrical components. They have a first digit of 2 and would be machined on an engine lathe. These components would be supported between centers or gripped at the headstock end by a chuck and supported by the tailstock center at the other end. If the *L/D* ratio is too large, the flexibility of the workpiece creates a problem because of the forces generated during machining. Thus, the designer should ensure that the workpiece, when supported by the work-holding devices, is sufficiently rigid to withstand the machining forces.

When a rotational component must be supported at both ends for machining of the external surfaces, internal surfaces of any kind cannot be machined at the same time. In any case, with slender components, concentric bores would necessarily have large length-to-diameter ratios and would be difficult to produce. Thus, the designer should try to avoid specifying internal surfaces for rotational components having large *L/D* ratios.

A common requirement on a long, cylindrical component is a keyway, or slot. A keyway is usually milled on a vertical-milling machine using an end-milling cutter (Figure 13.9a) or on a horizontal-milling machine using a side-and face-milling cutter (Figure 13.9b). The shape of the end of the keyway is determined by the shape of the milling cutter used and thus the designer, in specifying this shape, is specifying the machining process.

Before the ways of changing the basic nonrotational shapes by machining operations are discussed, some general points should be noted regarding undesirable features on rotational components. These undesirable design features can be categorized as follows:

1. Features impossible to machine
2. Features extremely difficult to machine that require the use of special tools or fixtures
3. Features expensive to machine even though standard tools can be used

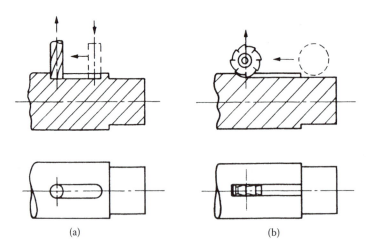

(a)   (b)

FIGURE 13.9 Machining of a keyway. (a) Vertical milling; (b) horizontal milling.

In considering the features of a particular design it should be realized that

1. Surfaces to be machined must be accessible when the workpiece is gripped in the work-holding device.
2. When the surface of the workpiece is being machined, the tool and tool-holding device must not interfere with the remaining surfaces on the workpiece.

Figure 13.10a shows an example of a component with external surfaces impossible to machine. This is because during the machining of one of the cylindrical surfaces, the tool would interfere with the other cylindrical surface. Figure 13.10b shows a component that would be extremely difficult to machine on a lathe because when the hole was drilled, the workpiece would have to be supported in a special holding device. Even if the workpiece were transferred to a drill press for the purpose of drilling the hole (in itself an added expense), a milled preparation would be required (Figure 13.10c) to prevent the drill from deflecting sideways at the beginning of the drilling operation. Figure 13.10 also illustrates the difficulties that may arise when nonconcentric cylindrical surfaces are specified.

Figure 13.11 shows two examples where the tool or tool holder would interfere with other surfaces on the workpiece. The small radial hole shown in Figure 13.11a would be difficult to machine because a special long drill would be required. The internal recess shown in the component in Figure 13.11b could not

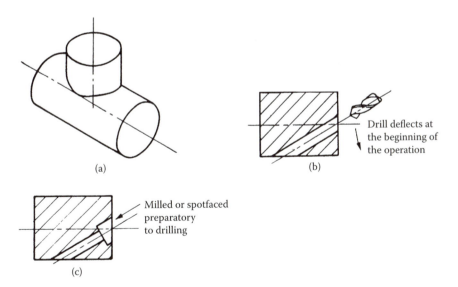

(a)

(b) Drill deflects at the beginning of the operation

(c) Milled or spotfaced preparatory to drilling

FIGURE 13.10 Difficulties arising when nonconcentric cylindrical surfaces are specified. (a) Impossible to machine; (b) difficult to machine; (c) can be machined on a drill press.

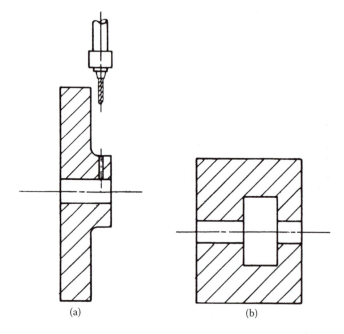

FIGURE 13.11 Design features to avoid in rotational parts. (a) Special drill required to machine radial hole; (b) impossible to machine internal recess.

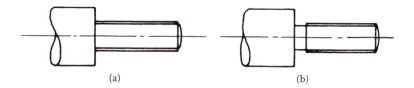

FIGURE 13.12 Provision of run-out groove for screw threads. (a) Wrong; (b) right.

be machined because it would be impossible to design a cutting tool that would reach through the opening of the bore.

Figure 13.12a shows a screw thread extending to a shoulder. Extending a screw thread to a shoulder would be impossible because when the lathe carriage is disengaged from the lead screw at the end of each pass, the threading tool generates a circular groove in the workpiece. Thus, it is necessary to provide a run-out groove (Figure 13.12b) so that the threading tool will have clearance and not interfere with the remaining machined surfaces.

Table 13.6 summarizes the rotational-workpiece features readily provided by machining. It should be kept in mind, however, that the relative difficulty of providing some of these features depends on the component's L/D ratio. For example, internal features most readily machined when the L/D ratio is small, become increasingly difficult to machine as this ratio is increased. Also, components

TABLE 13.6
Readily Provided Machined Features on Rotational Components

Machine	Type of surface	Orientation and nature of surface	Example
Lathe or vertical-boring machine	External surfaces (exposed when the workpiece is gripped in work-holding device)	Cylindrical concentric surfaces increasing in diameter from the exposed end of the workpiece	
		Plane surfaces normal to the workpiece axis	
	Internal surfaces (accessible when the workpiece is gripped in work-holding device)	Cylindrical, concentric surfaces decreasing in diameter from the exposed end of the workpiece	
		Plane surfaces normal to the workpiece axis	
Drill press	Auxiliary holes	Cylindrical holes radial or parallel to the workpiece axis and in accessible positions	
Milling machine or shaper	Auxiliary plane surfaces	Plane surfaces parallel to workpiece axis and in accessible positions	

with a thin section become flexible and difficult to grip firmly; hence, very small or very large L/D ratios should be avoided where possible. In general, the features listed in Table 13.6 are placed in increasing order of machining cost.

Table 13.7 summarizes rotational-workpiece features that are more expensive to machine; these features, too, are listed approximately in increasing order of machining cost. Features not in Table 13.6 and Table 13.7 should be examined carefully by the designer to see whether an alternative design would save manufacturing costs.

13.5.5 NONROTATIONAL COMPONENTS [$(A/B) \leq 3$, $(A/C) \geq 4$]

Nonrotational components having proportions such that the ratio A/B is less than or equal to 3 and the ratio A/C is greater than or equal to 4 are categorized as flat components and have a first digit of 6. Extremely thin, flat components should be avoided because of the difficulty of work holding in machining external surfaces. Many flat components are machined from plate or sheet stock and initially require machining of the outer edges. Outer edges are generally machined on either a vertical- or horizontal-milling machine. Figure 13.13 shows the simplest shapes that can be generated on the edge of a flat component. It can be seen that internal corners must have radii equal to the radius of the milling cutter used.

In general, the minimum diameter of cutters for horizontal milling [about 50 mm (2 in.) for an average machine] is much larger than the minimum diameter of cutters for vertical milling [about 12 mm (0.5 in.)]. Thus, small internal radii would necessitate vertical milling. However, as can be seen from Figure 13.13, a flat workpiece that must be machined around the entire periphery will generally be bolted to the machine worktable with a spacer beneath the workpiece smaller than the finished component. This means of work holding will require at least two bolt holes to be provided in the workpiece. In horizontal milling the workpiece can be gripped in a vise. The worktable feed motions in milling machines are generally parallel and perpendicular to the machine spindle and thus mutually perpendicular flat surfaces are the easiest to machine. The inclined surfaces shown in Figure 13.13 are more difficult to machine because the workpiece must be reclamped for the purpose.

With flat components required in reasonably large batch sizes, simultaneous machining of a stack of workpieces can often considerably reduce manufacturing costs. A shaper might be used economically for the machining of a stack of workpieces (Figure 13. 14a) and would allow a greater variety of outer shapes to be produced because of the less severe limitations on the radii of inside corners. Also, for flat workpieces that can be stacked for machining and that are required in large quantities, form-milling cutters for horizontal milling can provide almost any outer shape other than undercuts (Figure 13.14b).

Sometimes, large holes (principal bores) are required in nonrotational components. These principal bores are generally normal to the two large surfaces on the component and require machining by boring. This operation could be performed on a lathe (Figure 13.15a) where the workpiece would be bolted to a

TABLE 13.7
More Expensive Machined Features on Rotational Components

Machine	Type of surface	Orientation and nature of surface	Example
Lathe or vertical-boring machine	External surfaces (concentric)	Screw thread, functional groove, conical surface	
		Operating thread	
	Internal surfaces (concentric)	Screw thread, functional groove, conical surface	
		Operating thread	
Milling machine or broacher	Auxiliary plane surfaces	External spline or polygon section, internal through spline or polygon section	
Drill press	Auxiliary holes	Cylindrical holes not parallel or perpendicular to workpiece axis but in accessible positions	
Gear-cutting machine	Gear teeth	Concentric gear teeth	

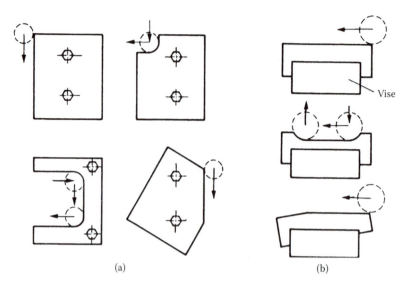

FIGURE 13.13 Milling external shape of flat components. (a) Vertical milling (plan view); (b) horizontal milling (front view).

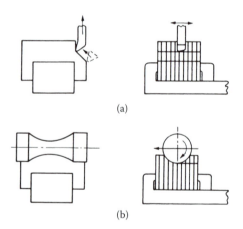

FIGURE 13.14 Machining of stacked workpieces. (a) Shaping; (b) horizontal milling.

faceplate, or on a vertical borer (Figure 13.15b), where the workpiece would be bolted to the rotary worktable. For small parts, however, where high accuracy is required, the bores would be machined on a jig borer. A jig borer is similar to a vertical-milling machine, but the spindle is fed vertically and can hold a boring tool (Figure 13.15c). From these examples it can be seen that where possible, principal bores should be cylindrical and normal to the base of the component. It can also be seen that a spacer is required between the workpiece and the work-holding surface.

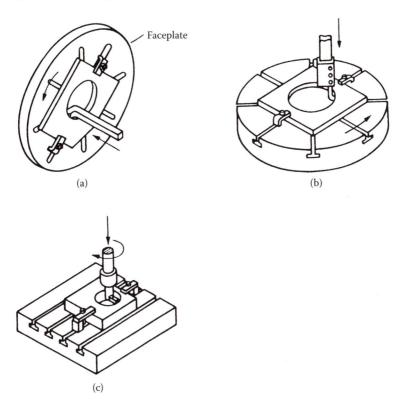

(a)

(b)

(c)

FIGURE 13.15 Machining of principal bores in nonrotational workpieces. (a) Lathe; (b) vertical borer; (c) jig borer.

The next type of secondary machining operation to be considered is the provision of a series of plane surfaces (represented by the fourth digit in the geometric code). These include the machining of steps, slots, and so on, in one of the large surfaces on the workpiece. If possible, plane-surfacing machining should be restricted to one surface of the component only, thus avoiding the need for reclamping the workpiece. Plane surfaces might be machined on milling machines, shapers, or, in very large workpieces (such as machine beds), on planing machines. Figure 13.16 shows a variety of plane-surface machining operations, and it can be seen that plane-machined surfaces should, if possible, be either parallel or normal to the base of the component. Also, internal radii need not be specified for the milling operations because the corners of the teeth on milling cutters are usually sharp.

Finally, auxiliary holes might be required in flat components, as in the case of disk-shaped, rotational components (Figure 13.7), and these auxiliary holes would generally be machined on a drill press. Similar requirements to those discussed for the machining of auxiliary holes in disk-shaped rotational components apply. Thus, auxiliary holes should, if possible, be cylindrical and normal

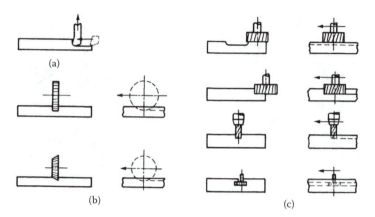

FIGURE 13.16 Plane-surface machining of flat components. (a) Shaping or planning; (b) horizontal milling; (c) vertical milling.

to the base of the component and preferably related by a pattern to simplify positioning of the workpiece for drilling.

13.5.6 NONROTATIONAL COMPONENTS [(A/B) > 3]

Nonrotational components having proportions such that the ratio A/B is greater than 3 are classified as long components, have a first digit of 7, and are often machined from rectangular- or square-section bar stock. Extremely long components should be avoided because of the difficulties of work holding. The most common machining operations are drilling and milling. For obvious reasons, bored holes should be avoided because generally the workpiece cannot be rotated about the required axis. Also, machined surfaces parallel to the principal axis of the component should be avoided because of the difficulties of holding down the entire length of the workpiece. Instead, the designer should, if possible, utilize work material preformed to the cross section required.

13.5.7 NONROTATIONAL COMPONENTS [(A/B) < 3, (A/C) < 4]

Nonrotational components having proportions such that the ratio A/B is less than 3 and the ratio A/C is less than 4 are classified as cubic components; they have a first digit of 8 and are often of quite complicated shape. Cubic components should be provided with at least one plane surface that can initially be surface-ground or milled to provide a base for work-holding purposes and a datum for further machining operations.

If possible, the outer machined surfaces of the component should consist of a series of mutually perpendicular plane surfaces parallel to and normal to its base. In this way, after the base has been machined, further machining operations can be carried out on external surfaces with minimum reclamping of the work-piece. Figure 13.17, for example, shows a cubic workpiece where the external

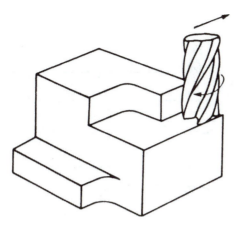

FIGURE 13.17 Milling outer surface of a cubic workpiece.

exposed surfaces can all be machined on a vertical-milling machine without reclamping. This figure shows that sharp, internal corners parallel to the base can be machined readily but that sharp, internal corners normal to base should be avoided.

The workpiece shown in Figure 13.17 is block like. However, many cubic components are boxlike, and some are split with flanges provided to allow the two split portions to be bolted together. When the split is provided at a main bore, this main bore is machined with the two sections bolted together. Main bores in cubic components are often machined on a horizontal-boring machine. For ease of machining, internal cylindrical surfaces should be concentric and decrease in diameter from the exposed surface of the workpiece. Also, blind bores should be avoided where possible because in horizontal boring, the boring bar must usually be passed through the workpiece. Internal machined surfaces in a boxlike cubic component should be avoided unless the designer is certain that they will be accessible.

With small cubic components, it is possible to machine pockets or internal surfaces using an end-milling cutter as shown in Figure 13.18. Again it can be seen that internal corners normal to the workpiece base must have a radius equal to that of the cutter. Also, the same cutter will be used to clear out the pocket after machining the outer shape, and the smaller the cutter diameter, the longer it will take to perform this operation. Consequently, the cost of the operation will be related to the radii of the vertical internal corners. Thus, internal corners, normal to the workpiece base, should have as large a radius as possible.

Finally, cubic components will often have a series of auxiliary holes. Auxiliary holes should be cylindrical and either normal or parallel to the base of the component; they should also be in accessible positions and have L/D ratios that make it possible to machine them with standard drills. In general, standard drills can produce holes having L/D ratios as large as 5.

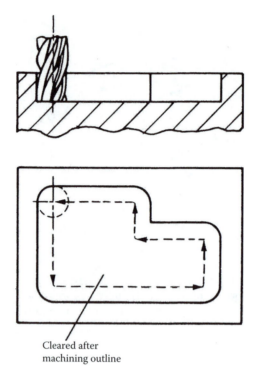

Cleared after
machining outline

FIGURE 13.18 Milling a pocket in a block-like cubic workpiece.

Figure 13.19a shows examples of features that would be difficult and expensive to produce in nonrotational components. In the first case the internal vertical corners are shown sharp; these features cannot be produced with standard tools. In the second case the through hole has an extremely large L/D ratio and would be difficult to produce even with special deep-hole drilling techniques. Figure 13.19b shows examples of machined features virtually impossible to produce because a suitable tool cannot be designed that would reach all the internal surfaces. Figure 13.20 shows the design of some blind holes. A standard drill produces a hole with a conical blind end, and therefore the machining of a hole with a square blind end requires a special tool. Thus, the end of a blind hole should be conical. If the blind hole is to be provided with a screw thread, the screw thread will be tapped, and the designer should not specify a fully formed thread to the bottom of the blind hole since this type of screw thread is impossible to produce.

Holes that have a dogleg, or bend, should be avoided if possible. A curved hole (Figure 13.20) is clearly impossible to machine; however, drilling a series of through holes and plugging unwanted outlets can often achieve the desired effect, although this operation is expensive.

Finally Figure 13.21 shows an example of a boxlike cubic component with internal features that cannot be machined because they are inaccessible.

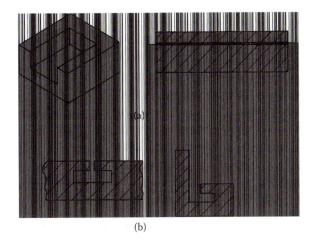

(b)

FIGURE 13.19 Design features to avoid in nonrotational components. (a) Difficult and expensive to produce; (b) virtually impossible to produce.

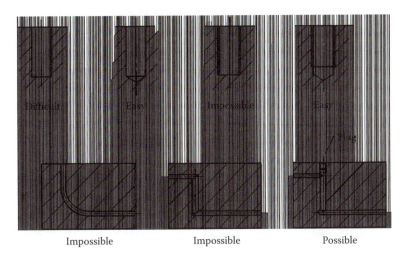

Impossible   Impossible   Possible

FIGURE 13.20 Design of blind holes.

Table 13.8 summarizes machined features that are relatively easy to provide on nonrotational workpieces in increasing order of cost; Table 13.9 presents machined features that would be more expensive to provide and should be avoided when possible. Again, it should be remembered that the relative difficulty of providing these features depends on the proportions of the component.

13.6 ASSEMBLY OF COMPONENTS

Most machined components must eventually be assembled, and the designer should give consideration to the assembly process. Design for ease of assembly

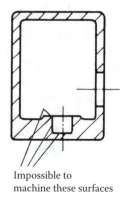

Impossible to
machine these surfaces

FIGURE 13.21 Inaccessible internal features in a boxlike cubic component.

is a separate subject [3]; however, there are one or two aspects of this subject that affect machining and can be mentioned here. The first requirement is, of course, that it should be physically possible to assemble the components. Obviously, the screw thread on a bolt or screw should be the same as the mating thread on the screwed hole into which the bolt or screw is to be inserted. Some assembly problems, however, are not quite so obvious. Figure 13.22 shows some impossible assembly situations, and it is left to the reader to decide why the components cannot be assembled properly.

A further requirement is that each operating machined surface on a component should have a corresponding machined surface on the mating component. For example, where flanges on castings are to be bolted together, the area around the boltholes should be machined perpendicular to the hole (spot-faced, for example) to provide proper seating for the bolt heads, nuts, or washers. Also, internal corners should not interfere with the external corner on the mating component. Figure 13.5 and Figure 13.12 were examples of how this interference can be avoided. Finally, incorrect specification of tolerances can make assembly difficult or even impossible.

13.7 ACCURACY AND SURFACE FINISH

A designer will not generally want to specify an accurate surface with a rough finish or an inaccurate surface with a smooth finish. When determining the accuracy and finish of machined surfaces, it is necessary to take into account the function intended for the machined surface. The specification of too-close tolerances or too-smooth surfaces is one of the major ways a designer can add unnecessarily to manufacturing costs. Such specifications could, for example, necessitate a finishing process such as cylindrical grinding after rough turning where an adequate accuracy and finish might have been possible using the lathe that performed the rough-turning operation. Thus, the designer should specify

TABLE 13.8
Readily Provided Machined Features on Nonrotational Components

Machine	Type of surface	Orientation and nature of surface	Example
Milling machine, shaper, or planer	External (exposed when workpiece is gripped in work-holding device)	Plane surfaces parallel and normal to base with no square internal corners normal to base	
	Plane-surface machining (on surfaces exposed when workpiece is gripped in work-holding device)	Rectangular slots or grooves parallel to base	
Drill press	Auxiliary holes	Cylindrical through holes normal to base Cylindrical blind holes with conical ends	
Lathe, vertical borer, jig borer, or horizontal borer	Principal bores	Cylindrical bores parallel or normal to base and decreasing in diameter from exposed surface of the workpiece	

TABLE 13.9
More Expensive Machined Features on Nonrotational Components

Machine	Type of surface	Orientation and nature of surface	Example
Milling machine, shaper, or planer	External	Accessible, inclined surfaces	
		Nonrectangular slots or grooves on accessible surfaces	
Vertical-milling machine	Plane surface machining	Accessible internal plane surfaces or pockets parallel or normal to base and with large radii specified for internal corners normal to base	
Drill press	Auxiliary holes	Accessible cylindrical holes inclined to base	
Lathe, vertical borer, jig borer, or horizontal borer	Principal bores	Cylindrical bores with functional grooves	
Gear-cutting machine	Gear teeth	In-line gear teeth on long components	

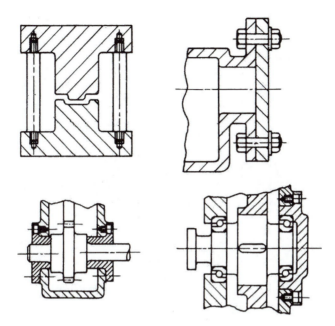

FIGURE 13.22 Components that cannot be assembled.

the widest tolerances and roughest surface that would give acceptable performance for operating surfaces.

As a guide to the difficulty of machining to within required tolerances it can be stated that:

1. Tolerances from 0.127 to 0.25 mm (0.005 to 0.01 in.) are readily obtained.
2. Tolerances from 0.025 to 0.05 mm (0.001 to 0.002 in.) are slightly more difficult to obtain and increase production costs.
3. Tolerances 0.0127 mm (0.0005 in.) or greater require good equipment and skilled operators and add significantly to production costs.

Figure 13.23 illustrates the general range of surface finish that can be obtained in different operations. It can be seen that any surface with a specified surface finish of 1 μm (40 μin.) arithmetical mean or better will generally require separate finishing operations, which substantially increases costs. Even when the surface can be finished on the same machine, a smoother surface requirement means increased costs.

To illustrate the cost increase as the surface finish is improved, a simple turning operation can be considered. If a tool having a rounded corner is used under ideal cutting conditions, the arithmetical-mean surface roughness R_a is related to the feed by

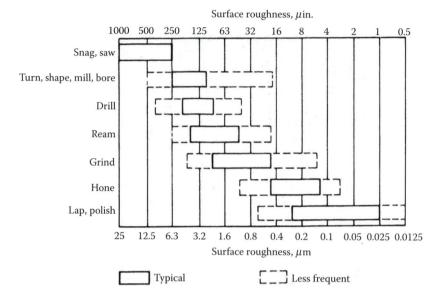

FIGURE 13.23 General range of surface roughness obtainable by various machining operations.

$$R_a = \frac{0.031 f^2}{r_\varepsilon} \tag{13.5}$$

where f is the feed, and r_ε is the corner radius.

The machining time t_m is inversely proportional to the feed f and related by the equation

$$t_m = \frac{l_w}{f n_w} \tag{13.6}$$

where l_w is the length of the workpiece, and n_w is the rotational frequency of the workpiece.

Substitution of f from Equation 13.5 in Equation 13.6 gives the machining time in terms of the specified surface finish:

$$t_m = \frac{0.18 l_w}{n_w \left(R_a r_\varepsilon \right)^{0.5}} \tag{13.7}$$

Thus, the machining time (and hence the machining cost) is inversely proportional to the square root of the surface finish. Figure 13.24 shows the relationship between production cost and surface finish for a typical turning operation.

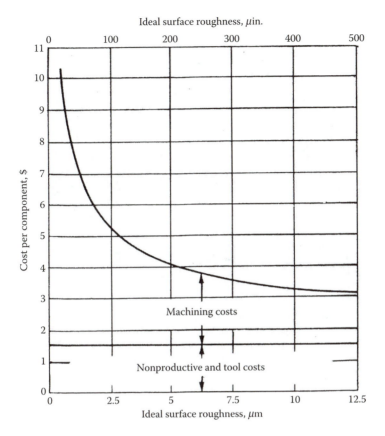

FIGURE 13.24 Effect of specified surface roughness on production costs in a turning operation, where the corner radius, $r_\varepsilon = 0.03$ in. (0.762 mm), the rotational frequency of the workpiece, $n_w = 200$ rpm (3.33 s^{-1}), and the length of the workpiece, $l_w = 34$ in. (864 mm).

It can be seen that the costs rise rapidly when low values of surface finish are specified.

For many applications a smooth, accurate surface is essential. Finish grinding can most frequently provide this smooth, accurate surface. When specifying finish grinding, the designer should take into account the accessibility of the surfaces to be ground. In general, surfaces to be finish-ground should be raised and should never intersect to form internal corners. Figure 13.25 shows the types of surfaces that are most readily finish-ground using standard-shaped abrasive wheels.

13.8 SUMMARY OF DESIGN GUIDELINES

This section lists the various design guidelines that have been developed. The section is intended to provide a summary of the main points a designer should keep in mind when considering the design of machined components.

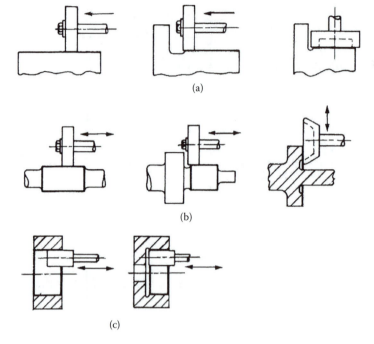

FIGURE 13.25 Surfaces that can readily be finish-ground. (a) Surface grinding; (b) cylindrical grinding; (c) internal grinding.

13.8.1 STANDARDIZATION

1. Utilize standard components as much as possible.
2. Preshape the workpiece, if appropriate, by casting, forging, welding, and so on.
3. Utilize standard preshaped workpieces, if possible.
4. Employ standard machined features wherever possible.

13.8.2 RAW MATERIAL

5. Choose raw materials that will result in minimum component cost (including cost of production and cost of raw material).
6. Utilize raw material in the standard forms supplied.

13.8.3 COMPONENT DESIGN

13.8.3.1 General

7. Try to design the component so that it can be machined on one machine tool only.
8. Try to design the component so that machining is not needed on the unexposed surfaces of the workpiece when the component is gripped in the work-holding device.

9. Avoid machined features the company is not equipped to handle.
10. Design the component so that the workpiece, when gripped in the work-holding device, is sufficiently rigid to withstand the machining forces.
11. Verify that when features are to be machined, the tool, tool holder, work, and work-holding device will not interfere with one another.
12. Ensure that auxiliary holes or main bores are cylindrical and have L/D ratios that make it possible to machine them with standard drills or boring tools.
13. Ensure that auxiliary holes are parallel or normal to the workpiece axis or reference surface and related by a drilling pattern.
14. Ensure that the ends of blind holes are conical and, in the case of a tapped blind hole, that the thread does not continue to the bottom of the hole.
15. Avoid bent holes or dogleg holes.

13.8.3.2   Rotational Components

16. Try to ensure that cylindrical surfaces are concentric and plane surfaces are normal to the component axis.
17. Try to ensure that the diameters of external features increase from the exposed face of the workpiece.
18. Try to ensure that the diameters of internal features decrease from the exposed face of the workpiece.
19. For internal corners on the component, specify radii equal to the radius of the rounded tool corner.
20. Avoid internal features for long components.
21. Avoid components with very large or very small L/D ratios.

13.8.3.3   Nonrotational Components

22. Provide a base for work holding and reference.
23. If possible, ensure that the exposed surfaces of the component consist of a series of mutually perpendicular plane surfaces parallel to and normal to the base.
24. Ensure that internal corners normal to the base have a radius equal to the tool radius. Also ensure that for machined pockets, the internal corners normal to the base have as large a radius as possible.
25. If possible, restrict plane surface machining (slots, grooves, etc.) to one surface of the component.
26. Avoid cylindrical bores in long components.
27. Avoid machined surfaces on long components by using work material preformed to the cross section required.
28. Avoid extremely long or extremely thin components.
29. Ensure that, in flat or cubic components, main bores are normal to the base and consist of cylindrical surfaces decreasing in diameter from the exposed face of the workpiece.

30. Avoid blind bores in large cubic components.
31. Avoid internal machined features in cubic boxlike components.

13.8.4 ASSEMBLY

32. Ensure that assembly is possible.
33. Ensure that each operating machined surface on a component has a corresponding machined surface on the mating component.
34. Ensure that internal corners do not interfere with a corresponding external corner on the mating component.

13.8.5 ACCURACY AND SURFACE FINISH

35. Specify the widest tolerances and roughest surface that will give acceptable performance for operating surfaces.
36. Ensure that surfaces to be finish-ground are raised and never intersect to form internal corners.

13.9 COST ESTIMATING FOR MACHINED COMPONENTS

Designers normally have a reasonable knowledge of the factors to bear in mind when attempting to minimize manufacturing costs for machined components, and the previous section listed some design rules that might be followed. Ultimately, however, the designer needs to know the magnitude of the effects of design decisions on manufacturing costs. The need for a method of estimating these costs is highlighted when considering the design of a product for ease of assembly. Techniques have been available for some time for analyzing an assembly in terms of the handling and insertion costs incurred as each part is added [3]. As a result of such analyses, many suggestions for design simplifications arise — often involving the elimination of individual parts or components. What has been lacking, however, are methods for quickly estimating the cost of these parts and the cost of tooling so that the total savings in product costs can be quantified.

Before embarking on a discussion of how an estimating method for designers might be developed, we should consider how the requirements for such a method differ from conventional cost-estimating procedures. These conventional procedures are meant to be applied after the component has been designed and its production planned. Thus, every step in production is known and can be estimated with a high degree of accuracy. During the early stages of design, however, the designer will not wish to specify, for example, all the work-holding devices and tools that might be needed; most likely, detailed design will not yet have taken place. Indeed, a final decision on the specific work material might not have been made at this stage. Thus, what is wanted is an approximate method requiring minimum information from the designer and assuming that the ultimate design will avoid unnecessary manufacturing expense and that the component will be manufactured under reasonable economic conditions.

Perhaps the simplest approach would be to have the designer specify the shape and size of the original workpiece and the quantity of material to be removed by machining. Then, with data on typical material costs per unit weight, an estimate can be made of the cost of the material needed to manufacture the component. If an approximate figure were available for the average cost of removal of each cubic inch of the material by machining, an estimate could also be made of the machining cost.

Unfortunately, this very simple approach will not take adequate account of the nonproductive costs involved in a series of machining operations. For example, if one cubic inch of material were to be removed in one pass by a simple turning operation, the nonproductive costs would be quite small — the component need only be loaded into the lathe and unloaded once — and the cutting tool need only be set and the feed engaged once. Compare this with one cubic inch of the same material removed by a combination of turning, screwing, milling, and drilling. In this case the nonproductive costs accumulate and become a highly significant factor in the ultimate cost of the machined component, especially when the machined component is relatively small.

What is needed is a method that forms a compromise between this oversimplified approach and the traditional detailed cost-estimating methods used by manufacturing and industrial engineers.

13.9.1 MATERIAL COST

Often, the most important factor in the total cost of a machined component is the cost of the original workpiece. This material cost will frequently form more than 50% of the total cost and, therefore, should be estimated with reasonable care. Table 13.10 gives densities and approximate costs in dollars per pound for a variety of materials in the basic shapes normally available. Provided the designer can specify the volume of material required for the original workpiece, the material cost can easily be estimated. Although the figures in Table 13.10 can be used as a rough guide, the designer can obtain more accurate figures from material suppliers. Material prices fluctuate considerably with market conditions so the values given in Table 13.10 are included for the purposes of calculations in this chapter.

13.9.2 MACHINE LOADING AND UNLOADING

Nonproductive costs are incurred every time the workpiece is loaded into (and subsequently unloaded from) a machine tool. An exhaustive study of loading and unloading times has been made by Fridriksson [4]; he found that these times can be estimated quite accurately for a particular machine tool and work-holding device if the weight of the workpiece is known. Some of Fridriksson's results are presented in Table 13.11, which can be used to estimate machine loading and unloading times. To these figures must be added the times for turning coolant on and off, cleaning the work-holding or clamping device, and so on.

TABLE 13.10
Approximate Costs in Dollar per Pound for Various Materials[a]

	Density		Bar	Rod	Sheet <0.5 in.	Plate >0.5 in.	Tube
	lb/in.[3]	Mg/m[3]					
Ferrous							
Carbon steel	0.283	7.83	0.51	0.51	0.36	0.42	0.92
Alloy steel	0.31	8.58	0.75	0.75	1.20	—	—
Stainless steel	0.283	7.83	1.50	1.50	2.50	2.50	—
Tool steel	0.283	7.83	6.44	6.44	—	6.44	
Nonferrous							
Aluminum alloys	0.10	2.77	1.93	1.93	1.95	2.50	4.60
Brass	0.31	8.58	0.90	1.22	1.90	1.90	1.90
Nickel alloys	0.30	8.30	5.70	5.70	5.70	5.70	—
Magnesium alloys	0.066	1.83	3.35	3.35	6.06	6.06	3.35
Zinc alloys	0.23	6.37	1.50	1.50	1.50	1.50	—
Titanium alloys	0.163	4.51	15.40	15.40	25.00	25.00	—

Note: Market prices for metals fluctuate frequently. These are representative values for the purposes of calculations in this chapter.

[a] To convert to dollars per kg multiply by 2.2.

13.9.3 OTHER NONPRODUCTIVE COSTS

For every pass, cut, or operation carried out on one machine tool, further nonproductive costs are incurred. In each case the tool must be positioned, perhaps the feed and speed settings changed, the feed engaged, and then, when the operation is completed, the tool withdrawn. If different tools are employed, the times for tool engagement or indexing must also be taken into account. Some time elements for these tasks for different types of machine tools are given in Table 13.12. Also included are estimates of the basic setup time and additional setup time per cutting tool. The total setup time must be divided by the size of the batch in order to obtain the setup time per component.

13.9.4 HANDLING BETWEEN MACHINES

One of the costs to be considered is that incurred in moving batches of partially machined workpieces between machines. Fridriksson [4] made a study of this by assuming that stacks of pallets of workpieces were moved around the factory using forklift trucks. He developed the following expression for t_f, the transportation time for a round trip by a forklift truck

$$t_f = 25.53 + 0.95\left(l_p + l_{rd}\right) \text{ s} \tag{13.8}$$

TABLE 13.11
Loading and Unloading Times (s) for Different Workpiece Weights

Holding Device	0–0.2 0–0.4	0.2–0.4.5 0.4–10	4.5–14 10–30	14–27 (kg) 30–60 (lb)	Crane
Angle plate (two U-clamps)	27.6	34.9	43.5	71.2	276.5
Between centers, no dog	13.5	18.6	24.1	35.3	73.1
Between centers, with dog	25.6	40.2	57.4	97.8	247.8
Chuck, universal	16.0	23.3	31.9	52.9	—
Chuck, independent (four jaw)	34.0	41.3	49.9	70.9	—
Clamp on table (three clamps)	28.8	33.9	39.4	58.7	264.6
Collet	10.3	15.4	20.9	—	—
Face plate (three clamps)	31.9	43.3	58.0	82.1	196.2
Fixture, horizontal (three screws)	25.8	33.1	41.7	69.4	274.7
Fixture, vertical (three screws)	27.2	38.6	53.3	—	—
Hand-held	1.4	6.5	12.0	—	—
Jig	25.8	33.1	41.7	—	—
Magnetic table	2.6	5.2	8.4	—	—
Parallels	14.2	19.3	24.8	67.0	254.3
Rotary table or Index plate (three clamps)	28.8	36.1	44.7	72.4	277.7
"V" Blocks	25.0	30.1	35.6	77.8	365.1
Vise	13.5	18.6	24.1	39.6	174.2

Source: After Fridriksson, L., Non-productive Time in Conventional Metal Cutting, Report No. 3, Design for Manufacturability Program, University of Massachusetts, Amherst, February, 1979.

where l_p is the length of the pathway between machines, and t_{rd} is the distance the truck must travel to respond to a request (both lengths are measured in feet). Assuming that $(l_p + l_{rd})$ is 139 m on average and that for every trip with a load of full pallets, a trip must be made with empty pallets, the total time is

$$t_f = 315 \text{ s} \tag{13.9}$$

If a full load of pallets and workpieces is 2000 lb, the number of workpieces of weight W transported will be 2000/W, and the time per workpiece t_{tr} will be given by

$$t_{tr} = \frac{315}{2000 / W} = 0.156W \text{ s} \tag{13.10}$$

Thus, for a workpiece weighing 10 lb, the effective transportation time is only 1.6 s, which is small compared with the loading and unloading time of

TABLE 13.12
Some Nonproductive Times for Common Machine Tools

Machine Tool	Time to Engage Tool; etc.[a] (s)	Basic Setup Time (h)	Additional Setup per Tool (h)
Horizontal band saw	—	0.17	—
Manual turret lathe	9	1.2	0.2
CNC turret lathe	1.5	0.5	0.15
Milling machine	30	1.5	—
Drilling machine	9	1.0	—
Horizontal boring machine	30	1.3	—
Broaching machine	13	0.6	—
Gear hobbing machine	39	0.9	—
Grinding machine	19	0.6	—
Internal grinding machine	24	0.6	—
Machining center	8	0.7	0.05

[a] Average times to engage tool, engage and disengage feed, change speed or feed (includes change tool for machining center)

between 19 and 43 s for that size of workpiece (Table 13.11). However, allowances for transportation time can be added to the loading and unloading times and these will become significant for large workpieces.

13.9.5 MATERIAL TYPE

The so-called machinability of a work material has been one of the most difficult factors to define and quantify. In fact, it is impossible to predict the difficulty of machining a material from knowledge of its composition or its mechanical properties, without performing a machining test. Nevertheless, it is necessary for the purposes of cost estimating to employ published data on machinability. Perhaps the best source of such data, presented in the form of recommended cutting conditions, is the *Machining Data Handbook* [5] from MetCut.

13.9.6 MACHINING COSTS

The machining cost for each cut, pass, or operation is incurred during the period between the feed being engaged and finally disengaged. It should be noted that the tool is not cutting for the whole of this time because allowances for tool engagement and disengagement must be made — particularly for milling operations. However, typical values for these allowances can be found and are presented for various operations in Table 13.13 as correction factors to be applied to the actual machining time.

TABLE 13.13
Allowances for Tool Approach Engagement and Disengagement

Operations	Allowances	
Metric Units Turn, face, cut-off, bore, groove, thread	$t'_m = t_m + 5.4$	$d_a < 0.05$
	$t'_m = t_m + 2200d_a^2$	$d_a \geq 0.05$
Drill (twist) (approach)	$t'_m = t_m \left(1 + 0.5 d_a / l_w\right)$	
Drill (twist) (start)	$t'_m = t_m + \left(25.9/vf\right)d_a^{1.67}$	
English Units Turn, face, cut-off, bore, groove, thread	$t'_m = t_m + 5.4$	$d_a < 2$
	$t'_m = 1.35 d_a^2$	$d_a \geq 2$
Drill (twist) (approach)	$t'_m = t_m \left(1 + 0.5 d_a / l_w\right)$	
Drill (twist) (start)	$t'_m = t_m + \left(88.5/vf\right)d_a^{1.67}$	
Helical side and key slot milling	$l'_w = l_w + 2\left(a_e\left(d_t - a_e\right)^{0.5} + 0.066 + 0.011 d_t\right)$	
Face and end milling	$l'_w = l_w + d_t + 0.066 + 0.011 d_t$	
Surface grinding	$l'_w = l_w + d_t / 2$	
Cylindrical and internal grinding	$l'_w = l_w + w_t$	
All grinding operations	$a'_r = a_r + 0.004$	$a_r \leq 0.01$
	$a'_r = a_r + 0.29\left(a_r - 0.01\right)$	$0.01 < a_r \leq 0.024$
	$a'_r = a_r + 0.008$	$a_r > 0.024$
Spline broaching	$l_t = -5 + 15 d_a + 8 l_w$	
Internal keyway broaching	$l_t = 20 + 40 w_k + 85 d_k$	
Hole broaching	$l_t = 6 + 6 d_a + 6 l_w$	

Source: Adapted from Ostwald [6]

t_m = machining time, s; d_a = diameter of work surface, diameter of drilled hole, or diameter of machined surface, m (in.); l_w = length of machined surface in direction of cutting, m (in.); vf = rate of surface generation (speed x feed), m²/min (in.²/min) (Table 13.13); a_e = depth of cut or depth of groove in milling, in; d_t = diameter of cutting tool, in; w_t = width of grinding wheel, in; a_t = depth of material removed in rough grinding, in; l_t = length of tool, in; w_k = width of machined keyway; d_k = depth of machined keyway; ' indicates corrected value after allowance is applied.

For an accurate estimation of actual machining time, it is necessary to know the cutting conditions, namely, cutting speed, feed, and depth of cut in single-point tool operations, and feed speed, depth of cut, and width of cut in multipoint tool operations. Tables giving recommended values for these parameters for different work materials can fill large volumes such as the MetCut *Machining Data Handbook* [5].

Analysis of the selection of optimum machining conditions shows that the optimum feed (or feed per tooth) is the largest that the machine tool and cutting tool can withstand. Then, selection of the optimum cutting speed can be made by minimizing machining costs. The resulting combination of cutting speed and feed in a single-point tool operation gives a rate for the generation of the machined surface that can be measured in in.^{2}/min, for example. The inverse of such rates are presented by Ostwald [6] for a variety of workpiece and tool materials and for different roughing and finishing operations. A problem arises, however, when applying the figures for roughing operations. For example, Ostwald recommends a cutting speed of 500 ft/mm (2.54 m/s) and a feed of 0.02 in. (0.51 mm) for the rough machining of low-carbon steel (170 Bhn) with a carbide tool. For a depth of cut of 0.3 in. (7.6 mm) this would mean a metal removal rate of 36 in.^{3}/min (9.82 μm^3/s), The *Machining Data Handbook* [5] quotes a figure of 1.35 hp min/m^3 (3.69 GJ/m^3) (unit power) for this work material. Thus, the removal rate obtained in this example would require almost 50 hp (36 kW). Since a typical medium-sized machine tool would have a 5 to 10 hp motor (3.7 to 7.5 kW) and an efficiency of around 70%, it can be seen that the recommended conditions could not be achieved except for small depths of cut. Under normal rough machining circumstances, therefore, a better estimate of machining time would be obtained from the unit horsepower (specific cutting energy) for the material, the volume of material to be removed, and the typical power available for machining.

For multipoint tools such as milling cutters, finishing and roughing costs are not usually distinguished. The chip load (feed per tooth) and the cutting speed are usually recommended for given tool materials. However, in these cases the machining time is not directly affected by the cutting speed but by the feed speed, which is controlled independently of the cutter speed. Thus, assuming that the optimum cutting speed is being employed, the feed speed that will give the recommended feed per tooth can be used to estimate the machining time. Again, a check must be made that the power requirements for the machine tool are not excessive.

13.9.7 Tool Replacement Costs

Every time a tool needs replacement because of wear, two costs are incurred: (1) the cost of machine idle time while the operator replaces the worn tool, and (2) the cost of providing a new cutting edge or tool. The choice of the best cutting speed for particular conditions is usually made by minimizing the sum of the tool replacement costs and the machining costs since both of these are affected by changes in the cutting speed.

Neglecting nonproductive costs, the cost of machining a feature in one component on one machine tool can be expressed by:

$$C_m = Mt_m + \frac{Q\left(Mt_{ct} + C_t\right)t_m}{t}$$ (13.11)

where

M = machine and operator rate

t_m = machining time (time the machine tool is operating)

Q = proportion of t_m for which a point on the tool cutting edge is contacting the workpiece

t = tool life while the cutting edge is contacting the workpiece

t_{ct} = tool-changing time

C_t = cost of providing a new cutting edge

Now the machining time t_m is given by

$$t_m = \frac{k}{v}$$ (13.12)

where k is a constant for the particular operation, and v is the cutting speed. Also, the tool life t is given by Taylor's tool-life equation:

$$vt^n = v_r t_r^n$$ (13.13)

where v_r and t_r are the reference cutting speed and tool life, respectively, and n is the Taylor tool-life index, which is dependent mainly on the tool material. Usually, for high-speed steel tools n is assumed to be 0.125 and for carbide tools, 0.25.

If Equation 13.12 and Equation 13.13 are substituted into Equation 13.11, the resulting expression is differentiated to give the cutting speed for minimum cost, and finally the optimum cutting speed is substituted in Equation 13.11 to give the minimum cost of machining C_{min}, it can be shown that

$$C_{min} = \frac{Mt_{m_c}}{1-n}$$ (13.14)

where t_{m_c} is the machining time when the optimum cutting speed is used.

It can be seen that the factor $1/(1 - n)$ applied to the machining time will allow for tool replacement costs provided that the cutting speed for minimum cost is always employed. The factor would be 1.14 for high-speed steel tooling and 1.33 for carbides. Under those circumstances, where use of optimum cutting conditions would not be possible because of power limitations, it is usually recommended that cutting speed be reduced. This is because greater savings in tool costs will result than if the feed were reduced. When cutting speed has been reduced, with a corresponding increase in machining time, the correction factor given by Equation 13.14 will overestimate tool costs. If t_{m_p} is the machining time where the cutting speed v_{p_0} giving maximum power is used, the machining cost C_p will be given by

$$C_p = Mt_{m_p} + \frac{Q\left(Mt_{c_t} + C_t\right)t_{m_p}}{t_{p_0}} \tag{13.15}$$

where t_{p_0} is the tool life obtained under maximum-power conditions which, from Taylor's tool-life equation, is

$$t_{p_0} = t_c \left(\frac{v_{p_0}}{v_c}\right)^{(1/n)} \tag{13.16}$$

The tool-life t_c under minimum-cost conditions is given by

$$t_c = \left[\frac{1-n}{n}\right] Q\left(t_{c_t} + \frac{C_t}{M}\right) \tag{13.17}$$

Substituting Equation 13.16 and Equation 13.17 in Equation 13.15 and using the relation in Equation 13.12 gives

$$C_p = Mt_{m_p}\left[1 + \left(\frac{n}{1-n}\right)\left(\frac{t_{m_c}}{t_{m_p}}\right)^{(1/n)}\right] \tag{13.18}$$

Thus, Equation 13.18 can be used instead of Equation 13.14 when the cutting speed is limited by the power available on the machine tool.

13.9.8 MACHINING DATA

In order to employ the approach described above, it is necessary to be able to estimate, for each operation, the machining time t_{m_c} for minimum-cost conditions and t_{m_p} the machining time where the cutting speed is limited by power availability. It was shown earlier that machining data for minimum cost presented in handbooks can be expressed in terms of the rate at which machined surfaces can be generated or speed × feed (vf). Table 13.14 gives typical values of the surface generation rates vf for several material classifications selected and for rough-turning

TABLE 13.14
Machining Data for Turning and Drilling Operations

		Turning, Facing, Boring			Drilling, Reaming (1 in.)	
		vf (in.²/min)[a]		p_s	vf (in.²/min)[a]	p_s
Material	Hardness (BHN)	HSS	Brazed Carbide	(hp/in.³/min)[b]	HSS	(hp/in.³/min)[b]
Low carbon steel (free machining)	150–200	25.6	100	1.1	33.0	0.95
Low carbon steel	150–200	22.4	92	1.35	13.4	1.2
Medium and high carbon steel	200–250	18.2	78	1.45	15.1	1.4
Alloy steel (free machining)	150–200	23.7	96	1.3	16.4	1.15
Stainless steel, ferritic (annealed)	135–185	12.6	48	1.55	9.4	1.35
Tool steels	200–250	12.8	54	1.45	6.2	1.4
Nickel alloys	80–360	9	42	2.25	14.3	2.0
Titanium alloys	200–275	12.6	24	1.35	7.9	1.25
Copper alloys (soft) (free machining)	40–150	76.8	196	0.72	38.4	0.54
Zinc alloys (die cast)	80–100	58.5	113	0.3	51.1	0.2
Magnesium and alloys	49–90	162	360	0.18	75.2	0.18
Aluminum and alloys	30–80	176	352	0.28	79.8	0.18

[a] To convert in.²/min to m²/min multiply by 6.45×10^{-4}
[b] To convert hp min/in.³ to GJ/m³ multiply by 2.73

Source: Machinery Data Handbook, 3rd ed., Vol. 1 and 2, MetCut Research Associates, Inc., 1980..

operations with HSS tools or brazed carbide tools. These values were adapted from the data in the *Machining Data Handbook* [5]. Analysis of the handbook data shows that if disposable-insert carbide tools are to be used, then the data for brazed carbide tools can be multiplied by an average factor of 1.17.

When turning a workpiece of diameter d_a for a length l_m, the figures for vf given in Table 13.14 would be divided into the surface area ($A_m = (\pi)l_m d_a$) to give the machining time t_m. Thus

$$t_{m_c} = \frac{60 A_m}{vf} \text{ s} \qquad (13.19)$$

TABLE 13.14 (continued)
Machining Data for Turning and Drilling Operations

Factor	for	Turning, Facing, Boring	Milling
k_f	Finishing	0.60	0.89
k_i	Disposable insert	1.17	1.13

Factor	for	Tool Diameter (in.) (mm)							
	Drilling and	1/16	1/8	1/4	1/2	3/4	1	1.5	2 in.
	reaming	1.59	3.18	6.35	12.7	19.7	25.4	38.1	50.8 mm
k_h		0.08	0.19	0.35	0.60	0.83	1.00	1.00	1.23	1.47

Factor	for	Length/Diameter Ratio					
		<2	3	4	5	6	8
k_d	Deep holes	1.00	0.81	0.72	0.56	0.52	0.48

All data are for rough machining. For finish machining multiply by k_f.

For cutoff or form tool operations multiply by 0.2.

The term *carbide* refers to tools with brazed carbide inserts. For tools with disposable carbide inserts multiply by k_i.

Data for drilling are for 1.0 in. (25.4 mm) diameter tools with a hole depth/diameter less than 2.

For sawing, multiply the data for turning with HSS tools by 0.33.

For tap or die treading, multiply data for turning with HSS tools by 10 and divide by TPI (threads per inch); for standard threads TPI = 2.66 + 4.28/d_a.

For single-point threading, multiply result for die threading by number of passes, approximately 100/TPI, and add tool engagement for each pass.

If the operation is one of facing, it can be assumed that the conditions giving minimum cost at the start of the cut are used.

For an estimate of the machining time t_{m_p} for maximum power it is necessary to know the power available for machining and the unit power p_s (specific cutting energy) for the work material. Table 13.14 gives average values of p_s for the selection of work materials employed here.

When estimating the power available for machining P_m, it should be realized that small components will generally be machined on small machines with lower power available, whereas larger components will be machined on larger higher-powered machines. For example a small lathe may have less than 2 hp available for machining, whereas an average-sized lathe may have 5 to 10 hp available. A large vertical lathe will perhaps have 10 to 30 hp available. Typical power values for a selection of machines are presented in Figure 13.26, where horsepower available for machining P_m is plotted against the typical weight capacity of the machine.

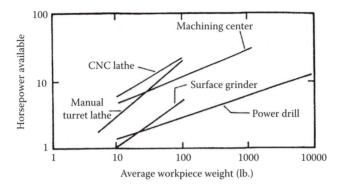

FIGURE 13.26 Relation between horsepower and workpiece weight for some machine tools.

The machining time for maximum power is given by

$$t_{mp} = 60 V_m p_s / P_m \text{ s} \qquad (13.20)$$

where V_m is the volume of material to be removed in the machining operation. If a_p is the depth of cut, then V_m is given approximately by $\pi a_p l_w d_m$. However for a facing operation or a cut-off operation carried out at constant rotational speed, the power limitations apply only at the beginning of the cut, and the machining time will be longer than given by Equation 13.20.

It was pointed out earlier that for milling operations, it is convenient to estimate machining time from knowledge of the feed speed v_f that will give the recommended feed per tooth. Data for milling selected materials are presented in Table 13.15.

The machining time t_m for recommended conditions is thus given by

$$t_{mc} = 60 l_w / v_f \text{ s} \qquad (13.21)$$

where l_w is the length of the feature to be milled. However, it is important to note that this result must be corrected for the approach and overtravel distances, which will often be as large as the cutter diameter.

The machining time for maximum power is given by Equation 13.20, but, again, corrections must be made for cutter approach and overtravel.

13.9.9 ROUGH GRINDING

Limitations on the rate at which operations can be carried out depend on many interrelated factors, including the work material, the wheel grain type and size, the wheel bond and hardness, the wheel and work speeds, downfeed, infeed, the type of operation, the rigidity of the machine tool, and power available. It appears

TABLE 13.15
Machining Data for Milling Operations

Material	Hardness (BHN)	Side and face HSS	Side and face Brazed Carbide	End (1.5 in.) HSS	End (1.5 in.) Brazed Carbide	ps (hp/in.^{3}/min)^b
Low carbon steel (free machining)	150–200	19.2	52.9	4.5	15.7	1.1
Low carbon steel	150–200	13.5	43.3	2.2	9.9	1.4
Medium and high carbon steel	200–250	10.8	37.3	1.8	8.9	1.6
Alloy steel (free machining)	150–200	13.7	40.2	2.7	10.5	1.3
Stainless steel, ferritic (annealed)	135–185	14.0	41.0	2.4	6.0	1.7
Tool steels	200–250	6.7	23.7	0.9	4.5	1.5
Nickel alloys	80–360	4.1	7.7	1.0	—	2.15
Titanium alloys	200–275	3.9	13.2	1.5	7.1	1.25
Copper alloys (soft) (free machining)	40–150	50.5	108.3	9.9	20.7	0.72
Zinc alloys (die cast)	80–100	28.0	60.1	9.8	16.0	0.4
Magnesium and alloys	49–90	77.0	240.6	27.7	55.0	0.18
Aluminum and alloys	30–80	96.2	216.5	20.4	36.7	0.36

Header: Milling v_f (in./min)^a

[a] To convert in./min to mm/s multiply by 0.423

[b] To convert hp/min/in.^{3} to GJ/m^3 multiply by 2.73

Source: Adapted from *Machining Data Handbook,* 3rd ed., Vols. 1 and 2, MetCut Research Associates, Inc. 1980.

that, assuming adequate power, these limitations can be summarized in terms of the maximum metal removal rate per unit width of grinding wheel Z_w/w_t. For example, the *Machining Data Handbook* [5] gives the following recommendations for the rough grinding of annealed free machining low carbon steel on a horizontal-spindle reciprocating surface grinder:

Wheel speed: 5500 to 6500 ft/min (28 – 33 m/s)
Table speed: 50 to 150 ft/min (0.25 – 0.75 m/s)
Downfeed: 0.003 in./pass (0.0076 mm/pass)
Crossfeed: 0.05-0.5 in./pass (1.25 – 2.5 mm/pass) (0.25 wheel width maximum)
Wheel: A46JV (aluminum oxide grain, size 46, grade J, vitrified bond)

If the wheel width w_t were 1 in (25.4 mm) and an average table speed (work speed) of 75 ft/min (0.38 m/s) is employed, then a downfeed of 0.003 in. (0.076 mm) and the maximum crossfeed of 0.25 in. (mm) would give a metal removal rate Z_w of 0.68 in.^{3}/ min (11.15 cm^3/min). In a plunge-grinding operation, the wheel width would be equal to the width of the groove to be machined and the rough grinding time t_{gc} for recommended conditions would be given by

$$t_{gc} = 60V_m / Z_w \qquad (13.22)$$

where V_m is the volume of metal to be removed, and Z_w is the metal removal rate (in.^{3}/min). If the groove depth a_d is 0.25 in. (6.25 mm) and the groove length l_w is 4 in. (101.6 mm), the grinding time would be given by

$$t_{gc} = 60a_d w_t l_w / Z_w = 60(1)(0.25)(4) / 0.68 = 88.2 \text{ s}$$

The *Machining Data Handbook* [5] also gives values of the unit power (specific cutting energy for surface grinding of various materials). The unit power p_s depends to a large extent on the downfeed, and for a downfeed of 0.003 in. (0.076 mm), 13 hp min/in.^{3} (35.5 GJ/m^3) would be required for carbon steel. In this example, the removal rate for a 2 in. (50 mm) wide groove would be

$$Z_w = 60(2)(0.25)4 / 88.2 = 1.36 \text{ in.}^3/\text{min (22.3 cm}^3/\text{min)}$$

and the power required P_m would be given by

$$P_m = p_s Z_w = 13(1.36) = 17.7 \text{ hp (13.2 kw)}$$

Clearly, for a particular rough-grinding operation, it will be necessary to check the grinding time t_{gp} when maximum power is used, and this will be given by

$$t_{gp} = 60V_m p_s / P_m \qquad (13.23)$$

The estimated rough-grinding time t_{gr} would be given by the grinding time t_{gc} for recommended conditions or the grinding time t_{gp} for maximum power, whichever is the largest. Table 13.16 gives recommendations for typical conditions for the horizontal-spindle surface grinding of selected materials. These recommendations are expressed in terms of Z_w/w_t, the metal removal rate per unit width of wheel in rough grinding, and the corresponding unit power p_s.

If the operation is one of plunge grinding, the width of the grinding wheel will be known. In a traverse operation, the width of the wheel will depend mainly on the grinding machine.

In a plunge-grinding operation, the depth of material to be removed will be specified by the geometry of the finished workpiece. In a traverse-grinding operation,

TABLE 13.16
Data for Horizontal-Spindle Surface Grinding

Material	Hardness (BHN)	Z_w/w_t (in./min)	p_s (hp/in.^{3}/min)
Low carbon steel (free machining)	150–200	0.68	13
Low carbon steel	150–200	0.68	13
Medium and high carbon steel	200–250	0.68	13
Alloy steel (free machining)	150–200	0.68	14
Stainless steel, ferritic (annealed)	135–185	0.45	14
Tool steels	200–250	0.68	14
Nickel alloys	80–360	0.15	22
Titanium alloys	200–275	0.9	16
Copper alloys (soft) (free machining)	40–150	0.89	11
Zinc alloys (die cast)	80–100	0.89[a]	6.5[a]
Magnesium and alloys	49–90	0.89	6.5
Aluminum and alloys	30–80	0.89	6.5

[a] Estimated values

For external cylindrical grinding, multiply Z_w/w_t by 1.24 and multiply p_s by 0.81. For internal grinding, multiple Z_w/w_t by 1.15 and p_s by 0.8.

Source: Adapted from *Machining Data Handbook,* 3rd ed., Vols. 1 and 2, MetCut Research Associates, Inc. 1980.

it is necessary to remove the rough-grinding stock left by the previous machining operation.

13.9.10 FINISH GRINDING

The time for a finish-grinding operation is usually determined by the desired surface finish. This means that the metal removal rate must be slow enough to generate an acceptable surface finish, and it therefore becomes independent of the parameters affecting the removal rate in rough grinding. From the *Machining Data Handbook* [5], typical average values of the removal rate per inch of wheel width are 0.16 in.^{3}/min (2.62 cm^3/min) for horizontal-spindle surface grinding, 0.08 in.^{3}/min (1.31 cm^3/min) for cylindrical grinding, and 0.06 in.^{3}/min (0.98 cm^3/min) for internal grinding. Recommended stock allowances for finish grinding range from 0.002 to 0.003 in. (0.05 – 0.076 mm) for horizontal grinding and 0.005 to 0.01 in. (0.127 – 0.254 mm) for cylindrical grinding.

13.9.11 ALLOWANCE FOR GRINDING WHEEL WEAR

In his analysis of the economics of internal grinding, Lindsay [7] shows that the cost per component associated with wheel wear and wheel changing are proportional to the metal removal rate during rough grinding and that the wheel costs due to dressing and finish grinding are negligible in comparison. Thus, the total cost C_g of a grinding operation will be given by

$$C_g = Mt_c + Mt_{gr} + C_w \qquad (13.24)$$

where

M = the total machine rate (including labor, depreciation, and overhead)

t_c = constant time that includes wheel dressing time (assumed to occur once per component), the loading and unloading time, the wheel advance and withdrawal time, and the finish grinding time

t_{gr} = is the rough-grinding time

C_w = wheel wear and wheel changing costs

If we substitute

$$C_w = k_1 Z_w \qquad (13.25)$$

where k_1 is a constant and Z_w is the metal removal rate during rough grinding, and

$$t_{gr} = k_2 / Z_w \qquad (13.26)$$

where k_2 is a constant, into Equation 13.24 we get

$$C_g = Mt_c + Mk_2 / Z_w + k_1 Z_w \qquad (13.27)$$

Differentiating with respect to Z_w and equating to zero for minimum cost, we find that the optimum condition arises when wheel wear and wheel changing costs (represented by the third term in Equation 13.27) are equal to the rough grinding costs (represented by the second term). This means that if optimum conditions are used in a grinding operation, wheel wear and wheel-changing costs can be allowed for by multiplying the rough grinding time by a factor of 2.

However, it was pointed out earlier that the recommended conditions may exceed the power P_m available for grinding. In this case, the metal removal rate must be reduced, resulting in a reduction in wheel wear and wheel-changing costs and an increase in rough-machining costs with a consequent increase in the total operation costs.

If Z_{wc} and Z_{wp} are the metal removal rates for optimum (recommended) and maximum power conditions respectively, the corresponding costs C_c and C_p are given by

$$C_c = Mt_c + 2Mk_2 / Z_{wc} \qquad (13.28)$$

$$C_p = Mt_c + Mk_2/Z_{wp} + k_1 Z_{wp} \qquad (13.29)$$

Also since the optimum conditions

$$k_1 Z_{wc} = Mk_2/Z_{wc} \qquad (13.30)$$

then after substitution and rearrangement, the following expression for the cost C_p, under maximum power conditions is obtained

$$C_p = Mt_c + \frac{Mk_2}{Z_{wc}}\left(\frac{Z_{wc}}{Z_{wp}} + \frac{Z_{wp}}{Z_{wp}}\right) = Mt_c + Mt_{gp}\left[1 + \left(\frac{t_{gc}}{t_{gp}}\right)^2\right] \qquad (13.31)$$

where t_{gc} and t_{gp} are the rough-grinding times for recommended and maximum power conditions given by Equation 13.22 and Equation 13.23, respectively, and where $t_{gp} > t_{gc}$.

This means that a multiplying factor equal to the term in square brackets in Equation 13.31 can be used to adjust the rough grinding time and thereby allow for wheel wear and wheel-changing costs. Under circumstances where the recommended grinding conditions can be used (i.e., when $t_{gp} < t_{gc}$) the multiplying factor is equal to 2. If, for example, because of power limitations the metal removal rate were 0.5 of the recommended rate, then t_{gp} would be equal to $2t_{gc}$ and the correction factor would be 1.25. Under these circumstances, the rough-grinding costs would be double those for recommended conditions, and the wheel costs would be one-half those for recommended conditions.

13.9.11.1 Example

Suppose the diameter d_w of a stainless steel bar is 1 in. and it is to be traverse ground for a length l_w of 12 in. If the wheel width w_t is 0.5 in., the power available P_m is 3 hp and the rough-grinding stock a_r left on the radius is 0.005 in., we get the volume of metal to be removed

$$V_m = \pi d_w a_r l_w = \pi(1)(0.005)(12) = 0.189 \text{ in.}^3$$

From Table 13.16 the recommended metal removal rate per unit width of wheel Z_w/w_t is 0.45 in.^{2}/min for horizontal-spindle surface grinding. Using a correction factor of 1.24 for cylindrical grinding, the rough-grinding time for recommended conditions is given by

$$t_{gc} = 60V_m/Z_w = 60(0.189)/(1.24)(0.45)(0.5) = 40.65 \text{ s}$$

However, Table 13.16 gives a unit power value of p_s of 14 hp in.^{3}/min for stainless steels and, therefore, with a correction factor of 0.81, the rough-grinding time for maximum power would be

$$t_{gp} = 60V_m p_s /P_m = 60(0.189)(14)(0.81)/(3) = 42.9 \text{ s}$$

Thus insufficient power is available for optimum grinding conditions and the condition for maximum power must be used. Finally, using the multiplying factor to allow for wheel costs, the corrected value t'_{gp} for rough-grinding time is

$$t'_{gp} = t_{gp}\left[1+\left(t_{gc}/t_{gp}\right)^2\right] = 42.9\left[1+\left(40.6/42.9\right)^2\right] = 81.3 \text{ s}$$

As explained earlier, the metal removal rate for finish grinding is basically independent of the material and is approximately 0.08 in.^{3}/min per inch of wheel width in cylindrical grinding. For the present example where the wheel width is 0.5 in., this would give a removal rate Z_w of 0.05 in.^{3}/min with a correction factor of 1.24 applied. Assuming a finish-grinding radial stock allowance of 0.001 in., the volume to be removed is

$$V_m = \pi(1)(0.001)(12) = 0.038 \text{ in.}^3$$

and the finish-grinding time is

$$t_{gf} = 60(0.038)/(0.05) = 45.6 \text{ s}$$

13.9.12 ALLOWANCE FOR SPARK-OUT

In spark-out, the feed is disengaged and several additional passes of the wheel or revolutions of the workpiece are made in order to remove the material remaining because of machine and workpiece deflections. Since the number of passes is usually given, this is equivalent to removing a certain finish stock. For estimating purposes, the finish grinding time can be multiplied by a constant factor of 2 to allow for spark-out.

13.9.13 EXAMPLES

Firstly the machining cost for a facing operation will be estimated. The workpiece is made from free machining steel bar 3 in. (76.2 mm) diameter and 10 in. (254 mm) long, where 0.2 in. (5.1 mm) is to be removed from the end of the bar using a brazed-type carbide tool. The surface area to be generated is

$$A_m = [\pi/4](3)^2 = 7.07 \text{ in.}^2 \text{ (4.5 mm}^2)$$

and the volume of metal to be removed is

$$V_m = [\pi/4](3)^2(0.2) = 1.41 \text{ in.}^3 (23.1 \text{ mm}^3)$$

For this work material-tool material combination, Table 13.13 gives a value of vf (speed × feed) of 100 in.^{2}/min (0.065 m^2/min), and from Equation 13.19 the machining time is

$$t_{mc} = 60(2)(7.07/100) = 8.5 \text{ s}$$

where the factor of 2 allows for the gradually decreasing cutting speed when constant rotational speed is used in a facing operation.

The weight of the workpiece is estimated to be

$$W = [\pi/4](3)^2(10)(0.28) = 20 \text{ lb } (9.07 \text{ kg})$$

From Figure 13.26 the power available for machining on a CNC chucking lathe would be approximately given by

$$P_m = 10 \text{ hp } (7.76 \text{ kW})$$

Table 13.13 gives a value of specific cutting energy (unit power) of 1.1 hp min/in.^{3} (3 GJ/m^3) and so, from Equation 13.20, the machining time at maximum power is

$$t_{mp} = 60(2)(1.1)(1.41/10) = 18.6 \text{ s}$$

Again, the factor of 2 has been applied for facing a solid bar.

It can be seen that, in this case, the conditions for minimum cost cannot be used because of power limitations and that a machining time of 18.6 s will be required. Now we can apply the factor given by Equation 13.18 to allow for tool costs.

For carbide tools the Taylor tool life index is approximately 0.25, and since the ratio t_{mc}/t_{mp} is 8.5/18.6, or 0.46, the correction factor is

$$\{1 + [0.25/(1 - 0.25)](0.46)^{(1/0.25)}\} = 1.01$$

and the corrected machining time is 18.6 (1.01), or approximately 18.9 s.

In this example the correction factor for tool costs is quite small because cutting speeds below those giving minimum costs are being employed. If optimum speeds could be used, the correction factor would be 1.33 and the corrected machining time would be 11.2 s.

Finally, in Figure 13.27, data are presented on the typical cost of various machine tools, where it can be seen that, for the present example of a CNC lathe,

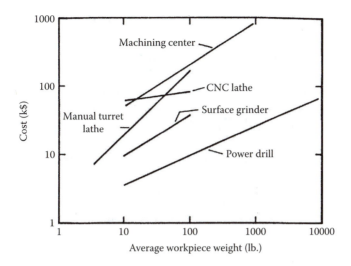

FIGURE 13.27 Relation between cost and workpiece weight for some machine tools.

a cost of about $80,000 would be appropriate. Assuming that the total rate for the operator and the machine would be $30 per hour, or $0.0083 per second, the machining cost for the facing operation would be $0.157.

Thus, using the approach described in this chapter, it is possible to estimate the cost of each machined feature on a component. For example, Figure 13.28 shows a turned component with the machining cost for each feature indicated. The small, nonconcentric hole and the keyway are relatively expensive features. This is because in order to machine them, the component had to be loaded on separate machine tools — significantly increasing the nonproductive costs. The designer who is able to make these estimates would clearly be encouraged to reconsider the securing operations that necessitated these features and thereby reduce the overall manufacturing costs of the product.

13.9.14 Machining Cost Estimating Worksheet

The various calculations required to estimate machining costs for turning and milling operations can be summarized in a worksheet as shown in Table 13.17. Each line on the worksheet is used for a specific cutting operation. The relevant tables and figures to be used for the required data are listed in Table 13.18. The following example illustrates the use of this worksheet for estimating the costs for a simple turning operation.

13.9.14.1 Example

Figure 13.29 shows a stepped rotational part that requires a cylindrical rough turning operation followed by a finish turning operation for completion of the part. The various entries in the worksheet are shown in Table 13.19, assuming

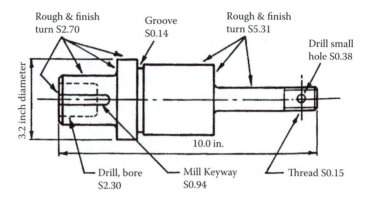

FIGURE 13.28 Turned component. Batch size, 1500; workpiece 3.25 in. dia × 10.25 in. long; material, low carbon free-machining steel.

that a CNC turret lathe is used to produce the part. The workpiece is carbon steel bar 3.00 in. diameter and 5.00 in long. From Table 13.10 the material density is 0.283 lb/in.³, and the material cost per unit weight is 0.51 $/lb. The workpiece weight is therefore 10.052 lb, and therefore the material cost per part is $5.13.

From Figure 13.27 for the workpiece weight the estimated cost of the machine tool is $80,000, and this gives with amortization and operator costs a machine rate of $30.00 per hour.

The estimated setup time is 0.65 hrs, and therefore the setup cost per part is

$$0.65 \times 30/1250 = \$0.016$$

The nonproductive time for workpiece loading and tool positioning is 31.9 + 3 = 34.9 s. Therefore the nonproductive costs per part are

$$34.9 \times 30/3600 = \$0.29$$

The total machining time including allowances for tool wear and tool approach is 105.63 from the worksheet. Therefore the machining cost is

$$30 \times 105.63/3600 = \$0.88$$

Therefore the total cost of the part is

$$\$5.13 + \$0.016 + \$0.29 + \$0.88 = \$6.316$$

13.9.15 APPROXIMATE COST MODELS FOR MACHINED COMPONENTS

During the initial conceptual stages, the designer or design team will be considering a variety of solutions to the design problem. Selection of the most promising design may involve trade-offs between the cost of machined components and components manufactured by other methods. However, the designer or design

TABLE 13.17
Machining Cost Analysis Worksheet

Machine Tool	Operation	Tool Type (HCD)*	Setup Time per Batch (hr) t_{su}	Load and Unload Time (s) t_l	Tool Positioning Time (s) t_{pt}	Dimension l_w	Dimension d_a	Dimension d_b	Volume v_m	Specific Cutting Energy p_s	Available Power P_m	Machining Time Max Power (s) t_{mp}	Rate of Surface Generation v_i	Milling Feed Speed v_f	Area A_m	Machining Time Recommended Conditions (s) t_{mc}	Time Corrected for Tool Wear (s) t_m	Time Corrected for Extra Tool Travel (s) t'_m
Part Name:																		
Material:																		
Density:																		
Workpiece weight:																		
Batch size (thousands):																		
	Totals																	Total

* H = high-speed steel, C = carbide (brazed), D = carbide disposable insert

TABLE 13.18
Summary of Data and Calculations for Cost Estimation Worksheet

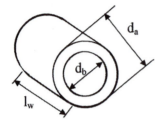

operation	area A_m	volume V_m
Turn, thread	$\pi l_w d_b$	$\dfrac{\pi}{4} l_w (d_a^2 - d_b^2)$
bore, drill, tap, ream	$\pi l_w d_a$	
face, thread**	$\dfrac{\pi}{2} d_a (d_a - d_b)$	$\dfrac{\pi}{2} l_w d_a (d_a - d_b)$

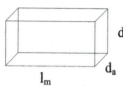

For milling:

$V_m = l_m d_a d_b$

$t_{m_p} = 60 p_s V_m / P_m$

$t_{m_c} = 60 A_m / vf$ or $60 l_m / v_f$

Variable	Source
Setup time, t_{su}	Table 13.12
Load/unload time, t_l	Table 13.11
Tool positioning time, t_{pt}	Table 13.12
Specific cutting energy, p_s	Tables 13.14 and 13.15
Power available, P_m	Figure 13.26
Rate of surface generation, vf	Table 13.14
Milling feed speed, v_f	Table 13.15
Machining time corrected for tool, t_m	Equations 13.14 and 13.18
Time corrected for approach allowance, t_m'	Factors from Table 13.13

team will not be in a position to specify all the details necessary to complete the type of analysis presented in the previous section. In fact, the information in the early stages of design may consist only of the approximate dimensions of the component, the material, and knowledge of its main features. Surprisingly, it is possible to obtain fairly accurate estimates of the cost of a component based on a limited amount of information. These estimates depend on historical data regarding the types of features usually found in machined components and the amount of machining typically carried out.

As an example consider a rotational component machined on a CNC turret lathe. In a study of the turning requirements for British industry [8] it was found that the average ratio of the weight of metal removed to initial workpiece weight was 0.62 for light engineering. Also, in light engineering, only 2% of workpieces

TABLE 13.19
Machining Cost Analysis Worksheet for Stepped Shaft[a]

Part Name: Stepped Shaft

Material: L. C. Steel

Density: 0.283 lb/in.³

Workpiece Weight: 10.052 lb

Batch Size (thousands): 1.2

Machine Tool	Operation	Tool Type (HCD)*	Setup Time per Batch (hr)	Load and Unload Time (s) t_l	Tool Positioning Time (s) t_{pt}	Dimension (in.) l_w	Dimension (in.) d_a	Dimension (in.) d_b	Volume (in.³) v_m	Specific Cutting Energy (hp min/in.³) p_s	Available Power (hp) P_m	Machining Time Max Power (s) t_{mp}	Rate of Surface Generation (in.²/min) vf	Milling Feed Speed v_f	Area (in.²) A_m	Machining Time Recommended Conditions (s) t_{mc}	Time Corrected for Tool Wear (s) t_m	Time Corrected for Extra Tool Travel (s) t'_m
CNC lathe	Rough turn	D	0.65	31.9	1.5	4	3	2.60	7.04	1.1	8	58	117	—	32.67	16.75	58.13	63.53
	Finish turn	D			1.5	4	2.60	2.57	0.05	1.1	8	0.4	70.2	—	32.30	27.6	36.7	42.1
		Totals	0.65	31.9	3												Total	105.63

[a] See Figure 13.29.

* H = high-speed steel, C = carbide (brazed), D = carbide disposable insert

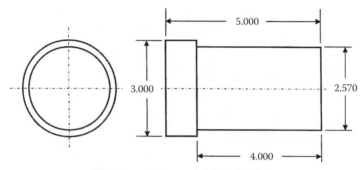

Workpiece: Carbon steel, 5.000 × 3.000 diameter

FIGURE 13.29 Simple stepped rotational part for cost estimating example.

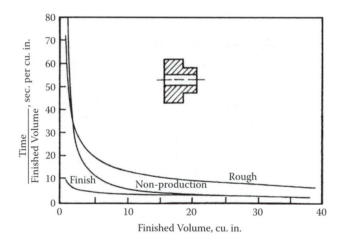

FIGURE 13.30 Effect of component size on rough machining, finish machining, and non-productive times per unit volume.

weighed over 60 lb (27 kg) and therefore required lifting facilities, and 75% of the workpieces were turned from bar. Usually, the proportion of initial volume of material removed by internal machining is relatively small for geometrical reasons.

The British survey [8] also showed a direct correlation between the length-to-diameter ratio and the diameter of turned components.

Using this type of data, Figure 13.30 shows, for a low-carbon-steel turned component, the effect of the finished size of the component on the rough machining, finish machining, and nonproductive times per unit volume. It can be seen that, as the size of the component is reduced below about 5 in.³ (82 μm³), the time per unit volume and hence the cost per unit volume increases dramatically, particularly for the nonproductive time. This increase is to be expected for the

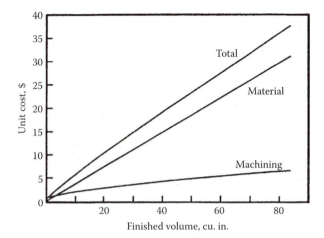

FIGURE 13.31 Effect of component size on total cost for machining a free-machining steel workpiece (cost 50 cents/lb) with an inserted carbide tool.

nonproductive time because it does not reduce in proportion to the weight of the component. For example, even if the component size were reduced to almost zero, it would still take a finite time to place it in the machine, to make speed and feed settings, and to start the cutting operations. For the rough-machining time, the higher times per unit volume for small components are a result of the reduced power available with the smaller machines used. The finish-machining time is proportional to the area machined. It can be shown that the surface area per unit volume (or weight) is higher for smaller components — thus leading to higher finish-machining times per unit volume.

These results have not taken into account the cost of the work material, and Figure 13.31 shows how the total cost of a finished steel component varies with component size. This total cost is broken down into material cost and machining cost, and it can be seen that material cost is the most important factor contributing to the total cost even though 62% of the original workpiece was removed by machining — a figure that results in relatively high rough-machining costs. In fact, for the larger parts, about 80% of the total cost is attributable to material costs.

From the results of applying the approximate cost models, it is possible to make the following observations:

1. For medium-sized and large workpieces, the cost of the original work-piece mainly determines the total manufactured cost of the finished component.
2. The cost per unit volume or per unit weight of small components (less than about 5 in.^{3} or 82 pm^3) increases rapidly as size is reduced because:
   a. The nonproductive times do not reduce in proportion to the smaller component size.

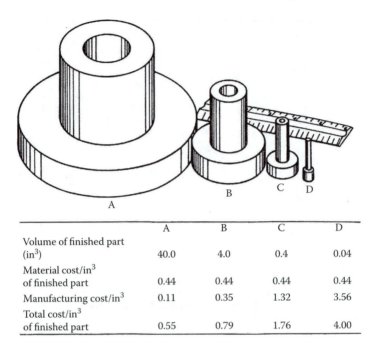

	A	B	C	D
Volume of finished part (in^3)	40.0	4.0	0.4	0.04
Material cost/in^3 of finished part	0.44	0.44	0.44	0.44
Manufacturing cost/in^3	0.11	0.35	1.32	3.56
Total cost/in^3 of finished part	0.55	0.79	1.76	4.00

FIGURE 13.32 Cost (dollars) for a series of turned components.

    b. The power available and hence the metal removal rate is lower for components.

    c. The surface area per unit volume to be finish-machined is higher for smaller components.

This is illustrated in Figure 13.32, where a series of turned components are shown, each being one-tenth the volume of the previous component. Although the cost per unit volume for the material is the same for all the components, the machining cost per unit volume increases rapidly as the components become smaller. For example, the total cost of the smallest component is $4.00 per in.^{3} whereas for the largest it is $0.55 per in.^{3}. Stated another way, it would clearly be much less expensive to machine one of the largest components rather than 1000 of the smallest components when using the same types of machines.

3. The choice of tool materials and optimum machining conditions only affects the finish-machining time. Since finish machining represents only about 25% of the total manufacturing cost, which in turn represents only about 20% of the total component cost for larger components, the effects of changes in tool materials or recommended conditions can be quite small under many conditions.

4. The factors to be taken into account in making early estimates of machining costs are:

    a. The amount of material to be removed. This factor directly affects the material costs per unit volume of the finished product and, less importantly, the rough-machining time.

    b. The rate for the machine tool and operator.

    c. The power available for machining and the specific cutting energy of the work material.

    d. The nonproductive times — especially for smaller components.

    e. The surface area to be finish machined.

    f. The recommended finish-machining conditions, which are in turn affected by the work material and tool material used.

5. The factors that affect the nonproductive times are:

    a. The number of times the component must be clamped in a machine tool — each clamping involves transportation, loading and unloading, and setup.

    b. The number of separate tool operations required — each operation requires tool indexing, and other associated activities and increases setup costs.

It was found in previous studies [9] that common workpieces can be classified into seven basic categories. Knowledge of the workpiece classification and the production data not only allows the cost of the workpiece to be estimated, but also allows predictions to be made of the probable magnitudes of the remaining items necessary for estimates of nonproductive costs and machining costs.

For example, for the workpiece shown in Figure 13.28, the total cost of the finished component was estimated to be $24.32, a figure obtainable from knowledge of the work material, its general shape classification and size, and its cost per unit volume. A cost estimate for this component, based on its actual machined features and using approximate equations based on the type of data listed above gave a total cost of $22.93, which is within 6%. A more detailed estimate obtained using the traditional cost-estimating methods presented in this chapter gave a total cost of $22.95. Thus, approximate methods, using minimum amounts of information, can give estimates surprisingly close to the results of analysis carried out after detailed design has taken place, at least for some types of parts.

PROBLEMS

1. A batch of 10,000 rectangular blocks $50 \times 25 \times 25$ mm is to be slab-milled on all faces. The material can be either mild steel costing $4,900/m^3 or aluminum costing $6,100/m^3. Estimate the total cost of the batch of components in each material if the average nonproductive time for each face = 60 s, the tool-changing time = 600 s, the cost of a sharp tool = $20.00, the machine and operator rate = $0.01/s, the tool-life index $n = 0.125$, the cutting speed for 60 s of tool life is 1 m/s for mild steel and 4 m/s for aluminum, the feed per revolution of the cutter = 1.25 mm, and the working engagement (depth of cut) = 2 mm.

(Assume that the distance traveled by the 50-mm-diameter, 38-mm-wide cutter for each face of the workpiece is 5 mm greater than the length of the face.)

2. Aluminum rod is available in diameters of 1-mm increments to 25-mm diameter and in 2-mm increments from 26- to 50-mm diameter. In the design of a particular shaft it has been determined that its finished diameter D should be within the range

$$33 - \frac{(L-229)^2}{84.75} > D > 27.43 + \frac{(L-254)^2}{127}$$

where L is the length of the shaft. If an allowance of 2 mm in diameter is to be allowed for machining, what finished diameter should be chosen for the shaft?

3. Write the geometric code number for each of the components shown in Figure 13.9, noting that run-out grooves for screw threads do not affect the coding.

4. Which of the two designs for a cylinder assembly shown in Figure 13.33 would be preferable from the manufacturer's point of view? Also, list other manufacturing and assembly aspects of both designs that could be improved. (Note that the omission of details of the bolting arrangements for the cylinder ends is not to be considered an error.)

5. The end covers shown in Figure 13.34 are to be milled from 10-mm-thick aluminum plate. Which of these two designs is preferable from the manufacturer's point of view? Also, what other aspects of the designs should be changed?

6. A designer has specified a surface finish on turned shafts of 0.4-pm arithmetical mean when a surface finish of 1.6 pm would suffice. Estimate the cost of this mistake when 2000 shafts are to be produced with a tool with a rounded corner if the machining time per component = 600 s, the number of components produced between tool regrinds = 4, the cost of a sharp tool = $2.00, the machine and operator rate = $00033/s, the tool-changing time = 120 s, and the nonproductive time per component = 240 s.

7. The part shown in Figure 13.33 is to be produced on a CNC turret lathe from free machining low carbon steel. Using the worksheet in Table 13.17 and the data in this chapter, estimate the processing time and costs for machining a batch of 500 parts. Assume that 0.015 in.

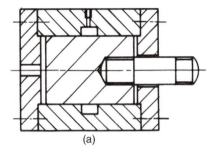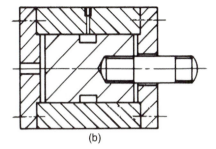

FIGURE 13.33
Alternative designs of cylinder assembly.

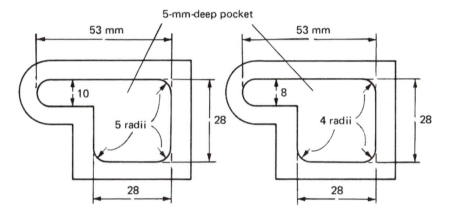

FIGURE 13.34
Alternative designs of an end cover.

of material is faced off from the end of the part prior to drilling the through hole. The center drilling operation prior to drilling the through hole can be assumed to have a machining time of 5 s. The turning tool is has a disposable carbide insert, and the drill is made from high speed steel.

REFERENCES

1. *Machinery's Handbook*, 19th ed., Industrial Press, Inc., New York, 1973.
2. Opitz, H., *A Classification System to Describe Workpieces*, Pergamon Press, New York, 1970.
3. Boothroyd, G., Dewhurst, P., and Knight, W.A., *Product Design for Manufacture and Assembly*, 2nd ed., Marcel Dekker, New York, 2002.
4. Fridriksson, L., Non-productive Time in Conventional Metal Cutting, Report No. 3, Design for Manufacturability Program, University of Massachusetts, Amherst, February, 1979.

5. *Machining Data Handbook*, 3rd ed., Vols. 1 and 2, MetCut Research Associates Inc., 1980.
6. Ostwald, P.F., *AM Cost Estimator*, McGraw-Hill, New York, 1985/1986.
7. Lindsay, R.P., Economics of Internal Abrasive Grinding, SME Paper MR 70-552, 1970.
8. Anon., Survey of Turning Requirements in Industry, Production Engineering Research Association, U.K., 1962.
9. Anon., Survey of the Machining Requirements of Industry, Production Engineering Research Association, U.K., 1963.

14 Nonconventional Machining Processes

14.1 INTRODUCTION

The recent increase in the use of hard, high-strength, and temperature-resistant materials in engineering has made it necessary to develop many new machining techniques. With the exception of grinding, conventional methods of removing material from a workpiece are not readily applicable to most of these new materials. Although the development of new cutting materials has enabled conventional machining of more difficult to process materials to be more widely carried out. Most of the new machining processes have been developed specifically for materials that are difficult to machine, but some of them have found use in the production of complex shapes and cavities in softer, more readily machined materials.

Descriptions of the main nonconventional machining processes are given in this chapter. More detailed descriptions of these processes and several others can be found elsewhere [1].

Nonconventional machining processes may be used as alternatives to the more traditional processes for a number of reasons, including the following [2]:

1. Machinability of the workpiece material: Many workpiece materials are difficult to machine by conventional methods because of high hardness, high thermal resistance, or high abrasive wear. For these materials many of the nonconventional machining processes are suitable.
2. Shape complexity of the workpiece: Many shape features are difficult or impossible to produce by conventional machining methods. For example, it is usually relatively simple to produce a cylindrical hole in a workpiece, but to produce a square, sharp-cornered hole of similar size is much more difficult. Nonconventional machining processes are often capable of machining complex shape features with comparative ease, even in materials that are normally difficult to machine.
3. Surface integrity: Conventional machining processes can result in surface cracks and residual stresses in the workpiece. Many nonconventional machining processes can remove material without causing these effects.
4. Precision: Many nonconventional processes are capable of high levels of precision which cannot be achieved by conventional machining.
5. Miniaturization: Some nonconventional processes are capable of producing very small shape features and fine detail not possible by conventional machining (e.g., the machining of deep holes with very small diameters).

6. Automatic data communication and computer integration: Many non-conventional processes can be computer-controlled and linked directly to CAD/CAM systems in an integrated manufacturing environment.

14.2 RANGE OF NONCONVENTIONAL MACHINING PROCESSES

The nonconventional machining processes utilize a variety of energy sources for removing the workpiece material. Some of the more important processes are shown schematically in Figure 14.1, grouped according to the energy source used [2]. The three main processes that utilize a mechanical action for removing material are ultrasonic machining, abrasive-jet machining, and water-jet machining, including abrasive water-jet machining. Four processes can be characterized as electro thermal processes; electro discharge machining, including wire-electro discharge machining; laser-beam machining, electron-beam machining, and plasma-beam machining. The remaining processes are basically chemical: chemical machining and electrochemical machining, including electrochemical grinding. There are several other less commonly used processes also available [1], but these will not be discussed here.

A number of the processes shown in Figure 14.1 remove material by means of a narrow jet or beam projected at the workpiece. A useful property of these processes is the ability to cut sheet or plate material with a very narrow "kerf" width. The kerf width is the width of the slit cut in the material, which determines the amount of waste material from the process and the intricacy of detail of the profile that may be cut by the process.

14.3 ULTRASONIC MACHINING

Ultrasonic machining is a valuable process for the precision machining of hard, brittle materials [1–4]. Material is removed by the abrasive action of a grit-loaded slurry circulating between the workpiece and a tool vibrating at ultrasonic frequencies. The tool is gradually fed into the workpiece to form a cavity corresponding to the tool shape. Frequencies of 20 to 40 kHz are most commonly used. A typical arrangement for ultrasonic machining is shown schematically in Figure 14.2. The main elements are:

1. A high frequency generator
2. A transducer, which utilizes the piezoelectric or magnetostrictive effect
3. A transformer, which is shaped to amplify the vibration output of the transducer
4. A tool holder and tool
5. Abrasive slurry, which is circulated into the small gap between the tool and workpiece

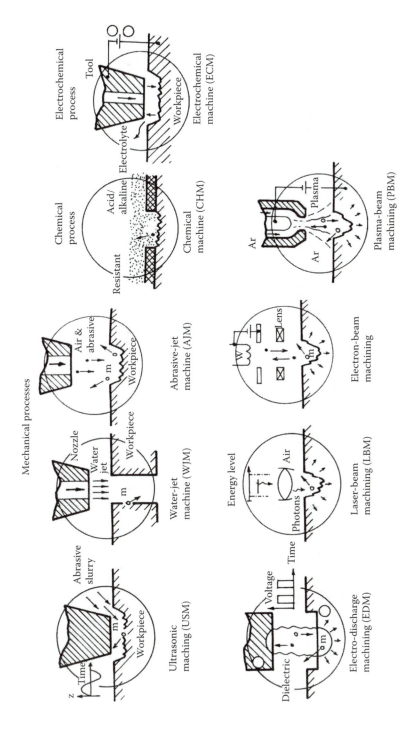

FIGURE 14.1 Schematics of various nonconventional machining processes. (Adapted from Snoeys [2])

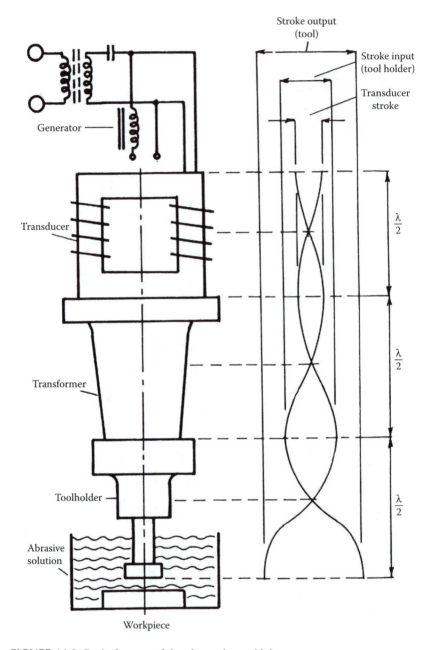

FIGURE 14.2 Basic features of the ultrasonic-machining process.

14.3.1 Transducers

A transducer converts energy from one form to another. Transducers used in ultrasonic machining convert electrical energy into mechanical motion, and they are based on piezoelectric or magnetostrictive principles. The amplitude of vibration possible depends on the length of the transducer and the strength of the material used, the limit being approximately 0.025 mm in practice.

14.3.1.1 Piezoelectric Transducers

Certain crystals such as quartz undergo small dimensional changes when subjected to a changing electrical current. Application of an alternating voltage to the crystal causes vibrations to occur and the amplitude is greatest at the resonant frequency of the crystal. For resonance, the length of the crystal must be equal to half the wavelength of sound in the crystal [4]. For quartz, Young's modulus $E = 5.2 \times 10^{10}$ N/m^2, the density $\rho = 2.6 \times 10^3$ kg/m^3, and hence the velocity of sound $C_s = E/\rho = 4480$ m/s. Finally, at a frequency f_s of 20 kHz, the wavelength = $C_s/f_s = 0.228$ m, and therefore the required crystal length is 110 mm.

Polycrystalline ceramics such as barium titanate are also used as transducers; they have a sandwich construction and lengths of 75 to 100 mm. A typical transducer (Figure 14.3) consists of a stack of ceramic disks, sandwiched between a high-density base and a lower-density block of material, which provides the radiating face of the transducer. The action of the transducer is based on the fact that, from momentum conservation, the velocity and amplitude of vibration are high in a lower density material. Such transducers are capable of converting electrical energy into vibrational energy at 96% efficiency and consequently do not require cooling.

14.3.1.2 Magnetostrictive Transducers

When ferromagnetic materials such as nickel and nickel alloys are magnetized, an expansion occurs. Application of an alternating current to coils surrounding the material (Figure 14.4) causes vibrations at the applied frequency. The magnetic material is laminated to reduce eddy current losses, but the energy conversion efficiencies are only 20 to 35%; consequently, water-cooling is required to remove waste heat from the transducer.

14.3.2 Transformer and Tool Holder

The purpose of the transformer in ultrasonic machining is to increase the vibration amplitude. The transformer's length is half the wavelength of sound in the transformer material. It is in effect a mechanical amplifier. The reduction in cross-sectional area causes amplification of the vibration, which increases in inverse proportion to the area ratio between the upper and lower ends of the transformer.

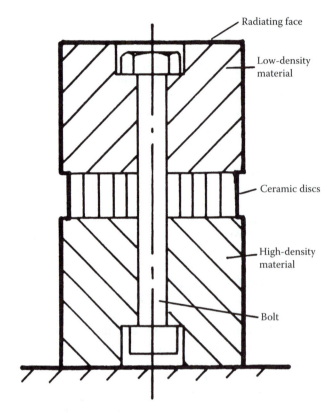

FIGURE 14.3 Piezoelectric ultrasonic transducer.

Amplitude may be increased by as much as 600% with a suitably shaped transformer. A similar increase in amplitude at the tool face can occur in the tool holder, and final amplitudes at the tool face are usually in the range of 0.013 to 0.1 mm.

14.3.3 TOOLS

The shape of the tool is the inverse of the cavity to be eroded. Relatively ductile tool materials, such as stainless steel, brass, and mild steel are generally used to minimize tool wear. The tools are usually brazed to the tool holder to reduce possible fatigue problems associated with screwed fasteners.

14.3.4 ABRASIVE SLURRY

The slurry consists of small abrasive particles mixed with water or sometimes oil, with the abrasive having a concentration of 30 to 60% by weight. Commonly used abrasive materials are boron carbide, silicon carbide, and aluminum oxide. The grit size is usually about the same as the amplitude of vibration being used.

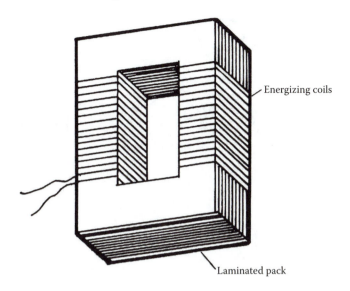

Energizing coils

Laminated pack

FIGURE 14.4 Magnetostrictive ultrasonic transducer.

14.3.5 Applications

Ultrasonic machining can be used to produce cavities in most materials whether conductive, nonconductive, metallic, ceramic, or composite. It is most effective on materials that are harder than $50R_c$. Material removal rates are low, usually less than 1.5 cm³/min. Holes as small as 0.076 mm can be drilled, and the upper limit on cavity size is approximately 75 mm. Machining depths of 50 mm or more are possible, but in practice depths are usually less than 5 mm. Tolerances of ±0.25 mm and surface finishes of 0.25 µm can routinely be obtained.

14.4 WATER-JET MACHINING

14.4.1 General Discussion

In water-jet machining, material is removed through the erosion effects of a high-velocity, small-diameter jet of water [1,5]. The process is primarily used for drilling and slitting operations on materials in sheet or slab form. Intricate shapes can be readily cut with a very narrow kerf width. The equipment (Figure 14.5) used for water-jet machining consists of a high-pressure intensifier and pumping unit, capable of delivering water at pressures up to 400 MPa, a delivery system to route the high-pressure water to the work area, a nozzle to form the jet, and a positioning system that moves either the nozzle or the workpiece to bring about the cutting action.

The nozzle conventionally contains a sapphire insert to form the small-diameter jet to resist wear, but some success has been achieved with metallic nozzle inserts made from tungsten carbide (Figure 14.6) [2,6]. Jet diameters are

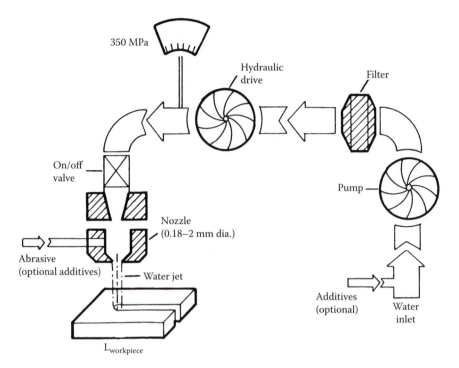

FIGURE 14.5 Basic features of water-jet and abrasive water-jet machining.

usually in the range 0.07 to 0.5 mm, and the typical life of the nozzle inserts is from 250 to 500 hours. The standoff distance of the nozzle from the workpiece is usually in the range 3 to 25 mm. The choice of positioning system used for the tool or workpiece depends on the application of the process. In many applications the nozzle or workpiece is moved manually and the process is then similar to conventional band sawing. Water-jet cutting machines are also available with numerically controlled workable motions or with the nozzle moved with an industrial robot.

Water-jet machining is mainly used for machining soft nonmetallic materials such as paper, wood, leather, and foam. High cutting rates are possible with no thermal damage to the work material. An adaptation of the basic process, which enables it to be used for machining hard materials, is the introduction of abrasive particles into the fluid stream after the jet; the process is then referred to as abrasive water-jet machining. The remainder of the equipment is essentially the same as in conventional water-jet machining. Commonly used abrasives for this process are garnet, silicon, and silicon carbide grains [1].

14.4.2 APPLICATIONS

Water-jet machining is used to cut soft, nonmetallic materials in sheet form, often in multilayer stacks. Table 14.1 shows some water-jet cutting rates for different

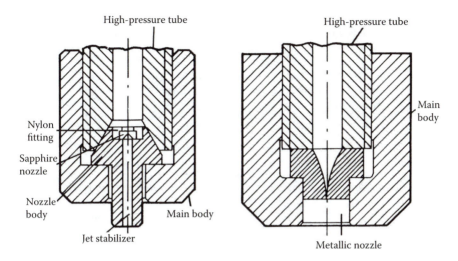

FIGURE 14.6 Nozzles for water-jet machining. (a) Conventional sapphire nozzle; (b) metallic nozzle.

TABLE 14.1
Water-Jet Cutting Rates

Material	Thickness (mm)	Pressure (Mpa)	Nozzle Diameter (mm)	Speed (m/min)
ABS Plastic	2.2	379	0.15	1.0
	3.8	379	0.15	0.2
Cardboard	1.4	379	0.10	6.1
Cheese	100	379	0.20	22.8
Foam Rubber	76	207	0.10	300.0
Leather	1.5	275	0.13	76.2
Plexiglas	3	379	0.13	0.9
Wood	3.2	379	0.13	1.0

Source: Adapted from Benedict, G., *Nontraditional Manufacturing Processes*, Marcel Dekker, New York, 1987.

materials [1]. Force levels on the workpiece are low, allowing crushable materials to be cut effectively without damage. The kerf width is approximately 0.28 mm larger than the nozzle diameter, enabling close nesting of the shapes to be cut from the sheet materials. Predrilling of starting holes is not necessary. The jet can pierce the material at any location and then cutting can be carried out in any direction. The surface finish of the cut edges is good, with no burrs and no thermal damage. Tolerances are a function of the material type and thickness, and they are usually within ±0.1 to 0.2 mm.

TABLE 14.2
Abrasive Water-Jet Cutting Rates

Material	Thickness (mm)	Pressure (MPa)	Nozzle Diameter (mm)	Speed (mm/min)
Acrylic	3.5	NA[a]	NA	1200
	13	NA	NA	460
Aluminum	3.3	207	0.45/1.5	900
	25	207	0.45/1.5	51
	152	207	0.45/1.5	10
Boron	0.33	207	0.33/1.2	305
Cast iron	8.1	207	NA	150
Glass	10	207	0.45/1.5	915
Graphite	51	NA	NA	125
Steel	20	207	0.45/1.5	51
Titanium	3.4	NA	NA	300

[a] NA, not available.

Source: Adapted from Benedict, G., Nontraditional Manufacturing Processes, Marcel Dekker, New York, 1987.

Abrasive water-jet machining is a considerable enhancement of the basic process, making it possible for a much wider range of materials to be cut [1,2,7]. Virtually all materials, regardless of hardness, reflectivity, and conductivity, can be successfully cut. Most metals can be readily machined, with thickness of up to 100 mm. Table 14.2 lists some cutting speeds for various materials machined by abrasive water-jet machining [1]. Kerf widths are usually between 1.5 and 2.3 mm and the cut surfaces reveal no thermal or mechanical damage. Some tapering of the cut slit does occur and its extent increases with the material thickness. Brittle materials such as acrylic, graphite, and silica glass can be cut without damage, leaving the cut edges free from cracks. Porous materials such as concrete can be cut at thicknesses up to 300 mm. A growing application of abrasive water-jet machining is the trimming of cured fiber composite and laminated materials because the materials do not delaminate during cutting and ragged fibers are not left protruding from the cut edges.

14.5 ABRASIVE-JET MACHINING

14.5.1 GENERAL DISCUSSION

Abrasive-jet machining removes material through the eroding action of a high-velocity stream of abrasive-laden gas [1,2,8]. An abrasive jet is formed by micro-abrasive particles being entrained in an inert gas stream and propelled through a small nozzle at pressures up to 0.7 MPa, which results in jet velocities of up to 300 m/s. Material is removed from the workpiece by a chipping action that is particularly effective on hard, brittle materials such as glass, silicon, tungsten, and ceramics.

In an abrasive-jet machining system, the gas is first compressed and mixed with the abrasive powder in a mixing chamber (Figure 14.7). The mixing chamber consists of a convergent-divergent inlet nozzle, followed by a chamber where the abrasive powder is drawn into the gas stream before passing through the outlet nozzle to the work surface. The delivery nozzle, made of either tungsten carbide or sapphire, has a useful life of about 30 hours. Jet diameters are usually 0.12 to 1.25 mm. The most commonly used abrasives are aluminum oxide, silicon carbide, and sodium bicarbonate, with particle sizes varying from 10 to 50 µm.

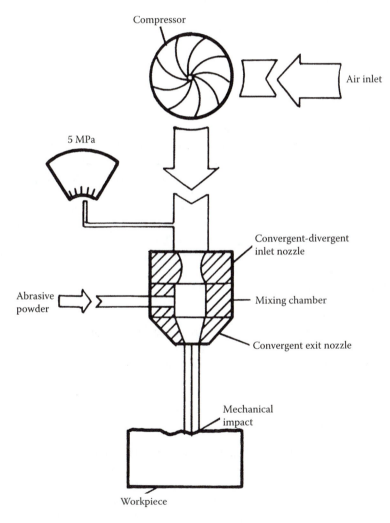

FIGURE 14.7 Basic features of abrasive-jet machining.

14.5.2 APPLICATIONS

The process can be used for performing intricate cutting, drilling, polishing, and etching operations on delicate workpieces because of the relatively low overall forces generated. The cooling action of the gas stream ensures that the workpiece material experiences no thermal damage. When abrasive-jet machining is not used as a cutting or drilling process, material removal rates are low — in the region of 0.015 cm³/min. Slots as small as 0.12 to 0.25 mm in width can be produced. For cutting and drilling operations, the process is restricted to metal foils and thin sections of ceramics and glass. Tolerances of ±0.12 mm are readily maintained, together with surface finishes ranging from 0.25 to 1.25 μm. Future development of the process is likely to be for micro grit-blasting applications such as deburring, cleaning, polishing, and etching. Typical applications are the etching of shallow, intricate holes in electronic components such as resistor paths in insulators and patterns in semiconductors. The engraving of characters on toughened-glass automobile windows is another significant application of the process.

14.6 CHEMICAL MACHINING

In chemical machining a strong acid or alkaline solution is used to dissolve materials selectively, resulting in controlled material removal [1,2,9]. A chemically resistant coating known as a *mask* is selectively applied to the workpiece to protect surfaces that are not to be machined. Chemical machining can be divided into two basic sub-processes: chemical milling and chemical blanking. In chemical milling, erosion of the material is used to produce pockets and channels in the workpiece surfaces or to remove material over large areas of the surfaces of a workpiece, for weight reduction, for example. In chemical blanking, holes and slots that penetrate entirely through the material are produced, usually in thin sheet materials.

The five steps in chemical machining are as follows:

1. Part preparation: cleaning
2. Masking: application of chemically resistant material
3. Etching: dip or spray exposure to the etchant
4. Mask removal: stripping the mask material and cleaning the part
5. Finishing: inspection and other processing

Three basic types of masks are used: cut-and-peel masks, screen-resist masks, and photoresist masks.

Cut-and-Peel Masks — A film of the chemically resistant material is applied to the workpiece by dipping, spraying, or flow coating, producing a dry film up to 0.13 mm thick. The materials most commonly used are vinyl, styrene, and butadiene. The rubbery film is then cut and peeled away selectively. The most common way of cutting the mask is by hand scribing using templates, but other methods, including laser cutting of the mask material, have been used [2]. Manual scribing of the mask material usually achieves an accuracy of ±0.13 to ±0.75 mm. The high mask thickness produced by the dipping or spraying of the mask material makes it possible for depths of as much as 13 mm to be cut.

Screen-Resist Masks — For a screen-resist mask, the mask material is applied to the workpiece surface by printing, using stencils and a fine polyester or stainless steel mesh screen. Screen-printing is a fast, economical method of applying the mask material. Relatively thin coatings with tolerances held to ±0.05 to ±0.18 mm are obtained by screen-printing. Etching depths are restricted to about 1.5 mm because of the thinness of the coating.

Photoresist Masks — Chemical machining using photoresist masks is quite widely used and is often referred to as photochemical machining. The process produces intricate and finely detailed shapes in materials by using a light-activated resist material. The workpiece is coated with the photoresist material and a master transparency is held against the workpiece, while exposure to ultraviolet light takes place (Figure 14.8). The

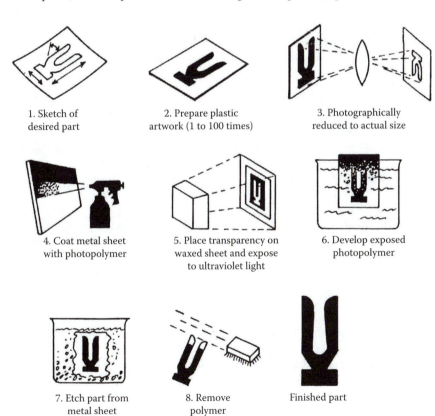

| 1. Sketch of desired part | 2. Prepare plastic artwork (1 to 100 times) | 3. Photographically reduced to actual size |

| 4. Coat metal sheet with photopolymer | 5. Place transparency on waxed sheet and expose to ultraviolet light | 6. Develop exposed photopolymer |

| 7. Etch part from metal sheet | 8. Remove polymer | Finished part |

FIGURE 14.8 Process steps for photochemical machining.

light activates the photoresist material in those areas corresponding to the transparent portions of the master negative. The areas corresponding to the opaque parts of the negative are dissolved away during the etching process. Photoresist is the most accurate of the available masking methods and can be used to produce shapes with tolerances of ±0.025 to ±0.005 mm, but typically feature sizes cannot be less than the material thickness.

14.6.1 ETCHANTS

A variety of materials are used for dissolving the workpiece material, and selection of the most appropriate chemicals depends on a number of factors [1,8], including:

1. The workpiece material
2. The mask material
3. The depth of etch required
4. The surface finish required, as some etchants attack the grain boundaries of some materials
5. The material removal rates required
6. The heat-treated condition of the material
7. The potential damage to the work material and the metallurgical properties required
8. Economic considerations

14.6.2 APPLICATIONS

Chemical machining using cut-and-peel masking is usually applied in the removal of material from large workpieces, for example, in the removal of material in selected areas for the lightening of aircraft wing panels and structural members.

Chemical blanking, using photoresists or screen-printed masks, is used for producing delicate shapes in thin materials. Figure 14.9 shows some intricate parts produced by photochemical blanking [1]. The process can be used to produce burr-free shapes for which conventional blanking die costs would be very high. Another advantage of the process is that the workpiece material can be hard and brittle, so that it could not be blanked by conventional methods. In addition to sheet metal applications, photochemical machining is used routinely for the production of printed-circuit boards. The unwanted copper deposited on the board is removed photochemically, leaving the required pattern of connections on the nonmetallic backing material of the board.

14.7 ELECTROCHEMICAL MACHINING

Electrochemical machining (ECM) refers to the process of metal removal in which electrolytic action is utilized to dissolve the workpiece metal. It is, in effect, the reverse of electroplating. A modification of the process is electrolytic grinding, which will be discussed later.

The ECM process is illustrated in Figure 14.10. The workpiece (which must be a conductor of electricity) is placed in a tank on the machine table and

FIGURE 14.9 Some part profiles produced by photochemical blanking (From Benedict [1]).

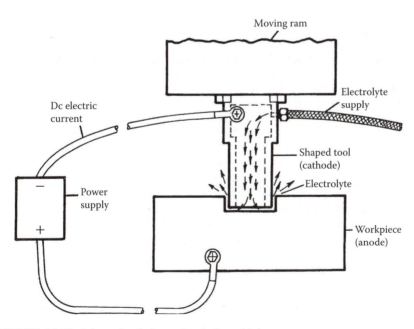

FIGURE 14.10 Schematic of electrochemical machining.

connected to the positive terminal of a high-current, low-voltage dc supply. The tool electrode, shaped to form the required cavity in the workpiece, is connected to the negative terminal of the supply. An electrolyte flows through the gap between the tool and the workpiece and is then pumped back to the working zone, either through the tool or externally, depending on the application.

The action of the current flowing through the electrolyte is to dissolve the metal at the anode, that is, the workpiece. The electrical resistance is lowest (and hence the current is highest) in the region where the tool and workpiece are closest. Since the metal is dissolved from the workpiece most rapidly in this region, the form of the tool will be reproduced on the workpiece, as the tool is fed gradually toward the workpiece.

There is no mechanical contact between the workpiece and the tool. The fast-moving electrolyte removes the deplated material while it is in solution, before it can become plated on the tool (the cathode). Hence, there is neither tool wear nor plating of the workpiece material on the tool so that one tool can produce a large number of components during its life.

The electrolyte serves two important functions in ECM. First, the electrolyte provides the medium by which electrolysis takes place; and second, it removes the heat that is generated in the working zone as a result of the high current flow through the electrodes and the electrolyte.

The electrolyte most commonly used is brine; other salt solutions have been used in specific applications. The corrosive nature of the electrolytes used necessitates that the parts of the machine exposed to the electrolyte be manufactured from stainless steel or plastic or be coated with corrosion-resistant paint.

14.7.1 Metal Removal Rates

The rate at which material is removed from the workpiece in ECM can be predicted from knowledge of its electrochemical equivalent.

The metal removal rate may be expressed as follows:

$$Z_{we} = \frac{\eta_e eI}{\rho} \qquad (14.1)$$

where

Z_{we} = volume of metal removed electrolytically per unit time

e = electrochemical equivalent of the work material

I = current

η_e = current efficiency (i.e., fraction of the current actually effective in electrolysis)

ρ = density of the work material

The metals cobalt, iron, nickel, molybdenum, chromium, and tungsten, common constituents of engineering alloys, all have values of e/ρ in the range 32.5 to 37.4 p m³/As (0.119 to 0.137 in.³/min/l000 A). The current efficiency η_e is usually on the order of 0.75 to 0.90, so that a good approximate figure for metal removal by ECM is 27.3 p m^3/As (0.1 in.^{3}/min/1000 A) [10].

From Equation 14.1 the tool feed speed is given by

$$v_f = \frac{Z_{we}}{A} = \eta_e \frac{e}{\rho} \frac{I}{A}$$

(14.2)

where

v_f = tool feed speed
A = area of work surface exposed to electrolysis
I/A = current density

Thus, to obtain the highest machining rate, the highest permissible current density should be used. The current density is limited by the capacity of the supply as well as the capacity of the work material, the tool material, and the electrolyte to carry the current. The current should not be so high as to produce excessive heat in the tool and the work or to cause electrical breakdown or decomposition of the electrolyte. A figure of 1.5 MA/m^2 (1000 A/in.^{2}) can usually be safely assumed, while quite often a current density as high as 4.5 MA/m^2 (3000 A/in.^{2}) is permissible. Thus, when the maximum possible cutting rate is desired, the highest permissible current for the particular job must be determined by experiment. Table 14.3 shows the material removal rates for a range of different materials [1].

If the permissible current density for a particular job limits the total current to a level well below the machine and supply capacities, the possibility of two or more pieces being worked simultaneously may be considered. In this way the machine can be used to its full capacity and the overhead cost per component kept low.

14.7.2 Nature of the Machine Surface

The fact that metal removal in ECM is not achieved by mechanical shearing (as in conventional machining) or by melting and vaporization of the metal (as in EDM, section 14.9) means that no thermal damage occurs and no residual stresses are produced on the worked surface. For this reason, ECM is often selected for highly stressed or fatigue-sensitive applications in the aerospace industries. The only heat generated results from electrical resistance, and the temperature cannot be allowed to rise above the boiling point of the electrolyte.

14.7.3 Effect of Tool Feed Speed and Supply Voltage on Accuracy

The predominant resistance to the current flow in ECM is that due to the electrolyte, and hence the current may be expressed as follows:

TABLE 14.3
ECM Removal Rates

Material	Density (gm/cm³)	Removal Rate (cm³/min)
Aluminum	2.7	1.9
Beryllium	1.8	1.3
Cobalt	8.9	1.9
Copper	8.9	3.9
Iron	7.8	2.1
Lead	11.3	5.0
Molybdenum	10.1	1.8
Nickel	8.9	1.9
Titanium	4.5	1.9
Tungsten	19.3	0.8
Uranium	19.0	1.8
Zirconium	6.5	1.9

Source: Adapted from Benedict, G., *Nontraditional Manufacturing Processes*, Marcel Dekker, New York, 1987.

$$I_c = \frac{V_s}{R_e} = \frac{AV_s}{r_e a_g} \tag{14.3}$$

where

V_s = supply voltage
R_e = resistance due to the electrolyte
r_e = specific resistivity of the electrolyte
a_g = gap between tool and work surfaces
A = area of work surface exposed to electrolysis

An expression for the gap width may now be obtained from Equation 14.2 and Equation 14.3:

$$a_g = \frac{\eta_e V_s}{r_e \rho v_f} \tag{14.4}$$

Equation 14.4 shows that as the feed speed is increased, the gap width decreases. A smaller gap between the tool and the workpiece means higher accuracy of reproduction; therefore a higher feed speed (and hence a higher rate of removal) gives greater accuracy.

From Equation 14.4 it is also seen that an increase in the applied voltage V_s increases a_g and hence reduces the accuracy of reproduction. However, as V_s is progressively increased with a given feed speed v_f (hence a constant current, by Equation 14.2), the increase in gap width a_g causes an increase in resistance. The increased heat generation and consequent temperature rise lead to an increase in resistivity r_e so that the linear relationship between a_g and V_s no longer holds; the effect of this nonlinearity is to reduce the increase of a_g with V_s at constant v_f [11].

The supply voltage commonly used in ECM ranges from 5 to 20 V [11], the lower values being used for finish machining and the higher values for rough machining.

14.7.4 TOOLS FOR ELECTROCHEMICAL MACHINING

Tools for ECM are made from materials resistant to the electrolyte and also relatively easy to machine. Commonly used materials are brass, copper, stainless steel, and titanium. Tool design is often based upon experience with the process.

If a vertical-sided tool electrode is used to form a cavity in a workpiece, the sides of the cavity produced will not be vertical. This lack of conformity is a result of the current flowing between the sides of the tool and the sides of the cavity. To eliminate this effect it is usual to cover the sides of the tool with insulating material [2]. When the sides of the cavity required are not parallel to the direction of tool feed, such insulation may be unnecessary. The desired cavity shape may then be obtained by correcting the tool form, with the amount of the correction determined from practical experience with the process.

A most important factor in ECM tool design is provision of a suitable passage through the tool for efficient electrolyte flow through the cutting gap and to prevent stagnation areas.

14.7.5 APPLICATIONS

The main applications of ECM are in the machining of complex shapes in difficult-to-machine materials such as those used in high-temperature service. The removal rate for such materials can be much higher than that obtained with conventional machining, if conventional machining can be used at all. The removal rate for softer and more ductile materials is lower than that possible with conventional machining processes, but even for softer, more ductile materials, ECM can be useful for producing complex shapes accurately.

The ECM process has been applied successfully to the following:

1. Machining of through holes of any cross section, usually by trepanning
2. Machining of blind holes with parallel sides
3. Machining of shaped cavities (i.e., forging dies)
4. Wire cutting of large slugs of metal
5. Machining of complex external shapes (i.e., turbine blade root profiles)

14.8 ELECTROLYTIC GRINDING

Electrolytic grinding is a modification of the ECM process described above [12]. The tool electrode consists of a rotating abrasive wheel, usually a metal-bonded diamond wheel that can conduct electricity. The electrolyte is fed between the wheel and the work surface in the direction of movement of the wheel periphery so that it is carried past the work surface by the action of the wheel rotation. The abrasive particles help to maintain a constant gap between the wheel and the workpiece. In addition, the abrasive continually removes electrically resistant films from the work surface, which otherwise tend to impede the electrolytic action. (See Figure 14.11.)

Thus, the metal removal rate can be expressed as

$$Z_w = Z_{we} + Z_{wa} \qquad (14.5)$$

where

Z_w = total metal removal rate

Z_{we} = metal removed electrolytically per unit time

Z_{wa} = metal removed by mechanical abrasion per unit time

The value of Z_{we} can be obtained from a knowledge of the current flowing, the electrochemical equivalent of the work material, and the current efficiency. For current densities up to approximately 1 MA/m^2 (700 A/in.^{2}), the current efficiency can be assumed to be 1.00. Usually in practice Z_{we} constitutes 90% and Z_{wa} 10% of the total metal removal rate Z_w [13].

In electrolytic grinding the predominance of the electrolytic action reduces wheel wear to a negligible amount and makes it possible to grind hard materials

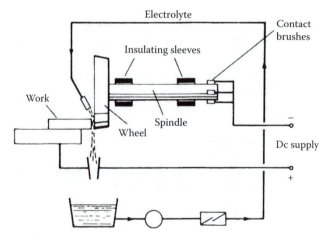

FIGURE 14.11 Electrolytic grinding.

rapidly. Moreover, the wheel can be used for long periods without dressing. The temperature rise and mechanical forces are very small, and hence thermal damage and surface cracks, common in conventional grinding, are eliminated [13]. The surfaces are also free from scratches and burrs.

In applying electrolytic grinding it is desirable to design the operation in such a way that the area of contact between the wheel and the work is as large as possible [10]. This large contact area gives the highest rate of removal for a given current density and allows full use to be made of the available current capacity. Thus, applications that would conventionally require multipass grinding can be performed with a single pass by electrolytic grinding.

Electrolytic grinding has been used very successfully in the grinding of hard materials, which must also be electrically conductive. One of the most successful applications of the process is in the grinding of carbide-tool inserts [1], where the absence of grinding cracks in the ground tip and the high rate of removal are a great advantage. The saving in wheel wear also constitutes a major economy. Surface finishes of the order of 0.1 μm (4 μin.) have been obtained when carbide inserts are ground.

14.9 ELECTRICAL-DISCHARGE MACHINING

Electrical-discharge machining (EDM), or spark machining, as it is also called, removes material with repetitive spark discharges from a pulsating dc power supply, with a dielectric flowing between the workpiece and the tool.

The simplified diagram in Figure 14.12 illustrates the principle of the EDM process. The tool is mounted on the chuck attached to the machine spindle whose motion is controlled by a servo-controlled feed drive. The workpiece is placed in a tank filled with a dielectric fluid; a depth of at least 50 mm (2 in.) over the work surface is maintained to eliminate the risk of fire. The tool and workpiece are connected to a pulsating dc power supply. Dielectric fluid is circulated under pressure by a pump, usually through a hole (or holes) in the tool electrode. A spark gap of about 0.025 to 0.05 mm (0.001 to 0.002 in.) is maintained by the servomotor.

In power supplies for EDM conventional solid-state rectifiers first convert the input power into continuous dc power. A bank of power transistors that are switched by a digital multivibrator oscillator circuit then controls the flow of this dc power. The high-power pulsed output is then applied to the tool and workpiece to produce the sparks responsible for material removal.

Each spark generates a localized high temperature on the order of 12,000°C in its immediate vicinity. This heat causes part of the surrounding dielectric fluid to evaporate; it also melts and vaporizes the metal to form a small crater on the work surface. Since the spark always occurs between the points of the tool and work that are closest together, the high spots of the work are gradually eroded, and the form of the tool is reproduced on the work. The flowing dielectric fluid carries the condensed metal globules, formed during the process, away. As the

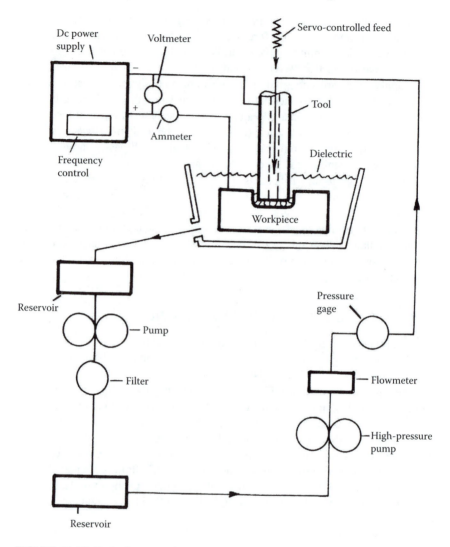

FIGURE 14.12 Basic features of electro-discharge machining.

metal is eroded, the tool is fed toward the workpiece by a servo-controlled feed mechanism.

Each pulse in the EDM cycle lasts for only a few microseconds. Repeated pulses, at rates up to 100,000 per second, result in uniform erosion of material from the workpiece and from the electrode. As the process progresses, the electrode is advanced by the servo drive toward the workpiece to maintain a constant gap distance until the final cavity is produced.

The servomechanism that controls the infeed of the tool into the workpiece, shown in its simplest form in Figure 14.13, consists of a reversible servomotor connected between one terminal of the condenser (D) and the slider of a variable

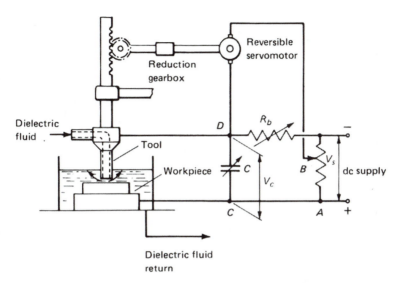

FIGURE 14.13 Basic features of EDM tool-feed control.

resistance (B). A and C are at the same potential; if the mean value of the voltage V_c (Figure 14.13) is equal to the preset voltage between A and B, the potentials at B and D are equal, and the servomotor remains stationary. If, however, the mean value of the voltage V_c changes because of changes in gap width, the potential at D will change, and the servomotor will rotate until the required gap is again obtained. In this way the servo system maintains the mean spark-gap voltage (and hence the gap width) at a constant value. The motor must be reversible because if the servomechanism overshoots or short-circuits as a result of the accumulation of swarf in the gap, the tool must be withdrawn. A constant gap of about 0.025 to 0.05 mm (0.001 to 0.002 in.) is usually maintained [1]; the exact gap value may be varied by changing the setting of the variable resistance. Other types of control used in practice usually operate by sensing either the gap voltage or the gap capacitance.

14.9.1 Tool Materials and Tool Wear

The EDM electrode shape determines the shape of the hole or cavity that is sunk into the workpiece. Many materials have been successfully used as EDM electrodes. The main requirements of a tool material for EDM are as follows:

1. Electrically conductive
2. Exhibits low wear rates
3. Is easily machinable
4. Provides good surface finish on the workpiece

Each spark discharge removes material from both the workpiece and the electrode, and consequently the electrode gradually wears during the machining

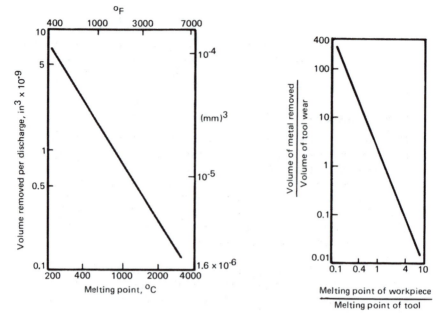

FIGURE 14.14 EDM metal removal and tool wear. (a) Volume removed per discharge as a function of melting point; (b) wear ratio as a function of workpiece to tool melting-point ratio.

process. The wear ratio, defined as the volume of metal lost from the tool divided by the volume of metal removed from the work, depends on the tool and work materials used. For example, with a brass tool and a brass workpiece the ratio is approximately 0.5, whereas for a brass tool and a hardened-carbon-steel workpiece it is approximately unity, and for a brass tool and a tungsten carbide workpiece the ratio may be as high as 3. With suitably chosen materials wear ratios as low as 0.1 can be achieved [1,14].

The volume of material removed per discharge decreases rapidly as the melting point of the material increases (Figure 14.14), thus high values of the wear ratio are obtained by having a high melting point for the tool material relative to the melting point of the work material [14]. Table 14.4 gives some properties of commonly used EDM tool materials. Graphite and copper graphite are by far the most common materials used. Graphite is easily machined and, because it vaporizes at very high temperatures relative to most metals, the wear rates of graphite electrodes are very low. Copper is added to graphite to form copper graphite, mainly to increase the conductivity of the electrode material. Copper and brass are also used as tool materials, but exhibit relatively high wear rates. Copper tungsten provides greatly improved wear rates but is a much more difficult material to machine. Tungsten wire is often preferred over other materials for drilling deep holes of small diameter because it can better withstand the tendency of the wire to buckle as a result of shock waves produced by the spark discharge or physical contact with the workpiece in the case of overshoot.

TABLE 14.4
Some Physical Properties of Electrical Discharge Machining (EDM) Electrode Materials

Property	Copper	Graphite	Tungsten	Iron
Melting point °C	1,083	a	3,395	1,535
Boiling point °C	2,580	>4,000	>5,930	>2,800
Heat to vaporize from room temperature, cal/cm³	12,740	About 20,000	About 22,680	About 16,900
Thermal conductivity, Ag = 100%	94.3	30.3	29.6	16.2
Electrical conductivity, Ag = 100%	96.5	0.1	48.1	16.2
Strength, N/mm²	241	34	3,134	275
Modulus of elasticity, N/mm² × 10³	124	5.9	351	186

a Sublimes or boils before melting at atmospheric pressure.

Source: From Kalpakjian, S. and Schmid, S.R., *Manufacturing Processes for Engineering Materials,* 4th ed., Prentice-Hall, New York, 2002.

14.9.2 DIELECTRIC FLUID

The dielectric fluid used must remain nonconducting until breakdown occurs; when the critical voltage is reached, it must break down rapidly and then deionize rapidly as each spark is discharged. The latent heat of vaporization of the dielectric fluid must be high so that only a small quantity vaporizes and the spark channel is confined to a small area. The dielectric fluid must have sufficiently low viscosity to flow easily and remove the metal globules effectively from the working zone. The dielectric fluid also acts as a coolant to carry away the heat generated by each spark.

The dielectric fluids commonly used are paraffin oil and transformer oil. Both these fluids are hydrocarbons, and it has been found [15] that the hydrogen in these fluids provides the deionizing action necessary for the fluid to become an effective insulator after each discharge.

The dielectric fluid may be fed through nozzles near the working zone, but where possible, and especially when deep holes are drilled, the fluid is fed or drawn through the tool itself. In electrode design it is necessary to provide sufficient passages through the tool to ensure adequate flushing of the tool-workpiece gap and prevent debris buildup, which can result in short circuits. Before the dielectric fluid is recirculated to the working zone, it is necessary to filter it to remove the metal particles produced in the process.

14.9.3 PROCESS PARAMETERS

Metal removal rates, surface finish, and accuracy are influenced mainly by the choice of electrical parameters. As the current is increased, each individual spark removes a large crater of workpiece material (Figure 14.15a), which increases the

(a)

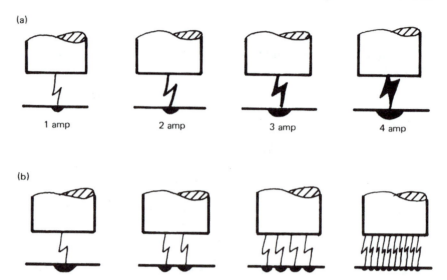

(b)

FIGURE 14.15 Influence of current and pulse frequency on metal removal and surface finish. (Adapted from Benedict [1].)

metal removal rate but also increases the surface roughness. Similar effects occur with increased spark voltage. Increasing the spark frequency, while keeping the other parameters constant, results in a decrease in surface roughness (Figure 14.15b) because the energy available is shared between more sparks, and smaller-sized surface craters are produced in the workpiece. The frequency range of modern EDM machines is from 180 Hz, for roughing cuts, to several hundred kilohertz, for fine finishing. When the sparking frequency becomes very high, the dielectric fluid cannot deionize at a sufficiently high rate, placing an upper limit on the frequencies possible. Volumetric removal rates vary from 0.001 to 0.1 cm^3/hr [1].

The accuracy of the process is closely related to the spark-gap width; the smaller the gap, the higher the accuracy, but a smaller gap results in a lower working voltage and a lower removal rate. A compromise is thus necessary. Typical spark gaps range from 0.0 12 to 0.05 mm. As the gap decreases, efficient flushing of the gap becomes more difficult to achieve.

Other factors affecting the accuracy of the process are the accuracy of the tool electrode itself and the allowance made on it for the spark gap. A cause of inaccuracy when deep parallel-sided cavities and holes are produced is the taper produced by sparks occurring between the sides of the electrode and the sides of the cavity. This sparking is assisted by any eroded particles that pass between the electrode and the sides of the cavity and effectively reduces the breakdown voltage of the dielectric fluid. The taper produced in this manner, on the order of 0.05 mm (0.002 in.) per side on the total depth of the cavity, becomes an asset in the machining of molds, blanking dies, and so on, where a small draw angle is required [2]. If the taper is not desirable, the tool form can be corrected by suitably relieving it behind the tip.

The spark-machined surface, having a matte appearance similar to a shot-blasted surface, consists of very small spherical craters as a result of the metal being removed by individual sparks. The finish is therefore nondirectional and very suitable for holding a lubricant [16]. Surface finishes of 0.25 μm (10 μin.) and better have been obtained. A layer of melted and resolidified material, known as *recast*, is left on the surface produced by EDM. This layer tends to be very hard and brittle (greater than $65R_c$ hardness). This recast layer is usually from 0.0025 to 0.05 mm in thickness and may have to be removed by other processes if high levels of fatigue resistance are required.

14.9.4 APPLICATIONS

Electrical-discharge machining can be used for all electrically conducting materials regardless of hardness. The process is most suited to the sinking of irregularly shaped holes, slots, and cavities. Fragile workpieces can be machined without breakage. Holes can be of various shapes and can be produced at shallow angles in curved surfaces without problems of tool wander [1].

The EDM process finds greatest application at present in tool making, particularly in the manufacture of press tools, extrusion dies, forging dies, and molds. Graphite electrodes produced by computer numerical control (CNC) machining or by copy milling from patterns are often used.

A great advantage of EDM is that the tool or die can be machined after it is hardened and hence great accuracy can be achieved. Tools of cemented carbide can be machined after final sintering, which eliminates the need for an intermediate partial sintering stage, thus eliminating the inaccuracies resulting from final sintering after holes, slots, and so on, are machined [15].

Electrical-discharge machining can be used effectively to drill small high-aspect-ratio holes. Diameters as small as 0.3 mm in materials 20 mm or more in thickness can be readily achieved. With efficient flushing, holes with aspect ratios as high as 100:1 have been produced. The process has been used successfully to produce very-small-diameter holes in hardened fuel-injector nozzles. Varying numbers of holes in a precise pattern can be drilled around the injector tip [1].

14.10 WIRE ELECTRICAL-DISCHARGE MACHINING

Wire electrical-discharge machining (wire-EDM) is an adaptation of the basic EDM process, which can be used for cutting complex two- and three-dimensional shapes through electrically conducting materials. Wire-EDM utilizes a thin, continuously moving wire as an electrode (Figure 14.16). It is a relatively new process and applications have grown rapidly, particularly in the tool making field [16].

The wire electrode is drawn from a supply reel and collected on a take-up reel. This continuously delivers fresh wire to the work area. The wire is guided by sapphire or diamond guides and kept straight by high tension, which is important to avoid tapering of the cut surface.

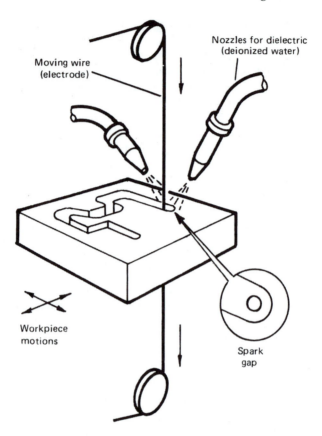

FIGURE 14.16 Basic features of wire-EDM.

High-frequency dc pulses are delivered to the wire and workpiece, causing spark discharges in the narrow gap between the two. A stream of dielectric fluid is directed, usually coaxially with the wire, to flood the gap between the wire and the workpiece. The power supplies for wire-EDM are essentially the same as for conventional EDM, except the current-carrying capacity of the wire limits currents to less than 20 A, with 10 A or less being most normal. In addition, the spark frequencies are higher, up to 1 MHz, to give a fine surface finish on the workpiece.

The workpiece is moved under CNC (Figure 14.17) relative to the wire, and this enables complex-shaped profiles to be cut through sheet and plate materials. Many machines incorporate further angular positioning of the wire, thus allowing varying degrees of taper on the cut surface to be obtained. Adaptive control, based on gap–voltage sensing, is necessary to avoid contact between the wire and the work material. Short circuits must be sensed and the wire backed off along the programmed path to reestablish the correct gap for efficient cutting.

Deionized water is the dielectric used for wire-EDM because it has low viscosity, presents no fire hazard, and results in high cooling rates and high

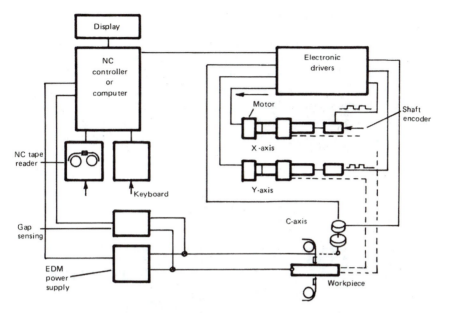

FIGURE 14.17 Schematic of wire-EDM system.

material removal rates [1]. A low viscosity is important to ensure adequate flow through the wire-workpiece gap to ensure efficiency of flushing. Copper and brass wires are commonly used for the electrodes, when the wire diameter is relatively large (0.15 to 0.30 mm). If very fine wire is required for high-precision cutting (0.03 to 0.15 mm), molybdenum-steel wire is used for increased strength. The wire-workpiece gap usually ranges from 0.025 to 0.05 mm, and consequently the kerf width of cut made by wire-EDM is less than 0.1 mm plus the wire diameter.

Wire-EDM is most commonly used for the fabrication of press stamping dies, extrusion dies, powder compaction dies, profile gages, and templates. Complicated cutouts can be made in difficult-to-machine metals without the need for high-cost grinding or expensive shaped EDM electrodes. Linear cutting rates are relatively low, ranging, for example, from 38 to 115 mm/hr in 25 mm thick steel. However, the linear speed is dependent on the material being cut and not on the shape of the cut. Although wire-EDM is a relatively slow cutting process, this is compensated for by the complexity and accuracy of the profiles that can be cut. Machines for wire-EDM are designed to operate unattended for long periods of time (often for several days).

Accuracies in the cut-out profile of ±0.007 mm can be routinely obtained with wire-EDM, and this figure can be further reduced with special care, including ensuring uniformity of the wire diameter. The minimum internal corner radius possible is limited only by the wire diameter being used. External corners can be produced with radii as small as 0.038 mm.

14.11 LASER-BEAM MACHINING

Laser-beam machining can be used for the cutting, drilling, slotting, and engraving of most materials. Material is removed rapidly by means of a high-energy coherent beam of light of a single wavelength. A typical 1 cutting-head configuration is shown in Figure 14.18. The lens to a short distance from the nozzle of the cutting head focuses the beam. The high, concentration of light at the focus produces energy densities as high as 106 W/mm² and lasers of less than 100 W power are capable of vaporizing or melting most materials [1]. Cutting and drilling

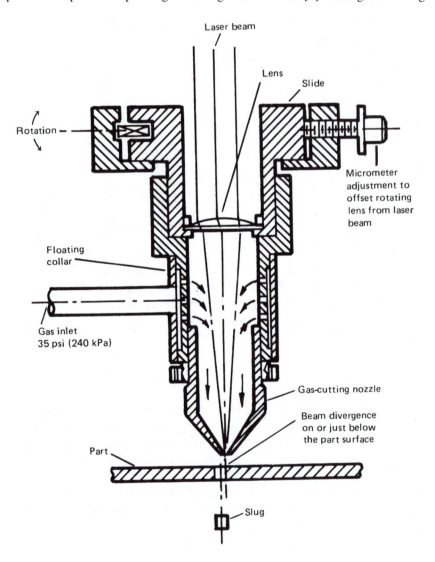

FIGURE 14.18 Typical laser cutting-head configuration.

TABLE 14.5
Industrial Laser Characteristics

Laser Type	Mode of Operation	Power (W)
Ruby	Pulsed	50
Nd:glass	Pulsed	50
Nd:YAG	CW/pulsed	600
Alexandrite	CW/pulsed	100
CO_2	CW/pulsed	25,000

Source: From Benedict, G., *Nontraditional Manufacturing Processes*, Marcel Dekker, New York, 1987.

can be achieved with no force applied to the workpiece and with only a small heat-affected zone.

14.11.1 TYPES OF LASERS

Table 14.5 lists the main types of lasers used for drilling and cutting [1]. They differ mainly in the power output and wavelength of light used. A high proportion (over 70%) of industrial applications use either Nd:YAG or CO_2 lasers, mainly because of the higher power capabilities. The higher power lasers are capable of both pulsed and continuous generation. High-rate pulsing or continuous operation is required for cutting applications.

14.11.2 PERCUSSION DRILLING OF SMALL HOLES

Percussion drilling is the simplest technique for producing small holes. The laser beam is focused near to the workpiece surface and a short duration pulse from the laser causes a small volume of material to melt and partially vaporize. The explosive escape of the vapor removes most of the melted material in the form of a spray of molten droplets. Multiple pulses can be used to penetrate thick materials that cannot be drilled with one pulse.

Holes produced by this process are generally tapered (Figure 14.19) and are only approximately round, but the hole size is highly repeatable. The depth of recast layer on the surface of the hole produced is very small. Holes can be drilled in times varying from 1 ms to 2 s dependent on the material thickness; the maximum hole diameter that can be drilled is about 1.3 mm. Larger holes require defocusing of the beam, which decreases the illumination intensity and produces uneven melting of the workpiece material.

14.11.3 TREPANNING OR CUTTING OPERATIONS

In the cutting or trepanning mode of operation, the laser pulses rapidly or operates continuously and a jet of gas is introduced through the cutting head around the

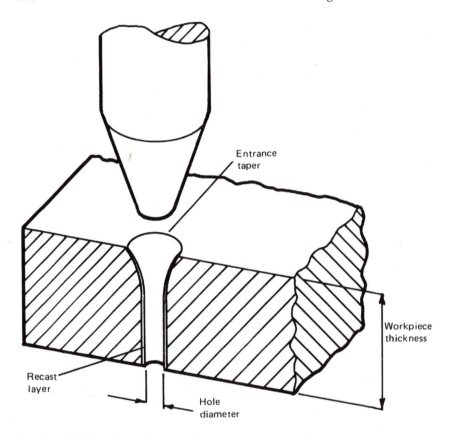

FIGURE 14.19 Hole configuration for laser percussion drilling.

focused beam (Figure 14.18). The jet of gas can assist the melting of the material and facilitate the clearing of molten material from the cut, together with keeping debris from contaminating the focusing lens. The most commonly used gases are oxygen, air, and argon. Oxygen is used for materials that oxidize such as carbon steels. The oxygen assists the melting process and results in high cutting speeds, particularly for thin materials (Figure 14.20). The cut edges left are oxidized and have a significant heat-affected zone. Air is much less expensive to use than oxygen, but cutting rates are reduced because of the much lower oxygen content. Argon is used when shielding of the cut surfaces from oxidation is required but results in cutting speeds about 25% lower than when using air.

In general there is a maximum recommended cutting speed for laser machining that depends on the material thickness and the laser power. Lower cutting rates should be used for better quality of the cut surfaces. Examination of the empirically determined recommended cutting conditions for different materials enables the following relationship for laser cutting speeds to be determined:

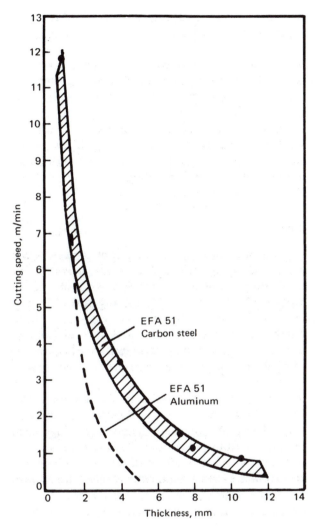

FIGURE 14.20 Cutting speeds for aluminum and carbon steel for a 100 W CO_2 laser. (Adapted from Benedict [1].)

$$v_{ls} = \left(q_1 Ln(T_s) - q_2\right)\left(q_3 Ln\left(P_l / q_4\right) + q_5\right) \qquad (14.6)$$

where

v_{ls} = cutting speed (m/min)

T_s = sheet or plate thickness (mm)

P_l = laser power (watts)

$q_1, \ldots, q_5$ = constants dependent on material and surface quality required

This expression can be used for the cutting of sheet and plate materials, where plate material is defined as having a thickness of greater than 6.25 mm. Different values of the constants q_1, ..., q_5 are applicable for the cutting speed for best surface quality and the maximum recommended cutting speed. Representative values of these constants are given, for different types of materials, in Table 14.6 for sheet materials and Table 14.7 for plate materials.

If the start of the cut path is not at the edge of the sheet, a starting hole must first be pierced, and the time for piercing is given by:

$$t_{ps} = q_{ps} / v_{ls_{max}} \qquad (14.7)$$

where

t_{ps} = piercing time (min)

q_{ps} = constant dependent on material

$v_{ls_{max}}$ = maximum recommended cutting speed obtained using Equation 14.6

The values of the constant q_{ps} for different material types are given in Table 14.6 and Table 14.7.

Thus the total machining time for a laser cutting operation is given by:

$$t_{ls} = t_{ps} + l_w / v_{ls} \qquad (14.8)$$

where

t_{ls} = total laser cutting time (min)

l_w = the length of the cut path

Laser cutting machines are often fitted with numerical control for moving the workpiece relative to the workhead. These machines are able to cut complex profiles and holes with high precision and high cutting speeds; without the requirement for complex tooling, kerf widths as low as 0.64 mm can be achieved in thin materials.

14.11.4 APPLICATIONS

Lasers are used for a variety of drilling and cutting applications. A wide range of materials can be cut with high precision and a small heat-affected zone. A typical drilling application is the formation of a pattern of small holes in gas turbine blades that provide cooling airflow through the blade surface. The small-diameter holes are readily produced in the high-temperature-resistant materials used for the blades. In addition, the holes can be drilled at shallow angles to the surface, without the problems of tool breakage and slippage that result with other mechanical methods of drilling.

TABLE 14.6
Data for Estimating Laser Cutting Speeds for Sheet Materials (Thickness less than 6.25 mm)

Material	Piercing Time Constant, q_{ps}	Best Surface Quality					Maximum Cutting Speed				
		q_1	q_2	q_3	q_4	q_5	q_1	q_2	q_3	q_4	q_5
Carbon and alloy steels	0.18	−3.92	−9.59	0.53	2200	1.00	−4.55	−11.1	0.61	2200	1.16
Stainless steels	0.22	−3.33	−8.15	0.53	2200	1.00	−3.87	−9.47	0.61	2200	1.16
Copper alloys	0.37	−1.96	−4.79	0.53	2200	1.00	−2.28	−5.57	0.61	2200	1.16
Aluminum alloys	0.46	−1.57	−3.83	0.53	2200	1.00	−1.82	−4.41	0.61	2200	1.16
Magnesium alloys	0.37	−1.96	−4.79	0.53	2200	1.00	−2.28	−5.57	0.61	2200	1.16
Titanium alloys	0.37	−1.96	−4.79	0.53	2200	1.00	−2.28	−5.57	0.61	2200	1.16
Nickel alloys	0.30	−2.85	−6.97	0.45	2200	0.86	−3.31	−8.10	0.52	2200	1.00

TABLE 14.7
Data for Estimating Laser Cutting Speeds for Plate Materials (Thickness more than 6.25 mm)

Material	Piercing Time Constant, q_{ps}	Best Surface Quality					Maximum Cutting Speed				
		q_1	q_2	q_3	q_4	q_5	q_1	q_2	q_3	q_4	q_5
Carbon and alloy steels	0.18	-1.80	-5.63	0.86	2200	1.00	-2.09	-6.55	1.00	2200	1.16
Stainless steels	0.22	-1.53	-4.79	0.86	2200	1.00	-1.78	-5.57	1.00	2200	1.16
Copper alloys	0.37	-0.90	-2.82	0.86	2200	1.00	-1.04	-3.28	1.00	2200	1.16
Aluminum alloys	0.46	-0.72	-2.26	0.86	2200	1.00	-0.84	-2.62	1.00	2200	1.16
Magnesium alloys	0.37	-0.90	-2.82	0.86	2200	1.00	-1.04	-3.28	1.00	2200	1.16
Titanium alloys	0.37	-0.90	-2.82	0.86	2200	1.00	-1.04	-3.28	1.00	2200	1.16
Nickel alloys	0.30	-1.31	-4.10	0.74	2200	0.86	-1.52	-4.77	0.86	2200	1.00

Laser cutting machines are now used quite extensively for cutting complex profiles in sheet and plate materials. Hard materials can be cut without the problems of tool wear. A typical application is the high-speed blanking of radial saw blades from plate stock. This can be achieved without the need for complex tooling. Numerous different profiles can be stored in the CNC positioning systems used and recalled as required [18].

14.12 ELECTRON-BEAM MACHINING

In electron-beam machining, material is removed by means of a focused beam of high-velocity electrons that strike the workpiece. The kinetic energy of the electrons is converted into heat that is sufficient to cause rapid melting and vaporization of the workpiece material. Figure 14.21 illustrates the main elements of an electron-beam machine. To eliminate scattering of the beam of electrons, the work is done in a high-vacuum chamber (10^{-5} mm of mercury or less). Electrons are emitted from an electron gun and are accelerated to speeds of about 75% of the speed of light, by voltages as high as 50 kV between the anode and cathode. A magnetic lens focuses the electron beam onto the workpiece, and energy densities of the order of 10^8 W/cm^2, which is sufficient to melt and vaporize any material, are produced.

The action of electron-beam drilling is illustrated in Figure 14.22 [1]. The focused beam from an electron pulse strikes the workpiece and melting and vaporization of some of the material occurs. The pressure of the escaping vapor is sufficient to form and maintain a small capillary channel into the work material.

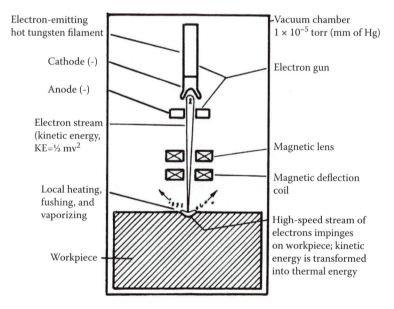

FIGURE 14.21 Basic features of electron-beam machining.

   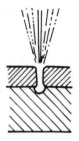

The sharply focused electron beam strikes the material to be drilled: local heating and fusion of the surface	The high power density of the electron beam produces a deepening metal vapor capillary encircled by molten material	The electron beam has drilled through the material being pierced and penetrates up to a certain depth into the auxilliary material	As a result of the high vapor pressure of the auxilliary material, the molten part of the material being drilled is expelled

FIGURE 14.22   Action of electron-beam drilling.

The beam and capillary channel rapidly penetrate the workpiece until an organic or synthetic backing material is reached at which time the backing material vaporizes rapidly with an explosive release of vapor. The relatively high vapor pressure of the backing material vapor expels the molten walls of the capillary in a shower of sparks leaving a hole in the work material.

A single pulse is often used to produce a single hole, but for thick materials multiple pulses may be required. Pulsing rates of up to 100 pulses per second are possible and deflection coils under program control can be used to move the beam around to cut holes of any shape.

Electron-beam machining can process a wide range of materials. As a result of the extremely high energy density of the beam and the short duration of beam-workpiece interaction, thermal effects on the workpiece material are limited to a heat-affected zone that seldom exceeds 0.025 mm in depth. The high beam-power density also enables high aspect ratio holes to be drilled, often as large as 15 to 1. For example, hole diameters of 0.1 to 1.4 mm can be drilled in materials up to 10 mm thick. Because no force is applied to the workpiece, brittle and fragile materials can be processed without fracture. Holes can be drilled at shallow angles to the work surface, as low as 20 degrees if required. A minimum spacing between holes of twice the diameter of the hole is normal, but this restriction still allows up to 1000 holes per square centimeter to be drilled, if required.

The most suitable application of electron-beam machining is for workpieces requiring large numbers of simple small holes to be drilled, or for drilling of holes in materials that are hard and difficult to machine by other processes. A typical example of the application of electron-beam machining [1] is the drilling of turbine engine combustor domes in Cn Nc Co Mo W steel. These parts have a wall thickness of 1.1 mm, and almost 4000 holes of 0.9 mm diameter are drilled in 60 mm by electron-beam machining.

The equipment for electron-beam machining is very expensive, and the necessity to work in a vacuum adds considerably to floor-to-floor machining time.

However, for certain highly specialized workpieces, electron-beam machining may be the only feasible method of production.

14.13 PLASMA-ARC CUTTING

14.13.1 General Discussion

Plasma-arc cutting is used mainly for cutting thick sections of electrically conductive materials. A high-temperature plasma stream interacts with the workpiece causing rapid melting. Any gas heated to extremely high temperatures dissociates into free electrons, ions, and neutral atoms and this condition is known as plasma. In the plasma state the material has electrical conductivity, and the condition can be sustained by the application of electrical power.

The basic configuration of a plasma-arc torch is shown in Figure 14.23. The electrically conductive workpiece material is positively charged and the electrode is negatively charged. An electric arc is maintained between the electrode and the workpiece. Gas flowing coaxially is heated to form the plasma stream, which has a peak temperature of approximately 33,000°C. Air and nitrogen are the most commonly used gases for the plasma stream. The gas flow is delivered to the nozzle at pressures up to 1.4 MPa, producing a plasma stream that flows toward the workpiece at several hundred meters per second. The secondary gas flow around the nozzles helps constrain the plasma flow into a narrow jet and cleans the cut of molten material. The highest quality cuts can be obtained by injecting water around the main plasma jet. Some of the water vaporizes to produce a thin layer of steam that constricts the plasma. The benefits of this water constriction include a smaller heat-affected zone and increased nozzle life.

14.13.2 Applications

Using plasma-arc cutting, materials up to 200 mm in thickness can be cut [1]. However, most practical applications are for materials in the thickness range of 3 to 75 mm. Relatively high cutting speeds can be achieved, depending on the thickness and type of material being cut. Table 14.8 shows some typical cutting speeds for different materials. Tolerances of ±0.8 mm can be achieved in materials of thickness less than 25 mm, and tolerances of ±3 mm are obtained for greater thickness. The heat-affected zone for plasma-arc cutting varies between 0.7 and 5 mm in thickness.

In general there is a maximum recommended cutting speed for plasma-arc machining that depends on the material thickness and the plasma current. Lower cutting rates should be used for better quality of the cut surfaces. Examination of the empirically determined recommended cutting conditions for different materials enables the following relationship for plasma cutting speeds to be determined:

$$v_{pl} = p_1 T_s^{p_2} \left(p_3 Ln \left(I/p_4 \right) + p_5 \right) \tag{14.9}$$

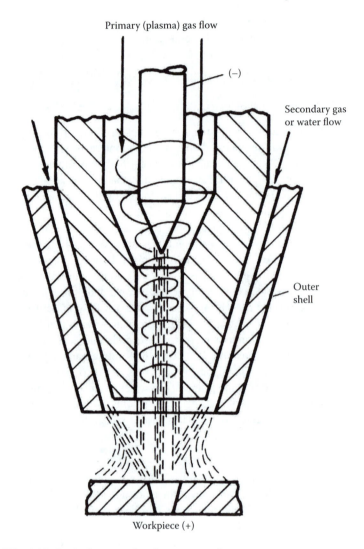

FIGURE 14.23 Basic features of a plasma-arc tool.

where

v_{pl} = cutting speed (m/min)

T_s = sheet or plate thickness (mm)

I = plasma current (A)

$p_1, ..., p_5$ = constants dependent on material and surface quality required

Different values of the constants $p_1, ..., p_5$ are applicable for the cutting speed for best surface quality and the maximum recommended cutting speed. Values of these constants are given for different types of material in Table 14.9 for sheet materials.

TABLE 14.8
Plasma-Arc Cutting Speeds

Material	Thickness (mm)	Arc Current (A)	Cutting Speed (mm/min)
Stainless steel	75	800	380
	130	1000	150
Aluminum	75	900	760
	180	1000	180
Brass	13	400	1780
Titanium	13	400	2285

If the start of the cut path is not at the edge of the sheet, a starting hole must first be pierced and the time for piercing is given by:

$$t_{ps} = q_{ps}/v_{pl_{max}} \qquad (14.10)$$

where

t_{ps} = piercing time (min)

q_{ps} = constant dependent on material

$v_{pl_{max}}$ = maximum recommended cutting speed obtained using Equation 14.9

The values of the constant q_{ps} for different material types are given in Table 14.9. Thus the total machining time for a plasma cutting operation is given by:

$$t_{pl} = t_{ps} + l_w/v_{pl} \qquad (14.8)$$

where

t_{pl} = total plasma-arc cutting time (min)

l_w = the length of the cut path

Most applications of plasma-arc cutting are in the metal fabrication and metal plate industries. A wider variety of materials can be cut than for flame cutting with an oxyacetylene torch. Plasma-arc nozzles can be fitted to cutting machines with both numerical and optical-line-following controls so that relatively complex profiles can be readily cut.

14.14 COMPARATIVE PERFORMANCE OF CUTTING PROCESSES

As described in this chapter, a variety of cutting processes are available in addition to the more conventional hard-tool and abrasive machining processes. Some of

TABLE 14.9
Data for Estimating Plasma Arc Cutting Speeds for Different Materials

Material	Piercing Time Constant, q_{ps}	Best Surface Quality					Maximum Cutting Speed				
		p_1	p_2	p_3	p_4	p_5	p_1	p_2	p_3	p_4	p_5
Carbon and alloy steels	0.01	21.9	−0.95	0.52	400	1.00	37.2	−1.09	0.45	400	1.00
Stainless steels	0.01	108.5	−1.38	0.56	400	1.00	226.5	−1.55	0.57	400	1.00
Copper alloys	0.01	21.4	−0.95	0.51	400	0.99	36.2	−1.09	0.51	400	0.99
Aluminum alloys	0.01	60.0	−1.10	0.53	400	1.00	91.4	−1.14	0.56	400	1.00
Magnesium alloys	0.01	23.3	−0.95	0.52	400	1.07	39.4	−1.09	0.52	400	1.07
Titanium alloys	0.01	23.3	−0.95	0.52	400	1.07	39.4	−1.09	0.52	400	1.07
Nickel alloys	0.01	21.4	−0.95	0.51	400	0.99	36.2	−1.09	0.51	400	0.99

these processes are only suitable for the machining of particular types of workpieces and applications, but some general comparison of the relative performance of the available processes is useful.

Figure 14.24 compares the surface finish and tolerances obtainable by various material removal processes. However, the rate at which material can be removed and the costs of removing material are also important considerations in the selection of processes. Table 14.10 [19] compares typical metal removal rates for some of the available processes, together with the approximate power consumption per unit volume of material removed. It can be seen that conventional machining processes in general remove material faster and with less power consumption than the less conventional processes. Table 14.10 also shows values of the ratio of the capital equipment costs to the metal removal rates, relative to conventional machining. Again it can be seen that the capital costs associated with removing material by nonconventional processes are much higher than for conventional machining. However, many of the nonconventional processes may be the only viable way to produce some workpieces and shape features, particularly in hardened materials.

PROBLEMS

1. Calculate the feed speed possible when drilling a 25-mm-diameter hole with electrochemical machining when a supply current of 5 kA is available. Assume that the electrochemical equivalent of the work material is 0.2 m.g/As, that its density is 8 M . g/m^3, and that a current efficiency of 80% is obtainable.

2. Estimate the time taken to grind a 1-mm layer from the face of a carbide insert 15 mm square by electrolytic grinding. Assume that the current efficiency is 0.95, that the mass of metal removed mechanically is 10% of the total mass of metal removed, that a current density of 1 MA/m^3 is used, and that the ratio of the electrochemical equivalent to the density of the work material is 26 p . m^3/As.

3. A 2000 W CO_2 laser cutting machine is to be used to cut some parts out of sheet and plate materials. The parts are all disks of 100 mm diameter with three holes of 15 mm diameter arranged equally spaced on a pitch circle of 50 mm diameter. Estimate the total cutting time for the following parts:
   a. A disk part cut from stainless steel sheet material 3 mm thick using cutting speeds that give the best surface quality.
   b. A disk part cut from stainless steel sheet material 3 mm thick using the maximum recommended cutting speeds.
   c. A disk part cut from aluminum alloy sheet material 3 mm thick using cutting speeds that give the best surface quality.
   d. A disk part cut from aluminum alloy plate material 8 mm thick using cutting speeds that give the best surface quality.

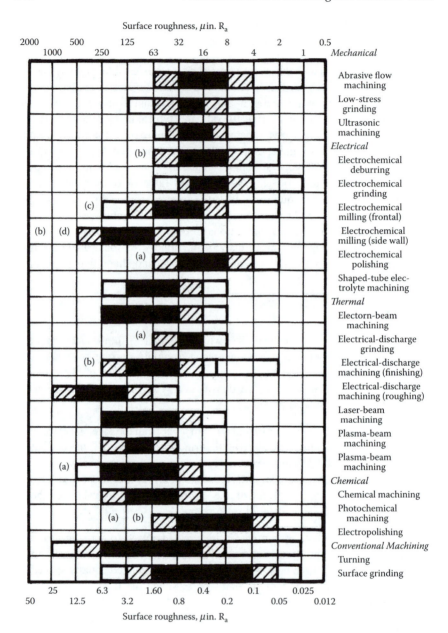

FIGURE 14.24 Surface roughness and tolerances for various material removal processes. (From Kalpakjian and Schmid [14].)

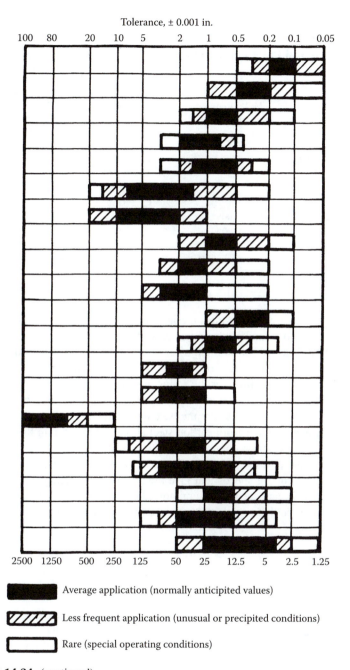

Tolerance, ± 0.001 in.

FIGURE 14.24 (continued).

TABLE 14.10
Typical Removal Rates and Capital Costs for Various Machining Processes

Process	Metal Removal Rate (cm³/s)	Approximate Unit Power (Specific Cutting Energy) (GJ/m³)	Relative Capital Equipment Cost Per Unit Material Removal Rrate
Edge-tool machining	14	3	1
Abrasive machining	8	30	2
Plasma jet	1.5	10	—
ECM	1.0	500	60
EDM	0.1	150	40
Ultrasonic machining	0.005	150	600
Electron beam	0.001	1,500	–40,000
Abrasive jet	0.0001	10,000	—
Laser	0.0001	150,000	160,000

Source: Adapted from Wager, J.G., *Proc. 2nd Int. Conf. on Prod. Res.*, Copenhagen, 1973, 231.

4. A plasma-arc cutting machine is to be used to cut some parts out of sheet and plate materials using a plasma current of 800 amps. The parts are all disks of 200 mm diameter with four holes of 20 mm diameter arranged equally spaced on a pitch circle of 100 mm diameter. Estimate the total cutting time for the following parts:

   a. A disk part cut from carbon steel sheet material 5 mm thick using cutting speeds that give the best surface quality.

   b. A disk part cut from carbon steel sheet material 5 mm thick using the maximum recommended cutting speeds.

   c. A disk part cut from aluminum alloy sheet material 5 mm thick using cutting speeds that give the best surface quality.

   d. A disk part cut from aluminum alloy plate material 15 mm thick using cutting speeds that give the best surface quality.

REFERENCES

1. Benedict, G., *Nontraditional Manufacturing Processes*, Marcel Dekker, New York, 1987.

2. Snoeys, R., Stuchens, I., and Dekeyser, W., Current Trends in Non-conventional Material Removal Processes, *Ann. CIRP*, Vol. 35, 2, 467, 1986.

3. Bellows, G. and Kohls, J.B., Drilling without Drills, *American Mach.*, Special Report No. 743, 187, March, 1982.

4. Bradford, J.D. and Richardson, D.B., *Production Engineering Technology*, 3rd ed., Macmillan, London, 1980.

5. Hashish, M. and Reichmen, J.M., Analysis of Waterjet Cutting at High Traverse Rates, *Proc. 5th Int. Symp. on Jet Cutting Technol.*, BHRA Fluid Engineering, Cranfield, England, 57, 1980.

6. Koenig, W. and Wulf, C., Wasserstrahlschneiden, *Industrie Anzeiger*, Vol. 106, no. 92, 1984.

7. Hoogstrate, A.M. and Lutterfelt, C.A. van, Opportunities for Abrasive Water Jet Machining, *Ann. CIRP*, Vol 46/2, 1997.

8. Domrowski, T.J., The How and Why of Abrasive Jet Machining, *Modern Machine Shop*, 76, January, 1983.

9. Dini, J.W., Fundamentals of Chemical Milling, *American Mach.*, Special Report 7.68, 1984.

10. Williams, L.A., How to Apply Electrolytic Machining, *Tool Eng.*, 43, December, 1959.

11. Opitz, H., Electrical Machining Processes, *Int. Res. in Prod. Eng.*, 1963, 225, presented at the Int. Prod. Eng. Res. Conf., Pittsburgh, 1963.

12. Phillips, R.E., Electrochemical Grinding — What is it? How does it work?, *SME Technical Paper*, MR 85-383, 1985.

13. Krahacher, E.J., Haggerty, W.A., Allison, C.R., and Davis, M.F., Electrical Methods of Machining, *Int. Res. in Prod. Eng.*, 1963, 232, presented at the Int. Prod. Eng. Res. Conf., Pittsburgh, 1963.

14. Kalpakjian, S. and Schmid, S.R., *Manufacturing Processes for Engineering Materials*, 4th ed., Prentice-Hall, New York, 2002.

15. Rudorff, D.W., Principles and Applications of Spark Machining, *Proc. Inst. Mech. Eng.*, London, Vol. 171, 495, 1957.

16. Anon., *Spark Machining*, Machinery Publishing Co., Brighton, England, 1960.

17. Foster, R.F., Evolution of wire-EDM, *Tooling and Prod.*, 90, March, 1982.

18. Wilson, M.J., Laser Speeds Saw Cutting, *Production*, 52, June, 1984.

19. Wager, J.G., New Metal Shaping Processes-Product Demands, Research Avenues and Assessment Criteria, *Proc. 2nd Int. Conf. on Prod. Res.*, Copenhagen, 1973, 231.

Nomenclature

A	area of the work surface exposed to electrolysis in electrochemical machining (ECM); constant in tool damage equation (Equation 4.5)
A_a	apparent area of contact between two surfaces
A_c	cross-sectional area of the uncut chip; i.e., the cross-sectional area of the layer of material being removed by one cutting edge measured normal to the resultant cutting direction
A_g	relative contact area of grains during grinding, percent
A_m	area of machine surface
A_r	real area of contact between two surfaces
A_s	area of shear or area of shear plane
A_α	tool flank; i.e., the surface over which the surface produced on the workpiece passes
A_γ	tool face; i.e., the surface over which the chip flows
a	characteristic linear dimension
a_c	undeformed chip thickness; i.e., the thickness of the layer of material being removed by one cutting edge at the selected point measured normal to the resultant cutting force direction
$a_{c_{av}}$	mean undeformed chip thickness; i.e., the mean value of a_c
$a_{c_{max}}$	maximum undeformed chip thickness; i.e., the maximum value of a_c
a_{cr}	critical depth of cut (back engagement) for chip breaking
a_d	depth of dress for grinding wheel; i.e., the amount of penetration of the dressing tool; groove depth to be machined
a_e	working engagement; i.e., the instantaneous engagement of the complete tool with the workpiece, measured, measured in the working plane P_{fe}, and perpendicular to the direction of feed motion (previously known as depth of cut in a slab-milling operation)
a_f	feed engagement; i.e., the instantaneous engagement of the tool cutting edge with workpiece, measured in the working plane P_{fe} and in the direction of feed motion (in single-point machining operations it is equal to the feed f; in multipoint tool operations, it is equal to the feed per tooth)
a_g	gap between the tool and workpiece in ECM
a_h	in-phase component of harmonic receptance
$a_{h_{min}}$	maximum negative in-phase value of operative receptance
a_i	inclination angle of regulating wheel in centerless grinding
a_m	material constant relating recommended feed to depth of cut
a_o	chip thickness, i.e., the thickness of the chip produced during machining
a_p	back engagement, i.e., the instantaneous engagement of the complete tool with the workpiece, measured perpendicular to the working plane P_{fe} (previously known as depth of cut in a single-point tool operation and width of cut in a slab-milling operation
a_{p0}	reference depth of cut (back engagement) for recommended cutting speed estimation

a_r	ratio of the inner to the outer radius of surface to be faced; critical back engagement (depth of cut) for chip breaking under representative cutting conditions; rough grinding stock on radius of rotational workpiece
a_t	total depth of material to be removed in a machining operation
a_v	amplitude of vibration
$a_{v_{max}}$	maximum amplitude of vibration, i.e., the maximum value of a_v
a_w	width of chip, i.e., the width of the chip produced during machining or the width of the uncut chip
$a_{w_{max}}$	maximum chip width, i.e., the maximum value of a_w
B	constant in tool damage equation (Equation 4.5)
B_s	batch size for one set-up
b	index in uniaxial true stress true stain relationship; constant
b_h	out-of-phase component of harmonic receptance
b_w	width of the machined surface; width of the workpiece
b_{w_m}	width of cut along asymptote to stability chart
b_{w_o}	width of cut for unconditional stability
C	cutting speed for 1 min of tool life (in feet per min); constant in empirical stress strain relationship
C_b	cost of setting up and preparing for the machining of a batch of components
C_c	ratio of k_2/k_1, assumed a constant; grinding cost when recommended conditions are used
C_g	number of active grains per unit area on a grinding-wheel surface; production cost of grinding operation
C_m	cost of machining, neglecting nonproductive costs
C_{mat}	cost of material for one workpiece
C_{mim}	minimum cost of production; i.e., the minimum value of C_{pr}
C_{mt}	total machining cost
C_p	cost of machining when maximum power is used; grinding cost when maximum power is used
C_{pr}	production cost; i.e., the average cost of producing each component on one machine tool
C_s	velocity of sound
C_t	cost of sharp tool; i.e., the average cost of providing the machine operator with a sharp tool or cutting edge, including regrinding costs or the cost of the insert and the tool or tool-holder depreciation
C_w	cost per lb of workpiece material; wheel wear and wheel changing cost in grinding
c	specific heat capacity
c_d	damping force per unit velocity; i.e., the viscous damping constant
D	constant in tool damage equation (Equation 4.4)
dF_r	variation in resultant cutting force
dF_1	force variation in-phase with tool displacement
dF_2	force variation out-of-phase with tool displacement
d_a	outer diameter of workpiece
d_b	inner diameter of workpiece
d_e	equivalent wheel diameter in grinding
d_g	average diameter of the grains in a grinding wheel
d_m	diameter of machined surface

d_r	diameter of regulating wheel in centerless grinding
d_t	diameter of the cutting tool or abrasive wheel
d_{tr}	diameter of the transient surface
d_w	diameter of a cylindrical workpiece; diameter of a work surface
E	Young's modulus; process activation energy
E_g	proportion of total energy flowing into the workpiece during grinding
e	electrochemical equivalent of the work material
e_n	chip-breaker land width (Figure 8.9)
F	instantaneous value of the external harmonic force
F_c	cutting component of the resultant tool force, F_r
F_c'	cutting component of the resultant force F_r', acting on the chip-tool interface region
F_f	frictional force on the tool face; frictional force between sliding surfaces
F_{max}	maximum or peak value of external force per unit mass
F_n	normal force on the tool face; normal load between surfaces
F_{ns}	normal force on shear plane
F_0	maximum or peak value of external force per unit mass
F_p	plowing force
F_r	resultant tool force
F_r'	resultant force acting on the chip-tool interface region
F_s	force required to shear the work material on the shear plane
F_t	thrust component of the resultant force F_r
F_t'	thrust component of the resultant force F_r' acting on the chip-tool interface region
F_{r_0}	threshold thrust force in grinding; i.e., the minimum value of the thrust force F_t to give grinding by cutting
F_u	harmonic force in the u direction
F_v	harmonic force in the v direction
f	Feed; i.e., the displacement of the tool relative to the workpiece, in the direction of feed motion, per stroke or per revolution of the workpiece or tool
f_{cr}	critical feed for chip breaking
f_d	feed during the dressing of a grinding wheel
f_r	critical feed for chip breaking under representative cutting conditions
f_s	frequency (Hz)
f_0	reference or known recommended feed
G	plan-setting angle (Figure 7.14); constant in tool damage equation (Equation 4.4)
G_r	grinding ratio; i.e., the ratio of the volume of metal removed to the volume of wheel removed during grinding
H	elevation setting angle (Figure 7.14)
H_n	grinding-wheel hardness number
h	chip-breaker height (Figure 8.3)
I	current in ECM; current in plasma-arc machining
i	$\sqrt{-1}$
K	constant for a machining operation; can be regarded as the distance traveled by the tool in relation to the workpiece during the machining time t_m
K_{dm}	workpiece material effect coefficient for critical depth of cut for chip breaking
K_{dT}	cutting tool (insert) effect coefficient for critical depth of cut for chip breaking
K_{dv}	cutting speed effect coefficient for critical depth of cut for chip breaking

K_f	specific friction pressure
K_{fm}	workpiece material effect coefficient for critical feed for chip breaking
K_{fT}	cutting tool (insert) effect coefficient for critical feed for chip breaking
K_{fv}	cutting speed effect coefficient for critical feed for chip breaking
K_n	specific normal pressure
K_p	constant of programming per unit machining time
KT	crater depth (Figure 4.3)
k	thermal conductivity
k_m	factor to allow for machine overheads
k_o	factor to allow for operator overheads
k_1	chip thickness coefficient or cutting stiffness; constant representing wheel wear and wheel changing costs per unit metal removal rate in grinding
k_2	out-of-phase force coefficient; constant representing rough grinding time multiplied by metal removal rate
L	roll-setting angle (Figure 7.14)
l_c	length of chip
l_f	contact length between the chip and tool
l_m	length of machined surface
l_n	chip-breaker distance (Figure 8.3)
l_p	length of pathway between machines
l_r	length to diameter ration of workpiece
l_{rd}	distance traveled by forklift truck in responding to a request
l_s	length of shear plane
l_{st}	sticking-contact length
l_t	length of tool or broach
l_w	length of workpiece or hole to be machined; length of cut path or cut surface
l_0	dimensionless length of contact between the chip and tool
M	total machine and operator rate (cost per unit time), including machine depreciation M_t , operator's W_o and machine and operator overheads
M_s	rate for each station on a transfer machine, including overheads (cost per unit time)
M_t	machine-tool depreciation rate (cost per unit time)
M_t'	machine-tool rate including overheads (cost per unit time)
m	number of revolutions of the workpiece during spark-out in plunge cylindrical grinding; friction factor during plastic flow
m_c	mass of chip specimen
m_e	equivalent mass of a machine-tool system
N	number of teeth on the cutting tool
N_a	number of automatic machines serviced by one operator
N_b	batch size; i.e., the number of components in the batch to be machined
N_c	number of chips produced per unit time in grinding
N_d	average number of cutting edges in cut
N_s	number of stations on a transfer machine
N_t	number of tool changes necessary during the machining of a batch of components
NB	wear on the tool flank measured normal to the cutting direction
n	constant in Taylor's tool life equation
n_o	number of operations

n_r	frequency of reciprocation; rotational frequency of regulating wheel in centerless grinding
n_s	rotational frequency of a machine-tool spindle; number of shifts
n_{s_c}	rotational frequency of a machine-tool spindle for minimum production cost
$n_{s_{ef}}$	rotational frequency of a machine-tool spindle for minimum efficiency (maximum profit rate)
n_{s_p}	rotational frequency of a machine-tool spindle for minimum production time
n_t	rotational frequency of the cutting tool or abrasive wheel; number of tools used
n_w	rotational frequency of workpiece
n_y	amortization period (years)
P	heat source
P_e	electrical power consumed by the machine tool during a machining operation
P_f	rate of heat generation in the secondary deformation zone; assumed working plane (Figure 7.10)
P_{fe}	working plane (Figure 7.10)
P_g	tool-face orthogonal plane
P_l	laser power (watts)
P_m	power required to perform the machining operation
P_n	cutting edge normal plane (Figure 7.9 and Figure 7.11)
P_o	tool orthogonal plane (Figure 7.6)
P_p	tool back plane (Figure 7.8)
P_{pe}	working back plane (Figure 7.10)
P_r	rate of profit (cost per unit time); tool reference plane (Figure 7.8 and Figure 7.9)
P_{re}	working reference plane (Figure 7.10 and Figure 7.11)
P_s	rate of heat generation in the primary deformation zone; tool cutting edge plane (Figure 7.9)
P_{se}	working cutting edge plane (Figure 7.11)
p	index for recommended cutting speed relationship to depth of cut
p_s	specific cutting energy; i.e., the work required to remove a unit volume of material
$p_1, ..., p_5$	constants for determining plasma-arc cutting speeds (Equation 14.9)
Q	proportion of machining time t_m during which the tool cutting edge is engaged with the workpiece
q	index for recommended cutting speed relationship to material hardness
q_n	chip-breaker groove radius (Figure 8.2)
q_{ps}	material constant for laser piercing (Equation 14.7); material constant for plasma-arc piercing (Equation 14.10)
$q_1, ..., q_5$	constants for determining laser cutting speeds (Equation 14.6)
R	thermal number; universal gas constant
R_a	arithmetical mean value of surface roughness
R_b	ballast resistance in electrical-discharge machining
R_c	Rockwell hardness number (C scale)
R_{c0}	reference material hardness for recommended cutting speed estimation
R_e	resistance of electrolyte in ECM
R_{max}	maximum height of surface irregularities
R_o	operator rate
R_{sg}	surface generation rate during machining

R_t	total machine and operator rate
R_{uu}	direct harmonic receptance in the u direction
R_{uv}, R_{vu}	cross harmonic receptance between u and v directions
R_{vv}	direct harmonic receptance in the v direction
r	radius at which cutting is taking place
r_c	cutting ratio
r_{chip}	radius of the chip curvature
r_e	specific resistivity of the electrolyte in ECM
r_g	grain aspect ratio in grinding
r_i	inside radius of the workpiece surface to be faced
r_l	chip breaking radius
r_{m_e}	ratio of workpiece volume removed externally by machining
r_{m_i}	ratio of workpiece volume removed internally by machining
r_o	outside radius of the workpiece surface to be machined
r_{sc}	radius of side curl chip
r_v	proportion of initial workpiece volume to be removed in machining
r_ε	corner radius; i.e., the radius of a rounded tool corner
S	amount received for machining one component; tool major cutting edge (Figure 7.1)
S'	tool minor cutting edge (Figure 7.1)
S_e	effective spring stiffness of the machine-workpiece-tool system (restoring force per unit displacement)
S_i	stiffness of the wheel and work in grinding
S_n	grinding-wheel-structure number
S_t	stiffness of the wheel support system in grinding
S_w	stiffness of the work support system in grinding
T	time between successive tool passes or revolutions of the workpiece
T_s	sheet or plate thickness
t	tool life; i.e., life of the tool cutting edge while the cutting edge is engaged with the workpiece
t'	time
t_c	tool life for minimum production cost; nonproductive time in grinding, includes wheel dressing time and time to load and unload workpiece
t_{c_t}	tool changing time; i.e., the average machine time to change a worn tool or to index (and, if necessary, replace) a worn insert
t_{ef}	tool life for maximum efficiency (maximum profit rate)
t_f	transportation time for a round trip with a forklift truck
t_{gc}	grinding time using recommended conditions
t_{gf}	finish grinding time
t_{gp}	grinding time under maximum power
t_{gr}	rough grinding time
t_l	nonproductive time; i.e., the average machine time to load and unload a component and to return the cutting tool to the beginning of the cut
t_{ln}	loading and unloading time
t_{ls}	total laser cutting time
t_m	machining time; i.e., the machine time to machine a component
t_{m_c}	machining time for minimum production cost

t'_{m_c}	machining time for minimum production cost corrected for tool replacement
t_{m_p}	machining time for minimum production time
t'_{m_p}	machining time for minimum production time corrected for tool replacement
t_p	tool life for maximum production rate (or minimum production time)
t_{ps}	piercing time for laser cutting; piercing time for plasma-arc machining
t_{pr}	production time; i.e., the average time to produce one component on one machine tool
t_{m_o}	tool life when maximum power is used
t_{pl}	total plasma-arc cutting time
t_{pt}	tool positioning time
t_r	tool life for a cutting speed of v_r
t_s	time for spark-out in grinding
t_{sa}	basic setup time per batch
t_{sb}	setup time per tool
t_t	transfer time for transfer machine
t_{tr}	transportation time per workpiece
t_w	thickness of the workpiece
V_b	percentage volume of bond material in a grinding wheel
V_c	condenser voltage in EDM
V_g	gap breakdown voltage in EDM
V_0	volume of chip produced in grinding
V_s	supply voltage in EDM and ECM
V_w	volume of tool material lost due to wear
VB	average width of flank wear land in the central portion of the active cutting edge (zone B in Figure 4.3)
VB_{max}	maximum width of flank wear land in the central portion of the active cutting edge (zone B in Figure 4.3)
$(VB)_m$	width of the flank wear land when tool must be reground
$(VB)_o$	increase in the flank wear land in zone B during the production of one component
VC	width of the flank wear land at the tool corner (zone C in Figure 4.3)
V_m	volume of material removed in machining
VN	width of the flank wear land at the wear notch (zone N in Figure 4.3)
v	cutting speed; i.e., the instantaneous velocity of the primary motion of the selected point on the cutting edge relative to the workpiece
v_{av}	mean cutting speed; i.e., the average value of v along the major cutting edge
v_c	cutting speed at minimum cost
v_e	resultant cutting speed; i.e., the instantaneous velocity of the resultant cutting motion of the selected point on the cutting edge relative to the workpiece
v_{ef}	cutting speed for maximum efficiency (maximum rate of profit)
v_f	feed speed; i.e., the instantaneous velocity of the feed motion of the selected point on the cutting edge relative to the workpiece; feed speed in ECM
v_i	machine infeed rate in grinding
v_{ls}	cutting speed in laser cutting
$v_{ls_{max}}$	maximum recommended cutting speed in laser machining
v_{max}	maximum cutting speed; i.e., maximum value of v
v_o	velocity of chip flow
v_p	cutting speed for minimum production time

v_{pl}	cutting speed for plasma-arc machining
$v_{pl\max}$	maximum recommended cutting speed for plasma-arc machining
v_{po}	cutting speed when maximum power is used
v_r	cutting speed giving a tool life of t_r
v_s	sliding velocity
v_t	surface speed of the wheel during grinding
v_{trav}	traverse speed in grinding
v_w	surface speed of the workpiece during grinding
v_1	recommended cutting speed at reference depth of cut (a_{p_0}) and reference material hardness (R_{c_0})
W	weight of workpiece
W_o	operator's wage (cost per unit time)
W_o'	operator's wage (wage plus overheads) (cost per unit time)
w	width of cut
w_n	width of chip breaking groove
w_o	dimensionless width of secondary deformation zone
w_t	width of grinding wheel
Z_t	wheel-removal rate in grinding
Z_w	metal-removal rate; i.e., the total volume of metal removed per unit time
Z_{wa}	metal-removal rate due to abrasive in electrolytic grinding
Z_{wc}	metal removal rate for recommended conditions in grinding
Z_{we}	metal-removal rate due to electrolysis in electrolytic grinding
$Z_{w\max}$	maximum metal-removal rate; i.e., the maximum value of Z_w
Z_{wp}	metal removal rate using maximum power in grinding
α	half angle between excitation directions u and v
α_c	proportion of chip thickness to neutral surface
α_n	tool normal clearance (Figure 7.12)
α_{n_e}	working normal clearance (Figure 7.13)
β	mean angle of friction on tool face
β_c	phase angle between successive waves cut on the workpiece surface
β_e	instantaneous cutting direction during oscillatory cutting
β_n	normal wedge angle (Figure 7.12)
β_R	angle between resultant displacement direction and vertical direction
Γ	proportion of heat generated in primary deformation zone conducted into workpiece
γ_{AB}	strain on shear plane
γ_f	tool side rake (Figure 7.5)
γ_g	tool geometric rake (Figure 7.4)
γ_m	proportion of real contact area over which metallic contact occurs
γ_n	tool normal rake (Figure 7.12)
γ_{n_e}	working normal rake (Figure 7.13)
γ_o	tool orthogonal rake (Figure 7.6)
γ_p	tool back rake (Figure 7.5)
γ_{set}	angle set on the vise to grind the tool back rake
δ	chip cross section related parameter
δ_c	work-surface slope; i.e., inclination of the work surface relative to the mean cutting direction

δ_R	resultant displacement in the R direction
δ_t	total relative deflection of the workpiece, wheel, and supports under steady-state grinding conditions
δ_u	response (deflection) in the u direction
δ_v	response (deflection) in the v direction
δ_w	amount to be removed from workpiece during grinding
ε	uniaxial true strain
ε_r	tool-included angle (Figure 7.12)
ε_{r_e}	working included angle (Figure 7.13)
η	resultant cutting speed angle; i.e., the angle between the direction of primary motion and the resultant cutting direction; factor to allow for additional plastic work after primary shear zone
η_e	current efficiency in ECM
η_m	overall efficiency of the machine-tool motor and drive systems
θ	temperature
θ_c	angle between resultant cutting force F_r and the normal to the cut surface
θ_f	average temperature rise of the chip owing to frictional heat source
θ_{int}	temperature at tool chip interface
θ_m	temperature rise of material passing through the secondary deformation zone
θ_{max}	maximum temperature along the rake face
θ_{mt}	melting temperature
θ_s	temperature rise of material passing through the primary deformation zone
θ_0	initial workpiece temperature
κ_r	tool cutting edge angle (Figure 7.12)
κ_r'	tool minor cutting edge angle (Figure 7.12)
κ_{r_e}	working cutting edge angle (Figure 7.13)
κ_{re}'	working minor cutting edge angle
Λ_t	wheel-removal parameter in grinding; i.e., the volume of wheel removed per unit time per unit thrust force
Λ_w	workpiece-removal parameter in grinding; i.e., the volume of metal removed per unit time per unit thrust force
λ	angle of resultant cutting force to shear plane
λ_s	tool cutting edge inclination (Figure 7.12)
λ_{s_e}	working cutting edge inclination (Figure 7.13)
λ_f	chip-flow angle (Figure 7.3)
μ	coefficient of friction
μ_c	overlap factor between successive cuts of the tool
ξ	damping ratio
ρ	density of work material
σ	chip-breaker wedge angle (Figure 8.3); uniaxial true stress
σ_f	normal stress on the tool face
$\sigma_{f_{max}}$	maximum normal stress on the tool face
σ_n	normal stress on tool face
σ_s	normal stress on the shear plane
σ_y	yield pressure of softer metal
σ_1	constant in uniaxial true stress true strain relationship

τ_f	shear stress on the tool face; shear strength of softer metal
τ_s	apparent shear strength of the work material (shear stress on the shear plane)
τ_s'	shear stress on the shear plane corrected for the tool-nose force
τ_{st}	shear stress on the tool face in the sticking region
τ_{s_o}	shear stress on the shear plane with zero normal stress applied
τ_1	shear strength of the softer metal in boundary lubrication
τ_2	shear strength of the lubricant layer in boundary lubrication
Φ_c	rate of heat transportation into the chip
Φ_t	rate of heat conduction into the tool
Φ_w	rate of heat conduction into the workpiece
ϕ	shear angle
ϕ_f	phase difference between disturbing force and resulting motion
ϕ_x	angle between normal to the cut surface and the vertical direction
ψ	angle between exciting force and vertical direction
Ω	rotational speed of the workpiece (rad/s)
ω	angular frequency of vibration
ω_f	angular frequency of external harmonic force
ω_n	natural angular frequency

Index